T0309764

Catalyst Engineering Technology

Catalyst Engineering Technology

Fundamentals and Applications

Jean W. L. Beeckman

This edition first published 2020

Registered Office
John Wiley & Sons, Inc., 111 River Street, Hoboken, NJ 07030, USA

Editorial Office
111 River Street, Hoboken, NJ 07030, USA

For details of our global editorial offices, customer services, and more information about Wiley products visit us at www.wiley.com.

Wiley also publishes its books in a variety of electronic formats and by print-on-demand. Some content that appears in standard print versions of this book may not be available in other formats.

Library of Congress Cataloging-in-Publication Data

Names: Beeckman, Jean W. L., author.
Title: Catalyst engineering technology: fundamentals and applications / Jean W. L. Beeckman.
Description: Hoboken, NJ : Wiley, [2020] | Includes bibliographical references and index.
Identifiers: LCCN 2020008260 (print) | LCCN 2020008261 (ebook) | ISBN 9781119634942 (hardback) | ISBN 9781119634966 (adobe pdf) | ISBN 9781119635017 (epub)
Subjects: LCSH: Catalysts. | Extrusion process. | Catalysts industry.
Classification: LCC TP159.C3 B44 2020 (print) | LCC TP159.C3 (ebook) | DDC 660/.2995–dc23
LC record available at https://lccn.loc.gov/2020008260
LC ebook record available at https://lccn.loc.gov/2020008261

Cover Design: Wiley
Cover Image: © Elsevier

Set in 9.5/12.5pt STIXTwoText by SPi Global, Pondicherry, India

Printed in the United States of America

SKY10019833_071720

Contents

About the Author

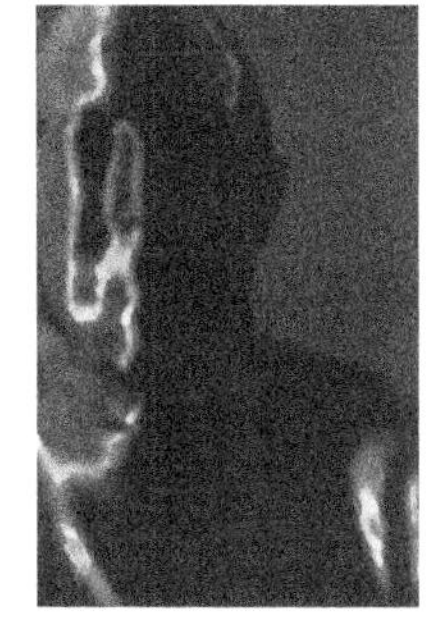
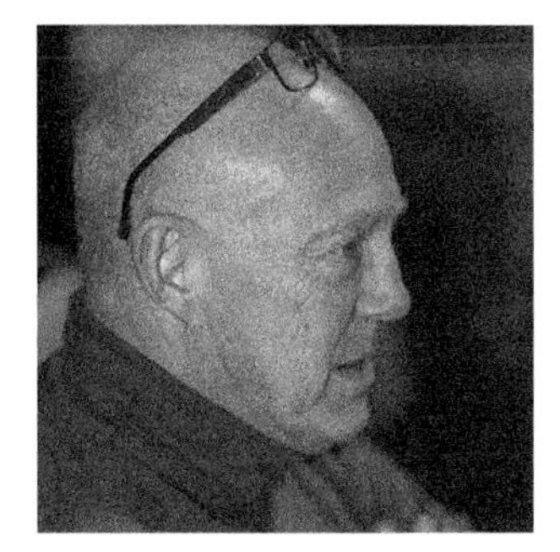

(Courtesy: Germain Beeckman)

Jean Willem Lodewijk Beeckman attended high school at the Koninklijk Atheneum in Zottegem, Belgium. He obtained his Chemical Engineering degree in 1975 and later his Doctorate in 1979 at the University of Ghent, Belgium, with Prof. Emeritus Gilbert F. Froment. Beeckman started his career as a technical contact engineer for the Gofiner and vacuum pipestill at the Esso refinery in Antwerp, Belgium. He then started a temporary foreign assignment on coal gasification at the Exxon Research and Development Laboratories in Baytown, TX, followed by a position as a chemicals coordinator at the Antwerp refinery. Thereafter, he spent a little over a decade at W. R. Grace and Co. in their corporate research laboratories in Columbia, MD. In 1997, Beeckman joined Mobil Oil Corporation in Paulsboro, NJ. Since 2001, Beeckman has been with ExxonMobil Research and Engineering, Process Technology Division, Catalyst Technology Department in Annandale, NJ. His entire career has been in the area of catalyst development, catalyst manufacturing, and mathematical modeling. As of February 1, 2020, Beeckman has retired and resides in Columbia, MD. The author can be reached by email at: jeanwlbeeckman@gmail.com

Acknowledgments - Dankwoord

It is my privilege and definite pleasure to acknowledge those who have worked with me in this area. Let me start with Theodore Datz, Natalie Fassbender, Eric Jezek, Nicole Vanderzee, Dana Mazzaro, Majosefina Cunningham, Patrick Hill, Glenn Sweeten, and Michael Hryniszak. I thank them wholeheartedly for their help with the experimental data gathering. I thank Michael Pluchinsky for his help with the high-speed photography work. I thank Joe Gatt for his help in collecting the commercial plant data that were needed to validate the theoretical approach on the breakage of extrudates. I thank Robert Mangene for many helpful discussions on laboratory equipment and pilot plant equipment. I thank Machteld Mertens for her expert help and discussions on proper analytical technique, sampling procedures, and sampling pitfalls. I thank Yohannes Soulages and Jaishankar Aditya for their helpful discussions on the rheology of extrusion pastes. I thank Arash Fathi for our discussion on stochastic differential equations. I thank Bello Laz for his kind help with this project. I also give a very special thank you to Cheryl Grimes and Charmaine Cooper Hussain for their help with the typing, editing, and correcting of the manuscript. I thank Melody Schottle for her counsel with the copyright permissions. I thank the late Jianxin (Jason) Wu for his most kind comments regarding the work on the breakage of extrudates during impact against a surface. Further, I would like to express my most sincere gratitude toward José Santiesteban, Chris Wright, J.J. Thiart, Kathy Keville, Beau Waldrup, Ivy Johnson, Marc Schreier, Ken del Rossi, and Vijay Swarup for reviewing this work and allowing me to bring this research into the open literature. I most sincerely thank Prof. Emeritus Thomas F. Degnan, Jr. for sharing his vast experience and invaluable insights in catalysis, catalyst modeling, and catalyst manufacturing. I thank the late Art Chester, one of the founding fathers of the Rutgers Catalyst Manufacturing Consortium, together with Prof. Ben Glasser, Prof. Fernando Muzzio, and Prof. Johannes Khinast for their unbridled enthusiasm and drive regarding helping to progress catalyst manufacturing technology toward it becoming an independent academic discipline (http://cbe.rutgers.edu/catalyst). I also thank Ben Glasser and Yangyang Shen for gathering and allowing me to use the compendium of research article references that have streamed from the Consortium's research. I thank Prof. Emeritus Gilbert F. Froment for teaching me the discipline of chemical reaction engineering and for his relentless drive toward improving, enhancing, and exploring the next steps in this arena.

I dedicate this book to my late parents and grandparents and to Barbara McKenna for their unconditional support of my research. I also give a warm thank you to my four brothers, Germain, Dirk, Renaat, and Marc Beeckman, for their quiet approval. I give a special thank you to Germain for his photographic artwork. I also thank Chris Van Malderghem, Els Beeckman, and Brenda, Gerd, Lena, and Ine Beeckman-Struyf for their kind support. I thank Gerd for the many interesting network discussions we have had – he is clearly bitten by the same bug, but works in such a clever way that his uncle now has difficulty following.

Last but not least, I thank Boris, who took me on many walks so that I could think on the problem du jour.

And finally, I take full responsibility for any errors that may arise in this book.

Foreword

Catalyst research and manufacturing is a major component of the chemical industry. It is often the backbone of process research and process commercialization. The catalyst industry is a fast-evolving industry where patenting new catalysts and bringing new catalysts to market are very important in order to become and remain competitive. The global catalyst business is a multibillion dollar industry. De Jong [1] cites US$12 billion per annum for solid heterogeneous catalyst sales in 2004, with the growth in catalyst sales foreseen to be at about 5% per year on average. The catalyst industry is supported by academic research worldwide, by research efforts and development efforts in numerous companies, and by a plethora of manufacturing sites.

The chemistry, the porous structure, and the mathematical modeling of catalysts are the subjects of many excellent textbooks and technical papers that have stretched over many decades. Mentioning only a few of these existing pearls of wisdom, however, does not give due respect to the enormous body of work and understanding that has been collected to date. An area of research and development that is part of catalyst know-how is the forming of extruded catalysts. This area is covered by patents and trade secrets, and it is typically very hard to obtain information on this topic in the open literature.

Catalysts are shaped porous material bodies and typically consist of an active phase held together with a binder. Such shaped bodies often come as extrudates, some are made in monolith form, some come as spheroids, and some are obtained by pelletizing. Most extruded catalysts and pelletized catalysts are at the millimeter scale, but monoliths can range from centimeter to meter scale. Some of the smallest monoliths are at the millimeter scale and are sometimes called Miniliths™. Once formed, the catalyst has to hold up under the typical stresses they encounter during their manufacture and use in process plants. Catalysts often must be and can be regenerated, and the stresses during unloading, regeneration, and reloading play an important role in the integrity of the catalyst's shape and average particle size. These stresses are induced through the variety of ways of handling the catalyst in a commercial plant.

Chapter 1 of this book will provide a background on the many different ways of forming catalysts in the laboratory and commercially. The method of manufacture is often determined by the available equipment in the plant and by the architecture of the plant. During handling, the catalyst may become damaged, and this is called natural breakage. It is often an objective in a plant to minimize this natural breakage because it hampers the production rate. At times, catalysts can be very resilient to natural breakage, but they still need to be sized appropriately. This sizing is then called forced breakage. Breakage of the catalyst is governed by the mechanical strength of the catalyst and by the severity of the handling. The mechanical strength of the catalyst is an important subject in this book, and I will demonstrate how physical–mechanical principles can be applied in the breakage arena.

Chapter 2 will provide a background on the technology of extrusion and the models that are in use. The open literature contains much information on the extrusion of plastics, but less so on the extrusion of ceramics. Extrusion technology regarding chemical and petrochemical catalysts is even less documented in the open literature.

Chapter 3 investigates in depth the breakage of catalysts due to collision and also the breakage of extrudates under the influence of stress in a fixed bed. It is shown that it is the bending strength of extrudates and especially the force of rupture in the bending mode that are the major strength properties of an extrudate when mathematically modeling breakage due to collision and stress. The laboratory measurement of catalyst strength is handled in detail herein. The method of mathematical modeling of breakage due to collision is based on finite-difference calculus. The mathematical model is then applied in order to quantitatively predict catalyst breakage in commercial plants. Modeling breakage in fixed beds is based on a mechanical force balance between catalyst strength and stress in the fixed bed. This phenomenon is then simulated with the help of the catalyst bulk crush strength test. A mathematical model is then derived that allows one to successfully predict catalyst breakage.

Chapter 4 will deal with the mathematical modeling of the porous structure of catalysts. This will be represented from the perspective of a collection of interconnecting nodes where mass flow combined with first-order reaction is solved for the entire network, satisfying all material balances in the individual nodes. The parameter space entails the network architecture, the rate of mass transfer, and the rate of reaction. The overall network architecture is looked at as "white noise" due to the abrupt changes in the interconnectivity and the properties of the nodes from point to point. The parameter space is threefold – structure, diffusivity, and reactivity – and in general leads to essentially no specifics. Limiting this parameter space to deep networks and perturbation-like reactivities in the nodes breaks a virtual logjam and opens up a plethora of theoretical results that are directly applicable to real catalysts. By applying this particular carving of the parameter space, here called VDNP or very deep network perturbation, it will be shown that arbitrary networks have the same characteristics as periodic or regular networks. These characteristics also apply for mass transfer alone or for a combination of mass transfer and chemical reaction. For VDNP, it will be shown that the solution of the stochastic finite-difference matrix equations converge to the classic solutions of reaction and diffusion in catalysts.

Reference

1 de Jong, K. (2009). *Synthesis of Solid Catalysts*. Wiley-VCH Verlag GmbH & Co. KGaA.

1

Catalyst Preparation Techniques and Equipment

1.1 Introduction

A catalyst is made up of its constituent components that are one or more active phases held together with one or more binder components. The components of a catalyst are initially in the form of loose powders. The forming of a catalyst entails using powders that serve as building blocks and making the catalyst hold together as a shaped particle. Excellent textbooks by Stiles [1] and Le Page [2] cover catalyst formation and show the typical equipment, methods, and materials used commercially. Adding catalytic components inside a particle by impregnation or by ion exchange with solutions of metals followed by using appropriate drying and calcination procedures makes catalysts useful for many processes.

The science and experience of impregnation, drying, and calcination is a vast area of knowledge represented by many excellent papers. Cited here, among a wealth of research on the subject, will be works by Neimark et al. [3], Chester and Derouane [4], Marceau et al. [5], Lekhal et al. [6, 7], and de Jong [8]. The Rutgers Catalyst Consortium headed by Professor Benjamin Glasser has been investigating manufacturing fundamentals for the last two decades. Many of their published contributions yield a solid basis for applied methods and numerical models for the many facets of catalyst manufacturing.

In the area of catalyst impregnation, I will cite Chester et al. [9], Chester and Muzzio [10], Liu et al. [11], Shen et al. [12], Romanski et al. [13–15], and Koynov et al. [16]. On the subject of metal profiles in catalyst particles, I will mention Kresge et al. [17] and Liu et al. [18–20]. For rotary calcination, I will mention Chaudhuri et al. [21, 22], Gao et al. [23], Paredes et al. [24], Emady et al. [25], and Yohannes et al. [26, 27]. Last but not least, this book cites a very interesting historical perspective of the many pioneers in heterogeneous catalysis by Davis and Hettinger [28].

Catalyst Engineering Technology: Fundamentals and Applications,
First Edition. Jean W. L. Beeckman.

The common structural properties of interest for catalysts include surface area, porosity, pore size distribution, particle shape, and particle size. A wonderful textbook in this area is by Thomas and Thomas [29].

Part of the catalyst-forming technology includes the development of the mechanical strength of the catalyst. This aspect is sometimes overlooked initially, but should not be underestimated or forgotten and taken for granted. Insufficient mechanical strength requires the changing or adjusting of recipes in order to yield the strength needed for handling and subsequent use. Changing recipes at the end of the game always brings the risk of an unforeseen detrimental impact on catalyst activity.

Catalysts come in many shapes, and the processes that allow their manufacture include extrusion, spray drying, bead forming by dripping, bead forming in granulation pans, fluid bed granulation, spheronization, and pelletizing. The process of use in the chemical and petrochemical industries dictates the size and the shape of the catalyst, while the manufacturing cost often dictates the choice of the catalyst-forming process. Extrusion and spray drying, Masters [30], yield economic manufacturing solutions when large quantities of catalyst are required, such as in the petrochemical industry.

In the automotive industry, monolith-type extruded catalysts are well established for their low pressure drop and hydrothermal stability properties. Monolith manufacturing processes have been described in the open literature and are available in textbooks, including those of Cybulski and Moulijn [31] and Satterfield [32].

Beyond the forming of a catalyst particle, many more operations are required to make a final catalyst. For instance, after the forming, drying followed by calcination at high temperatures set the matrix of a catalyst and yield the basic strength properties of the catalyst. Thereafter, the catalyst may require metal components to be introduced into its structure, followed by drying and calcination.

Catalytic metal impregnation, drying, and calcination can lead to further breakage of catalyst extrudates and a loss of catalyst and catalyst quality. Bringing an optimal recipe to fruition may require sacrificing many thousands of pounds of material together with precious plant manufacturing time.

Catalyst manufacturing plants are built with a combination of different pieces of production equipment in mind. The layout and the design of the production path needs to allow for product quality assurance and product quality control. Usually, new plants are designed based on experience with existing plants, but there is little guidance from first-principles methods based on the science of the phenomenon of breakage by collision or by stress in a fixed bed of catalyst. Often, a reference catalyst is tested in a piece of equipment and the decision to go ahead with that equipment or to change it depends on the outcome of that test. Engineers and chemists with many years of experience in the catalyst manufacturing industry are valuable resources when it comes to designing new plants or modifying existing ones.

1.1.1 What are Catalysts?

Catalysts are solid bodies with a porous structure that allows for the fantastic ingress of reagents and egress of products. This accessibility exposes the bulk fluid to enormous surface areas and densities of the catalytic sites located inside these catalytic bodies. Maximizing surface area and accessibility is always one of the goals, but it is necessary to balance this goal with the mechanical strength of the catalyst and the cost of the catalyst in order to obtain a commercially viable product.

Catalysts come in all shapes and sizes, from the micron scale in fluid catalytic cracking to the millimeter scale in hydrotreating applications to the meter scale for monoliths and coated catalysts in coal- and gas-fired applications. Catalysts are loaded in reactors as a fixed bed or as a moving bed, or are dispersed in and travel with the bulk fluid. It is important to consider the process pressure drop in reactor designs when it comes to catalyst selection.

Most if not all catalysts deactivate over time and have a limited life span that can range widely from fractions of a second to several years, depending on the application. Catalyst deactivation results in a loss in terms of conversion and/or selectivity, and also hampers throughput, which requires either a catalyst change-out or an *in situ* regeneration.

Catalyst abrasion and attrition are also factors that have a direct impact on both their practical use and economics.

Fouling of catalyst beds by scale also requires engineering solutions due to pressure drop considerations.

1.1.2 Catalyst Composition

Catalysts transform reagents into products, and catalyst activity dictates the rate of this transformation. It is often one of our objectives to let the products escape readily from the catalyst without undergoing secondary reactions, which enhances the selectivity of the catalyst. Needless to say, catalyst compositions are heavily guarded by patents and trade secrets. Although this book only mentions generic compositions, there are very good textbooks by Stiles [33] and Satterfield [32] that give a rich flavor of the many possibilities in catalyst composition.

A catalyst may consist of one or more active phases and one or more binders. The active phase is often γ-alumina (aluminum oxide) because of its stability, large surface area, and porosity. Other active phases include silica or zeolite (silicon oxide or crystalline silica alumina). Binders can vary, but may be alumina or silica.

Base metals include cobalt, nickel, molybdenum, etc., while noble metal catalysts often include platinum or palladium.

1.2 Forming of Catalysts

1.2.1 Catalysts Formed by Extrusion

1.2.1.1 Typical Materials

Standard formulations for catalysts entail adding an appropriate amount of binder to an active material and an appropriate amount of liquid (often water). The sequence of the addition of the components can be an important factor for certain catalysts.

1.2.1.2 Mixing, Mulling, Granulation, and Kneading

Mixing, mulling, granulation, and kneading occur before the extrusion of powders. It is necessary to blend powders and "work" them with a liquid – typically water – and to bring them to the desired rheological consistency before extrusion. The recipes are ad hoc, and it is difficult to predict which combination will bring success. A term often used when preparing a batch for extrusion is the "amount of work" put into a batch during mulling. Because this phrase has different meanings, it can lead to miscommunication among crews or teams. This process goes together with what is called the "peptization" of a batch, and both essentially go hand in hand when judging whether a batch is ready to extrude.

The appearance and response of a catalyst mixture in a batch will differ from one technique to another. Mill wheels will essentially go through stages: mixing and mulling followed by granulation and kneading. Mixing refers to both the mixing of dry powders and, later, the mixing of liquids. Mulling reduces the particle size of larger dried knurls to an appropriate size.

Granulation occurs early on in the mixing process and starts with the creation of different-sized spheroidal materials. This size distribution will get larger and larger with time. Once the size of the agglomerates increases to the centimeter scale, the term "kneading" more realistically depicts what happens to the batch.

The kneading stage induces a lot of shearing when the mill wheels smear out and reconstitute the agglomerates continuously. If not caught in time, the size of the clumps will increase to a size that no longer enables their transport to downstream equipment. It is important to catch this stage in time and not let it progress, for obvious reasons.

The period of time for working a mixture can vary from minutes to tens of minutes. Up-front technology transfer is crucial to alerting and advising the plant on operating time. Powders are typically mixed in large stationary tubs that allow vertically rotating mill wheels to roll around on the powders with scrapers inside the tub to help with the mixing. The scrapers are often multifold, which means that one scraper will keep the wheel surface clean, while other scrapers keep the bottom of the tub and sides clean. Agglomerated knurls dried in an earlier operation might be used instead of powders. The rotating mill wheels easily break up the knurls, given sufficient time. The mill wheels weigh a few hundred kilograms and are about 30 cm wide, thus producing a lot of shear on the mixture.

In catalyst manufacturing plants, mill wheels are not the only option. Manufacturers might use mixers that consist of large rotating tubs where the powders and liquid are charged to. These tubs are equipped with a fast-rotating studded pillar mounted vertically near the side of the tub, providing much kinetic energy to the mixture over a period of time. This kinetic energy is also felt as a gradual heating up of the mixture in the tub. Finally, sigma mixers, which depend on rotating arms, are also practical commercial means of making a mixture extrudable.

It is vitally important to practice “best safety practices and procedures” in the vicinity of these pieces of equipment (and others) and always to adhere to site safety policies. Often the equipment appears slow and sluggish, but you are no match for its power or its speed. Always check for loose garments that could get caught because most of this equipment has rotating parts. Always be aware of the locations of emergency stop devices for the equipment and, if possible, use the buddy system.

At times, the powders may be very similar in color (often white), and visual assessment is difficult when thorough mixing has been achieved. It is possible here to apply an elegant statistical method that the researchers at the Rutgers Catalyst Manufacturing Consortium use. Their method relies on statistics and measuring either a chemical component, the water content, or a physical attribute such as color (when applicable) to judge the degree of mixing. Section 3.8 at the end of Chapter 3 of this book touches on the subject of statistical methods.

Judging whether a batch is ready to extrude is critical, and this is where experienced operators and plant engineers play an important role. After mixing and mulling, the operator often has to fine-tune the mixture, and preferably this is done by adding more liquid and by further working the mixture until finally obtaining the correct rheology. I say “preferably” here because in the alternative case of a batch being too wet, one has to add the dry powder constituents, and this is much more labor-intensive and difficult to measure. In addition, uniformly mixing these extra ingredients into the batch at the end is a challenge.

As the mixture is being worked, there is often a noticeable change in its appearance, as well as in the sound that the equipment produces as the mixture becomes more granular. Initially, there are small satellites (very small spheres) being formed. These satellites tend to adhere to each other and/or attach to larger granules and grow by agglomeration; this phenomenon can give the mixture a fuzzy appearance.

As the granular material becomes larger, it may form large clumps the size of a human fist or even larger. The mixture often gets warmer and may even become hot to the touch. It is important to monitor the mixture's temperature, and it may be necessary to add moisture in order to prevent the mixture from drying out, which would ruin the chances of obtaining a workable recipe.

As the clumps get bigger, studded pillar tub-type mixers struggle to disperse any water that is added to the clumps, while mill wheels, due to their weight, have little difficulty smearing out clumps and mixing added moisture uniformly. It is difficult to decide when the mixture is finally ready to extrude. Without knowledge acquired during earlier trials, it is essentially impossible to determine this with any kind of certainty. If it is possible to prepare or drop a small quantity of the mixture and perform a test run with a commercial extruder, it is certainly advisable to do so, because dropping an entire batch too early may require heavy manual labor when the extrudability test fails, meaning the unit has to be cleared and the material has to be recycled.

In my opinion, "work" and "peptization" are terms that quantify the shearing of particles to the point where enough fines have been produced during mulling to make the batch ready for transport to the auger in the extruder, de-aeration in the auger, the production of a steady pressure at the die face, and finally extrusion through the die holes or die channels.

The amount of water that needs to be added to a particular batch depends on the material itself. Materials with high porosity and large surface areas require more water than those with little porosity. An estimate of the required amount of water can be obtained by performing a water adsorption test in the laboratory. The extrusion batch is then wetted to below the point of clumping. A water adsorption test takes just a few minutes and provides good quantitative information on how much liquid to add. Be aware of liquid density differences when using liquids other than water. Aim to stay 5–10% below the clumping level and avoid adding too much water, especially at the beginning of a commercial campaign. The particles soak up the liquid because of their porous structure (the available pore volume), and the mixture often appears very dry and free-flowing until the end (the point of clumping), and this can be misleading.

During the work and peptization phase, the particles shear and lose a certain percentage of their structure through breakage or attrition of their outer surface; these fines then become part of the mixture. The loss of material structure incurred during shearing is small, but raises the water-to-available pore volume ratio, even without physically adding water. As this ratio increases, the mixture may become what is called "over-worked" or "over-peptized," potentially leading to the formation of a slurry in which all of the material becomes liquid-like and can even be poured. Adding a small quantity of the powder components will produce a workable mixture again.

The shearing of the powder mixture is part of the recipe; it is a natural phenomenon and it is experienced during mulling and kneading. Some plant equipment such as the extruder may also induce shearing in the extruder itself. Sometimes over-shearing in the extruder happens and is unintentional, and in that case, the batch will need to be adjusted to be drier.

1.2.1.3 Extrusion

The extrusion of catalyst bodies is a very economical way to produce catalytic materials (or materials that will become catalytic after adding catalytic components such as metal). Typical sizes of extrudates are 1–4 mm in diameter with a constant cross section, but their lengths vary.

The extrudate bodies may be solid or hollow. Small-diameter extrudates are made solid, because their short diffusional path does not necessitate any extra channels for easy access of the bulk fluid. Large-diameter extrudates are often prepared hollow to enable adequate diffusion and a greater geometric surface area. Shape, size, and bed packing play key roles in the process pressure drop over the catalyst bed and can be determining factors in the selection of the catalyst.

The cross section is normally constant except for the typical wear of the extrusion channels. This wear may require changing of the dies when it is beyond the acceptable specification limits of the catalyst diameter or when the catalyst cross-sectional shape is compromised. Figure 1.1 shows typical catalyst cross sections, but many varieties exist.

Extrudates coming from the die face of the extruder will extrude as long strands, often several tens or hundreds simultaneously depending on the size of the die and the extruder equipment. Extrudate strands with lengths from several centimeters to several tens of centimeters are transported via conveyors to the drying equipment, where the material is spread out uniformly across a drying belt or a vibratory drier. During passage through the drier, the material loses the moisture it had acquired during its preparation (usually the moisture added during mulling)

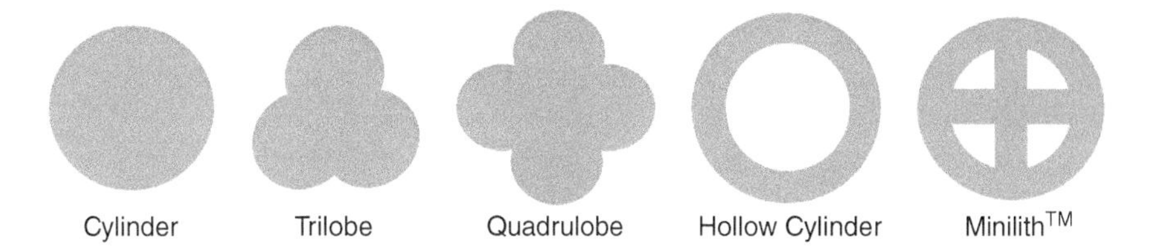

Figure 1.1 Typical extrudate cross-sectional shapes.

and the material gains mechanical strength – to a degree. The strands at the exit of the drier may still be long, measuring several centimeters.

As mentioned, the lengths of the extrudates vary, but some extrudate lengths are created by cutting the extrudate strands at the die face of the extruder as they are exiting, and those are typically much more uniform in length. Cutting at the die face is possible when the green extrudate body exiting the die face has sufficient strength to allow a cut to progress without deforming the extrudate. This practice is often done with some sort of spinning blade or wire.

The catalyst extrudates exit the drier and are transported via conveyors, bucket elevators, gravity falls, or alternative means to other parts of the plant. Several more steps such as sizing, sieving, impregnation, or ion exchange may be required before obtaining a commercial catalyst. It is during all of these handling steps that the catalyst breaks into shorter extrudates, either in a natural way or in a forced way, and becomes part of the final product. The science of catalyst breakage is where I hope to make a contribution to the literature concerning the fundamentals of catalyst scale-up and catalyst manufacturing in Chapter 3.

1.2.1.4 Extrusion Efficiency

Catalyst recipes are tried at the laboratory scale or at the pilot scale before large commercial campaigns. A large component of industry competitiveness is the rate at which materials can be extruded, because this rate has the largest impact on catalyst manufacturing costs. Acquiring data in pilot extruders or during commercial demonstrations is necessary in order to predict extrusion rates. Extrusion and predicting extrudability are still very much art forms and highly complex problems; trial and error and on-site assistance are still the most effective ways to gain information because no other reliable method is available.

Here are two basic methods that may help with understanding and quantifying some of the steps in extrusion.

The first step is to understand more of the rheological behavior of an extrusion paste where it pertains to extrusion efficiency. It is important to understand a mixture's slip behavior, meaning the point where a mixture under normal operating stress starts to slip under shear. A mixture under normal stress has a linear increase in torque when the strain rate increases. However, at some strain rate, the grip is lost and torque reduces dramatically. In commercial settings, the onset of slip behavior and the loss in production rate are often misinterpreted as being due to a plugged die because the extrusion rate drops precipitously. It is possible to investigate mixture behavior up front with a rheometer that measures torque versus strain rate, allowing the setting up of an operating window.

The second step is to understand the basics of extrusion and how the revolution rate of an auger inside a barrel, facing a die plate, translates to a particular extrusion rate.

I will discuss both aspects in more detail in Chapter 2.

1.2.2 Spheroidal Catalysts

1.2.2.1 Spray Drying

Spray drying of a liquid catalyst slurry is ideal when the objective is to make spherical catalysts in the 30 μm to ~120 μm diameter range; typical ranges are 50–80 μm. The process of use dictates the preferred size range, and the spray drier operating conditions and the slurry viscosity can be adjusted to obtain the right size. A short diffusional path length constitutes an advantage for this catalyst shape.

An excellent textbook by Masters [30] presents the many variations of and options for this technology. The components are prepared in a slurry of adequate viscosity, which is pumped to a spray nozzle that operates at a certain pressure or to a spray disk that runs at a few thousand revolutions per minute (rpm). The spray drier yields larger particles when the viscosity increases (by increasing the solids in the slurry, for instance) or when the pressure of the nozzle or the rotation rate of the spray disk drops. The slurry droplets come into contact with hot gas injected into the spray drier, lose their moisture by evaporation, and gain mechanical strength. Droplets that are too large often require too much time to dry and can reach the drier wall while still containing moisture. As the large mixture droplets touch the wall, they create an undesirable mud-like adhesion, which may lead to a shutdown of the spray drier.

Small droplets have a very high surface-to-volume ratio; hence, their drag with the gas leads to a strong deceleration. Larger droplets have less drag and hence less deceleration, and they will penetrate and overtake the cloud of slower-moving smaller droplets. Contact between larger and smaller droplets can lead to multisphere particulates. Slowing down the slurry feed rate and creating a less dense particle cloud can help reduce this phenomenon.

The dried material is gathered in the main chamber, while small particulates are gathered in a cyclone and are often combined with the main chamber product in order to maximize product. The smaller particulates often enhance fluidization properties later. The fines of the process of spray drying are often collected in a baghouse and discarded or reused in slurry making, depending on the size of the campaign and the purity of the fines.

1.2.2.2 Bead Dripping

Dripping a slurry of adequate viscosity (either in a hot-oil column or in a medium of adequate chemistry) will make spherical catalyst bodies in a size range between ~0.8 mm and ~1.5 mm. Immiscibility during first contact with either the oil or other medium is very important. Initially, the beads of slurry can be formed in a true spherical shape due to their immiscibility with the bulk fluid and due to the proper viscosity and surface tension of the slurry. The objective is for the beads to quickly gain strength along their surface and to harden and maintain the spherical shape either by drying in a hot-oil column or chemically via gelation.

In some forming processes, a second bulk fluid layer exists, strengthening the beads throughout by applying more intense conditions. After their travel through the column, the beads are collected and separated from the bulk fluid. In the event of an aqueous bulk fluid of proper chemistry (controlled by the operator), the beads have enough strength and can be dried on a belt drier. The devices used to drip the beads need the correct slurry viscosity. Slurries with a high degree of viscosity tend to clog the drip devices, while slurries with a degree of viscosity that is too low tend to form long slurry trails and non-spherical granules.

As droplets of slurry form, they must separate cleanly from the drip box, and surface tension and viscosity play key roles herein. Establishing an operating window requires laboratory or pilot testing to identify the properties of the slurry that allow the drip device to operate for a prolonged period of time. The viscosity of a slurry batch once prepared is unfortunately not a constant. Slurry viscosity changes with time due to aging, slowly drifting toward a higher level of viscosity – often too high to continue the dripping process.

It is therefore important to analyze the slurry for viscosity aging as time progresses in order to plan for viscosity thinning. Thinning can sometimes be accomplished through shear thinning. Temporarily increasing the mixing blade's rotational rate can bring the viscosity back to the appropriate window. Another technique to slow viscosity creep is cooling the slurry vessel. The slurries are often prepared by milling the original particle sizes formed by precipitation. Milling tends to slightly increase the temperature, but that temperature increase can be enough to enhance slurry aging and viscosity creep. Cooling the slurry mixture by 5–20 °F either by external or internal means can substantially slow the aging.

While drying, a fair amount of bead shrinkage occurs because all slurry liquid (mostly water) is still present during the initial solidification of the boundary, and that slurry liquid sets the bead diameter. This substantial shrinkage makes the drying process critical. Depending on the recipe, bead shrinkage can lead to the formation of micro-cracks that standard quality assurance tests and specifications may not reveal. Further processing in downstream equipment such as a rotary calciner sets the matrix and turns the beads into a catalyst support with appropriate properties. It is important to test the beads for their final use because these micro-crack defects can lead to greater-than-expected chemically induced attrition rates during their final process of use. Bead dripping is expensive compared to extrusion and it is labor-intensive, especially because the drip devices tend to clog and need to be cleaned out manually, which also reduces the rate of production.

1.2.2.3 Granulation Pans

Granulation pans are also used to make spherical catalysts. Commercial granulation pans are ~1–3 m in diameter; they rotate at ~5–20 rpm and are operated on an incline. Catalyst material is administered continuously together with liquids, and spheres start to form and collect at the bottom part of the pan. The spheres vary in size; the smaller spheres tend to accumulate below the larger spheres. As more and more material is formed, the larger spheres at the top of the bed fall freely over the edge of the pan onto a conveyor belt. After forming, the beads go through the standard methods of drying and calcination to remove moisture and set the matrix of the support during calcination in order to finally become a finished catalyst support.

The sphericity of granulation beads does not match the quality of beads obtained by dripping. However, although the beads do show a rougher surface, they are very well formed nonetheless. Granulation beads can be used in fixed beds, but are less attractive for moving beds because of the higher attrition.

1.2.2.4 Fluid Bed Granulation

Fluid bed granulation essentially sprays a bed of catalyst support spheres or granules or even extrudates from above with a slurry or a coating liquid of interest while being fluidized by a hot gas stream. This operation forms a coating on the beads and also makes them larger.

An interesting phenomenon occurs in fluid beds with granules of different sizes. Because of their size, it is often assumed to be impossible for the large-diameter beads to fluidize since they have a much higher minimum fluidization velocity compared to the smaller beads in the size distribution. However, this is an argument based on fluidization by gas and momentum transfer by gas. In a fluidized bed of small 100-μm particulates, which have far lower fluidization velocities than, say, 1–2-mm-diameter beads, larger beads still fluidize. Fluidization of large beads is actually based on momentum transfer with the small particulates, which have a much greater density than the fluidizing gas, and hence the small particulates can enable fluidization of the large beads. There is a threshold, however, for the concentration of the large beads in the fluidized bed, above which the large beads do collect and accumulate at the bottom of the bed. This threshold concentration can be fairly large, at 5–20% or even higher depending on conditions.

1.2.2.5 Spheronization

An interesting way to prepare spheroidal pellets is by extrusion into cylinders that have a diameter that is roughly the desired size of the granules. After extrusion, dropping the strands onto a spinning plate with a rough surface breaks up the strands, which then naturally become spheroidal in shape before being fed to dryers and calciners.

1.2.3 Catalysts Formed by Pelletizing

The pelletizing method forms beautifully shaped particulates, but it has a high fabrication cost given its low throughput. This method is popular in the pharmaceutical industry as a niche operation.

1.2.4 Monolith-type Catalysts

Monolith catalysts prepared by extrusion have a very intricate shape, and producing them on a large scale is an art. Their size ranges from millimeters to meters. In the millimeter scale, they are often referred to as miniliths (Pereira, [34]). Large-scale monoliths are often used in the automotive industry (Kubsh [35]) and coal-fired power plant selective catalytic reduction catalytic reactors (Beeckman and Hegedus [36]). Methodologies for cutting, drying, and calcination are based on trial and error, and high-quality bodies are now possible after years of development.

1.3 Impregnation and Drying

The impregnation of catalyst supports is done to bring the active component of the catalyst onto the catalyst internal surface area. A solution is prepared that contains the desired concentration of metal, and this solution is sprayed onto the catalyst batch. Since one can only spray the top surface of the batch, it is necessary to create good turnover of the catalyst on the surface with the catalyst inside the core of the bed. This turnover can be accomplished by tumbling in vessels that allow rotation. Spraying too rapidly results in the formation of clumps of catalyst extrudates that then become over-wetted and stick together, and in addition to this being undesirable, it can also take long periods of time to redistribute the liquid evenly. Overall spraying times of 10–20 minutes are desirable in order to achieve good contact of the liquid with the entire catalyst bed. For vessels with a considerable spraying area, it is preferred to have multiple nozzles in order to obtain a good, even distribution of the active component. In order to cover the entire surface area of the catalyst with the active component, it is necessary to fill the pore volume of the catalyst with the liquid, nearly to its entirety. In order to prevent coating the wall of the vessel with wet catalyst that then no longer mixes with the bulk of the catalyst bed, it is advisable to stay a few percent below the total liquid adsorption capacity of the bed. A catalyst bed impregnated to just below the theoretical maximum then also remains free-flowing, which is another advantage.

In certain applications, the catalyst can be highly mass transfer limited because of its high activity. Avoiding secondary reactions in the core of the catalyst becomes very desirable. In such instances, I will cite Bai [37], who provides a methodology for depositing a thin coating of active metal on the surface of the catalyst by operating with ultrasonic nozzles.

Another method of impregnation is by dipping, where the operator dips the catalyst into excess liquid for a period of time. The catalyst adsorbs the active metal components and is then separated from the waste liquid and dried in baskets. This method guarantees full soaking of the catalyst with the liquid without moving the catalyst around (hence avoiding breakage). However, the method is labor-intensive, and the metal distribution in a batch needs to be checked for uniformity. Using this method with metals that do not adsorb readily onto the catalyst surface may be less desirable because of the loss of metals with the waste liquid.

Drying an impregnated catalyst is an art in many cases. Great progress in the science of drying of commercial 1–2-mm-diameter extrudates has been made by Liu et al. [18–20]. In the drying of monoliths, the size of the catalyst requires careful control of the humidity in the drying ovens. Controlling this driving force for drying is key to obtaining uniform activity profiles.

1.4 Rotary Calcination

1.4.1 Introduction

Drying and calcination of a catalyst in rotary tubular devices is a common practice in commercial catalyst manufacturing. Rotary calciners are essentially metal or ceramic tubes that rotate near horizontally, with only a slight downward slope toward the exit or the discharge of the tube. Calciners operate at single-digit rpm and often have a cam ring on the outside of the tube on the drive end. The cam ring enables a chain connected to a drive mechanism to rotate the tube. A set of support cylinders on both ends supports the horizontal tube. A set of support cylinders turned at a 90° angle keeps the tube from traveling along its axis.

A rotary calciner allows the catalyst to be heat treated for a specific duration at a specific temperature in a particular atmosphere. The heating, drying, and calcination procedures are developed either in a laboratory or at a pilot plant level. The time and temperature profile in this context refers to a period of time during which the catalyst extrudates experience a heating up temperature gradient then a line-out period of time at a certain constant temperature, followed by a cool-down gradient.

There may be a single or sometimes a dual line-out period, and in the case of a dual line-out period, each is at an individual temperature. A dual line-out period, however, if necessary, is essentially limited because of the available length and residence time of a typical rotary calciner. For empty tubes, the residence times at temperature range from 10 to 30 minutes. Using a dam forces the catalyst to spend more time in the heated zone, and this can increase the residence time in the rotary calciner to one to three hours depending on the height of the dam.

A dam is a steel annular flat ring spot-welded in place that forces the catalyst to build up behind it before spilling over. Placing and removing dams are lengthy procedures, so dams are rarely changed once they are in place. Some dams have a split design that allows dual operation, making it possible to operate the calciner either with or without the dam by simply changing the direction of the tube's rotation. The tube's slope is small: from 0.5% to ~2% of its horizontal distance. In other words, for a 1% slope in a 5-m-long horizontal tube, for example, the height difference between the discharge and the inlet is 5 cm, with the exit being at the lower elevation.

The slope of the rotary calciner is the true slope according to the gravitational field. The slope can be measured with a water-filled transparent hose that extends from one end of the rotary tube to the other end. The water level in the hose at both ends is perfectly horizontal and the difference in vertical distance from the water level to the rotary tube at each end then allows one to calculate the slope. Lasers can also be used to measure the slope. In addition, as the tube rotates over prolonged periods of time, the support bearings wear out, which will slightly impact the slope and residence time in the rotary calciner. This change is often not very large since the bearings wear out more or less evenly at the inlet and the discharge and the impact on residence time is small. Nonetheless, when working with calciners either commercially or at pilot plants, the correct slope is required in order to correctly apply well-known literature correlations that allow the calculation of the residence time in the rotary calciner. Hence, when possible, confirming the slope by measuring is preferred. In addition, it is essential to consider – and account for – the potential slope of the plant or laboratory floor.

When rotating even at high temperatures, the tubes remain straight and there are no issues. If they are not kept rotating, tubes that run at high temperatures will sag and fail. In the event of a power failure during a hot tube operation, backup power must be available or else the tube will need frequent manual rotation to avoid sagging. High temperatures in rotary calciners can vary. Ceramic tubes can withstand temperatures up to ~1200 °C, which is typically well above the ordinary operational temperatures used in catalyst manufacture.

Tubes in the commercial catalyst industry are ~30–150 cm in diameter, but those in the cement industry are substantially larger at 3–5 m in diameter, with lengths as long as several hundred meters. I will not discuss these larger cement kilns further, but the correlations presented later still apply.

Rotary calciners can handle many sizes of particulates, from micron-sized powders to granules of several millimeters in size for a variety of shaped extrudates. Granules are fed into the rotary through a sloped chute, a horizontal auger in a tube, vibratory feeders, or some other mechanism. It is important to keep the feed end clog-free, and so the device that is actually used to feed the rotary is typically based on the experience of the operating crew or engineer.

It is important to have a consistent and reliable feeder mechanism because when feed is lost in the tube, the rotary calciner quickly runs empty and requires a restart of the line-out. Product that comes out during this period may be compromised. Catalyst that is fed into the rotary calciner runs along the bottom of the tube until it reaches the exit. The cross-sectional area of the catalyst bed in a rotary calciner is typically ~5–15% of the calciner cross-sectional area. A good visual reference is that a bed that extends from the top of the bed to the bottom of the bed over a quadrant of the circular cross section of the tube covers 9% of the cross-sectional area.

Above the catalyst bed in the rotary, gas is often injected in a countercurrent mode. This gives rise to two distinct atmospheres in the rotary calciner, and they are free to mix along the length of the rotary. The first atmosphere is the gas injected into the rotary calciner above the bed of catalyst. This gas can range in composition from ambient air to dry air, to an oxygen-free gas, nitrogen, steam, or any combination of the above.

The second atmosphere constitutes the gas emitted by the catalyst during the heat treatment of the catalyst. The composition of gas in a rotary calciner varies widely and may contain any of the chemistries that make up the catalyst, such are volatile organic compounds (VOCs), water, sulfur oxides, chlorides, and nitrogen oxides.

The top of the bed is in contact with the bulk of the gas, and gas exchange with the bulk gas is obvious. This time of contact is limited, however, before the catalyst avalanches down from the top of the rotating bed and becomes stationary again below the surface, then travels upward with the tube and the rest of the bed. During the time that the catalyst spends in the bed below the surface, it is exposed to any of the gases, such as steam released during drying or calcination, accumulating inside of the bed. Excessive steam contact may result in poor catalyst quality. The rate of exchange of gas between the bulk gas phase and the catalyst bed is still being researched.

The heat treatment of the catalyst and the mixing of the two gas atmospheres alter the gas atmosphere along the tube. For instance, the burn-off of catalyst carbon components typically results in a high concentration of oxygen at the inlet of the injected gas stream (catalyst discharge end), while the discharge end of the gas stream (inlet or catalyst charge end) may be totally oxygen-free. It is important to calculate the heat load during such a burn-off up front, since it may be necessary to limit the catalyst feed rate in order to control the exotherm in the rotary.

From a safety standpoint, when heating a catalyst with a certain organic load in an oxygen-free atmosphere, there is a high likelihood of ending up with fines, dust, or particulates that are captured in a baghouse. The fines are laden with organics that potentially have been exposed to high temperatures in the absence of oxygen and hence are susceptible to oxygen adsorption when exposed to air during discharge from the baghouse. This potential exothermic chemical reaction can be a safety hazard. The organics may not necessarily be the originating organics in the catalyst per se, but rather are formed by secondary reactions while exposed to high temperatures in the absence of oxygen in the rotary calciner.

The gas and catalyst may be fed from the same end of the rotary calciner (cocurrent feeding) or from the opposite end (countercurrent feeding).

Countercurrent feeding enables control of the atmosphere in the hot zone of the rotary calciner while keeping the cooling and discharge ends of the catalyst well controlled. Any gas can be injected close to the hot zone of the rotary calciner by lancing. It is possible to keep the discharge end dry (by using dry air injection) or oxygen-free (by using nitrogen injection).

In cocurrent feeding, any of the gases released from the catalyst travel along with the catalyst toward the exit and can re-adsorb onto the cooled catalyst. This scenario makes cocurrent feeding not a very desirable option because it makes it difficult to control catalyst specifications regarding composition.

When steam is part of the atmosphere, controlling the rotary calciner operation can be particularly difficult due to the propensity of steam to condense in cold spots. Pure steam at 100 vol.% in the atmosphere of a rotary calciner is nearly impossible to manage because it condenses so readily against the colder surfaces of the inlet and outlet heads. This condensation essentially draws a vacuum and pulls even more steam toward the surface, resulting in the collection of even more condensate. The catalyst, dust, and fines will readily adsorb the condensate and get stuck on the surfaces. The slow buildup of material can clog the feed and discharge ends of the rotary and, after some time, this process ceases the operation.

In order to minimize the possibility of condensation, I recommend keeping the feed and discharge ends at or below a 60 vol.% steam level by injecting air or nitrogen.

A rudimentary setup for measuring the steam content in an operating rotary calciner is to extend a sampling lance all the way into the hot zone and connect it in series to an ice-filled flask, then to a wet test meter, and finally to a vacuum pump. Turning on the vacuum pump and sampling over a time interval of several minutes allows the steam to condense in the ice-filled flask and be weighed as an amount W_s, while the wet test meter measures the volume V_{nc} of non-condensable gas. With these two measurements, Eq. (1.1) then determines the approximate volumetric percentage steam content of the gas atmosphere in the rotary calciner:

$$C_s \cong \frac{100}{1 + 0.96 \times (\mathrm{V}_{nc}/22.4)/(\mathrm{W}_s/18)} \tag{1.1}$$

Equation (1.1) contains a small yearly average temperature correction (0.96).

As in any operation, always be aware of the necessary safety concerns during the measuring procedures. The sampling lance and the sampling line will be hot and, therefore, one must use the proper personal protective equipment. When collecting these measurements, follow any required safety procedures. For instance, one "what if" scenario that comes to mind is: when leaving the lance only dunked into a bucket of water, it is conceivable that steam after some time might make its way through the sampling lines and finally start to condense in the water. The steam will keep condensing in the water bucket and will draw more steam from the rotary calciner through the sample tubing without any need for a vacuum pump because it creates its own vacuum. This is perhaps unlikely, but it could happen.

On a different subject, the heating of a rotary calciner can be by gas firing – direct or indirect – or by electrical heating.

Indirect heating by gas firing occurs in a shell around the rotary calciner; several burners operate in parallel from the sides or underbelly of the calciner. Segregating these burners and their off-gases in sections enables better control of the axial temperature profile of the rotary calciner. In some cases, it is possible to recirculate the off-gases back to the rotary shell to enhance heat transfer and operational efficiency. Direct gas firing allows a flame inside of the tube, thus allowing contact between the catalyst and the combustion products of the flame, which is not usually a desirable event. It is also best to avoid exposing the top layer of the catalyst bed to the heat radiation of a flame. For these reasons, direct firing is not often practiced on catalysts, particularly in the later stages of preparation, such as when a catalyst has been impregnated with active metals.

Electrical heating is a good option – likely the best option among the three – from the standpoint of ease of control. With electrical heating, the temperature of the rotary calciner is directly derived from the shell temperatures. Thermocouples located on the top of the rotating tube inherently leave a small gap between the tube and the rotary. Because of their particular location – as close as possible to the shell but not close enough to cause wear – these thermocouples do not give an exact temperature measurement, but it is close to exact, and they do have a long and durable lifetime. The operating crew and engineers have very good knowledge of the typical temperature difference between the shell and the bed based on years of experience.

Besides shell thermocouples, some rotary calciners are equipped with internal thermocouples that measure the temperature directly in the bed. The bed temperature is the temperature that the catalyst experiences during lined-out heating, also known as the temperature in the lined-out hot zone. A target for this temperature is determined during pilot plant or laboratory experiments. It is this temperature that, when established for the appropriate amount of time in the bed of a large rotary calciner, will deliver the best possible catalyst.

Some rotary calciners have no internal thermocouples in place, necessitating the insertion of thermocouples. Temperature profiles measured with such types of thermocouples must occur at regular intervals, at least in the beginning, in order to line-out the unit at the correct bed temperature.

Some rotary calciners have internal thermocouples in place, but they are sparsely located in the bulk gas phase above the bed and in the bed itself. Depending on the number of thermocouples, it may be necessary to measure the axial temperature profile with a thermocouple inserted from the outside in order to determine the presence and effect of temperature gradients.

Some rotary calciners are equipped with a thermo-well that includes several bed and bulk gas-phase thermocouples. In those cases, combining the bed and gas readings with the shell readings will offer a very good understanding of the temperature and thermal history of a catalyst during its preparation. Because the walls of industrial thermocouples can be thick, make sure to insert thermocouples deep enough in the bed. Thick-walled thermocouples can dissipate heat away from the thermocouple tip (along its thick wall) and hence affect the temperature reading. If the bed is shallow and the thermocouple cannot be inserted deep, then try to use a thermocouple bent at an obtuse angle so that it can be inserted just below the bed surface without penetrating deep and possibly touching the rotating tube of the calciner.

In order to avoid loss in heat transfer from the shell to the bed, caking of catalyst fines and dust onto the tube wall must be avoided. For this, one sometimes resorts to the use of a hammer that impacts the shell externally after each rotation, which tends to clear the tube wall. Chains are also used for the same purpose, but they may result in attrition of the formed catalyst, and so they are more often used for powders.

The rotating tube mostly fits close, with little gap between either the stationary discharge head or the stationary feeder head. Downstream equipment creates a low-level sub-ambient pressure in both the discharge and feeder heads, and this creates a draft of ambient air into the rotary heads, thus avoiding causing any leakage of process gases onto the floor. The downstream equipment typically services all of the rotaries in a plant for off-gas abatement; hence, the draft setting for one rotary calciner can influence the draft setting for another rotary calciner.

There are a multitude of options concerning the catalyst discharge end of the rotary calciner depending on the plant design and customer requirements. For example, options include a drum open to ambient air, a drum with a dry air blanket to keep the catalyst moisture-free, a drum with a nitrogen blanket to keep the catalyst oxygen- and moisture-free, or, for intermediate stages, a bucket elevator to bring the catalyst to a different location in the plant or a sieving operation to eliminate fines.

The back plate of the rotary calciner refers to the plate at the catalyst feed end that keeps the catalyst from spilling over into the bottom of the feed end's head. The back plate has a circular opening through which the feed chute or feed screw is passed to deliver the catalyst into the rotary. There is a small gap between the back plate and the feeder chute through which all of the off-gas can exit the rotary calciner and be drafted away to the scrubber or thermal oxidizer. When feeding catalyst on top of an existing rotating bed, the catalyst will travel both forward and backward – with perhaps a small inclination forward according to the slope. The back plate allows the bed to build up at the feed end without spilling over, creating a bed depth gradient that finally forces all of the catalyst to move forward. A feed rate set so high that the bed depth must be higher than the bottom part of the back-plate hole in order to make the catalyst move forward will create wasteful spillover.

Figure 1.2 shows a snapshot of the dynamic behavior of the solids. The top of the bed of solids starts at the dark blue line. Upon a small clockwise rotation of the tube, the bed exceeds the dynamic angle of repose and starts to cascade down in the direction of the arrows shown above the bed. Once the solids travel down, they cover the previously exposed layer. This layer now becomes fixed, and particles no longer move relative to one another. The catalyst below the top layer now travels with the tube along the thin blue lines as indicated. Overall, the catalyst runs down the rotary calciner along flattened semi-spiral lines.

Some catalysts will run down the rotary in the pattern shown in Figure 1.2. However, depending on the operating conditions, shape, and size of the catalyst, a "snaking" phenomenon can occur as well. With a snaking phenomenon, the catalyst does not turn over or avalanche down, but rather is carried up the wall as the tube turns. At some point below the angle of repose, the catalyst slides back down to the bottom of the tube and the process begins again. Because the catalyst did not

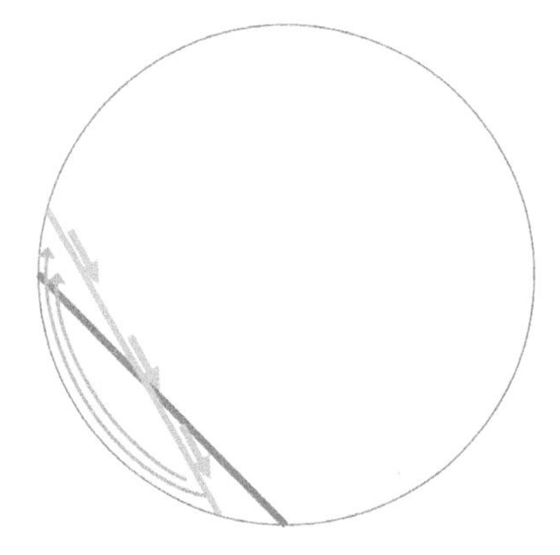

Figure 1.2 Dynamic behavior of solids in a rotary calciner.

turn over, the mixing in a cross section is poor, and the catalyst at the bottom likely remains at the bottom for longer than expected and hence heats up quicker. The top layer of the bed remains at a lower temperature until conduction and some poor mixing can finally raise the temperature. Direct observation of the way a bed of catalyst progresses through the tube can be achieved when the unit has viewing ports, which are sometimes present in commercial operations.

The proper abatement of any process off-gases is a requirement in all plants. VOCs in rotary calciner off-gas can be removed by passing the gas through a thermal oxidizer that oxidizes any organics to water, carbon dioxide, nitrogen oxides, etc. The thermal oxidizer is often direct gas-fired because direct contact with the flame is desired. After passing through the thermal oxidizer, the temperature of the off-gas is quenched by water injection.

Many acidic components in off-gas, such as nitrogen oxides and sulfur oxides, can be properly removed by passing the gas through a scrubber operating at the proper pH.

1.4.2 Residence Time in a Rotary Calciner without a Dam

Figure 1.3 shows, in a very simplified way, the material flow in a rotary calciner without a dam. The terms "rotary calciner" and "rotary kiln" are used interchangeably and both are used in the industry. The rotary kiln has a small slope downward and the shallow moving bed only occupies 5–15% of the cross section of the rotary kiln during typical operation.

Sullivan et al. [38] and Saeman [39] established a correlation shown in Eq. (1.2) for a rotary without a dam, with or without internal baffles/chains:

$$R_t = \frac{1.77\sqrt{\theta}\,L_r f}{\varphi D N} \tag{1.2}$$

where L_r is the length of the rotary measured from the feed point to the discharge point, f is a dimensionless flow factor equal to unity (1) for an empty tube, φ is the downward angle of the tube in degrees, θ is the dynamic angle of repose of the material in the rotary, D is the diameter of the tube, and N is the rate of rotation of the tube. Baffles or chains, here called internals, obstruct flow, making the flow factor larger than unity, hence increasing the residence time. The fractional slope of the tube s is here defined as the vertical height difference per linear axial

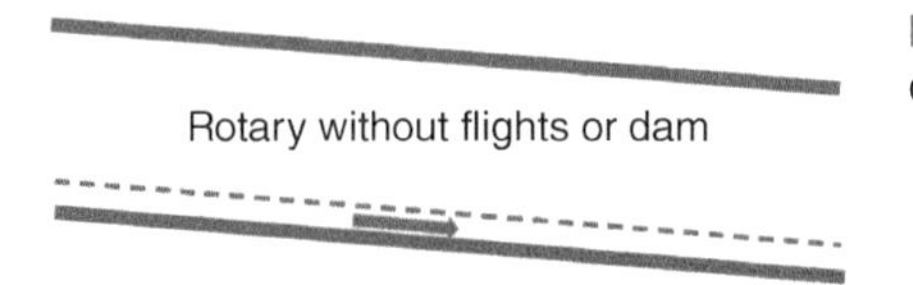

Figure 1.3 Flow of material in a rotary calciner without a dam.

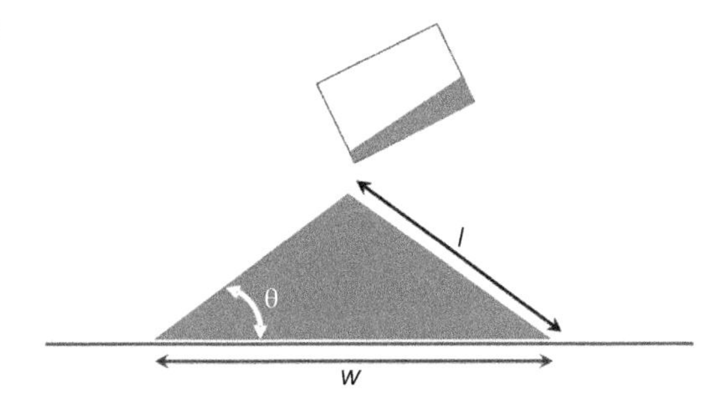

Figure 1.4 Measurement of the angle of repose.

distance. For small downward angles of the tube, the angle φ in degrees can be calculated from the fractional slope s using Eq. (1.3):

$$\varphi \simeq \left(\frac{360}{2\pi}\right) s \tag{1.3}$$

A typical fractional slope for a rotary calciner is 0.0104 (or ~1% or 1/8 in./ft or 0.6°).

An estimate of the amount of material prior to the feed point can be obtained from the distance of the back plate to the feed point assuming (i) the bed is uniform once lined-out and (ii) the feed point is not far from the back plate.

The angle of repose can be determined using the well-known method shown in Figure 1.4. Pour enough material gently and allow it to build up to an amount that allows for measurement of the diameter w of the footprint of the hill and the length l of the uphill with a reasonable degree of accuracy. Equation (1.4) then allows one to obtain the angle of repose θ in degrees:

$$\theta = \left(\frac{360}{2\pi}\right) \text{arcos}\left(\frac{w}{2l}\right) \tag{1.4}$$

It is understood that the function value of the "arcos" function in Eq. (1.4) is expressed in radians. This measurement and calculation yields the stationary angle of repose, which is the angle of repose when the material is at rest. The dynamic angle is slightly larger, but the stationary angle can serve as a good approximation.

The correlation shown in Eq. (1.2) calculates the average residence time that the catalyst spends traversing a section of length L_r and has been proven valid over the years. The residence time can be calculated with the same correlation for any axial distance of interest along the length of the rotary calciner. Gao et al. [23] was able to confirm these findings from his efforts on a pilot-scale rotary calciner, and also that the effect of the feed rate on the mean residence time is minimal and that the bed depth increases with a higher feed rate.

Measuring the residence time in a rotary calciner without a dam but with internals is straightforward, but can take several hours. Here are the steps:

1) Allow the rotary to line-out for a given feed rate.
2) When the rate of discharge stabilizes to a constant value, shut off the feed to the rotary calciner and begin collecting all of the discharge material into drums or appropriate containers from that time on.
3) When the entire rotary calciner is empty, weigh the net discharged inventory.
4) Correct the inventory for the material prior to the feed point.
5) Dividing the corrected inventory by the lined-out feed rate yields the residence time R_t of the rotary calciner.

Note that the entire tube also encompasses the cooling section, and the heated length is really only a fraction of the entire tube length. The residence time in the lined-out hot zone is of greatest interest, however, and this length is only a fraction of the heated length of the rotary calciner. The lined-out hot zone is here considered to be the length of the rotary where the bed can be considered at a uniform temperature. It is only possible to measure the temperature in the lined-out hot zone by using thermocouples and measuring the temperature profile along the axis of the rotary calciner. Equation (1.5) calculates the residence time R_{hz} in the lined-out hot zone:

$$R_{hz} = R_t(L_{hz}/L_r) \tag{1.5}$$

where L_{hz} is the length of the lined-out hot zone.

Other properties of interest for beds in rotary calciners are the cross-sectional area and the maximum bed depth. The cross section of the bed is here approximated by a segment of a circle with the same radius as the radius of the tube. Define the height h_s of the segment as the largest perpendicular distance from the chord to the circle. The area A_s of a segment of height h_s in a circle of radius R can then be approximated within 4% by Eq. (1.6):

$$A_s \cong 1.8\sqrt{Rh_s^3} \quad \text{for } 0 < h_s/R < 0.5 \tag{1.6}$$

With the aid of Eq. (1.6), the bed depth h_e for a calciner without a dam is then obtained from Eq. (1.7):

$$h_e = 0.85 \times (\mathrm{Q}R_t/L_r)^{2/3}/\mathrm{D}^{1/3} \tag{1.7}$$

where Q is the lined-out volumetric feed rate.

1.4.3 Residence Time in a Rotary Calciner with a Dam

Figure 1.5 shows, in a simplified way, the flow of material and mixing in a rotary calciner with a dam. The dams considered herein are shallow, and their heights amount to less than 20% of the diameter of the tube.

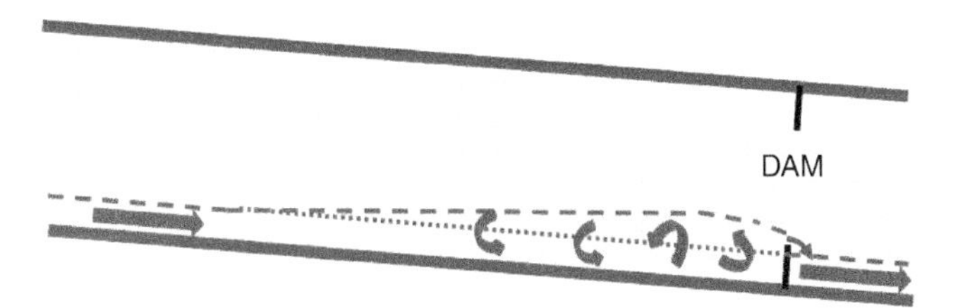

Figure 1.5 Flow of material in a rotary calciner with a dam.

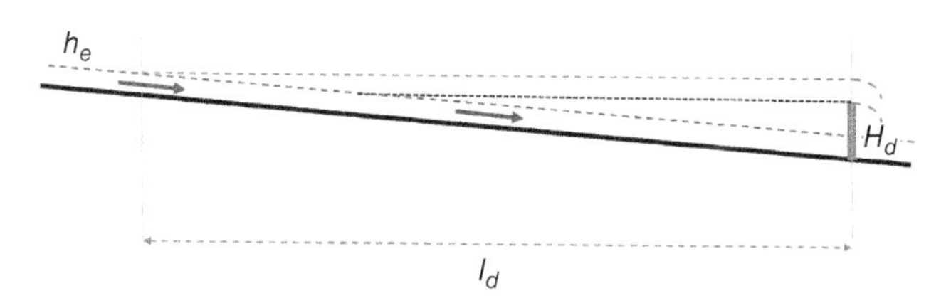

Figure 1.6 Geometry used to evaluate the residence time in a rotary calciner with a dam.

When a shallow dam is present, the material flow well ahead of the dam is the same as if the rotary calciner had no dam. Figure 1.6 shows the geometry used to calculate the residence time with a dam. As the material flows closer to the dam and reaches a critical distance l_d from the dam, it starts to build up and forms the extra inventory that the addition of the dam is designed to hold. Equation (1.8) calculates the critical distance l_d for a rotary with a dam of height H_d:

$$l_d = (H_d + h_e)/s \tag{1.8}$$

The axial distance l_d can be quite substantial at typical slopes for even very modest dam heights. For instance, for a rotary kiln with a 5-cm dam height and a typical fractional slope of 1/100th (i.e. 1%), the length over which the bed builds up is 5 m just to offset the dam itself. You still need to add the length required to offset the height of the bed for the rotary without a dam.

Once lined-out, the material at the dam has built up enough to clear the dam continuously, and the volumetric rate is the same as it would be without a dam. In fact, for the shallow dams considered here, it is assumed that the bed depth at the dam equals the height of the dam plus the bed depth for the calciner without a dam. As shown in Figure 1.5, it is the second half of the bed being the deepest and closest to the dam that is most important for obtaining the true residence time at temperature, as it is this section that contains the bulk of the catalyst.

The material in the inventory before the dam is more or less mixed and could, in the first approximation, be considered to be a continuous or complete stirred tank reactor (CSTR). Because of the small incline of the rotary, however, the material prior to the dam is likely acting more like two or three CSTRs in series because the effect of the dam extends quite far. Just after the dam, the catalyst flow quickly becomes the same as it would be in a rotary calciner without a dam; in fact, the dam functions much like a shallow back plate. The location of the dam should be at the end of the heated zone in order to maximize the residence time at temperature.

Equation (1.9) obtains the approximate volume of the slab of catalyst over the distance l_d before the dam by direct integration using Eqs. (1.6) and (1.8):

$$V = 0.51 \times \frac{D^3}{s}\left\{\left(\frac{h_e + H_d}{D}\right)^{2.5} - \left(\frac{h_e}{D}\right)^{2.5}\right\} \tag{1.9}$$

where h_e is the bed depth without a dam given by Eq. (1.7). As already mentioned, the length l_d is appreciable for small slopes and is likely to be greater than what is considered to be the length of the lined-out hot zone L_{hz}. In this latter case, calculate the volume of the catalyst prior to the dam from Eq. (1.10):

$$V = 0.51 \times \frac{D^3}{s}\left\{\left(\frac{h_e + H_d}{D}\right)^{2.5} - \left(\frac{h_e + H_d - sL_{hz}}{D}\right)^{2.5}\right\} \tag{1.10}$$

The length of the lined-out hot zone has to be determined from thermocouple readings, and it remains a judgment call as to what temperature variations are acceptable.

The residence time R_d in the lined-out hot zone for a rotary with a dam is then calculated from Eq. (1.11) while using either Eq. (1.9) or (1.10) for the volume of the catalyst:

$$R_d = V/Q \tag{1.11}$$

Table 1.1 shows a worked example for a rotary calciner with and without a dam. The length of the lined-out hot zone is assumed to be obtained from on-site thermocouple readings. It can be seen that the critical distance over which the bed builds up is very substantial when the dam is present. In addition, the deeper bed extends far beyond the lined-out hot zone. Hence, Eq. (1.10) must be used to calculate the catalyst volume exposed to the lined-out temperature in order to compensate for this. The increase in residence time from an empty rotary to a calciner with a modest dam is substantial (about threefold here for the conditions shown).

Be aware that rotaries with a dam have a much wider residence time distribution than rotaries without a dam. The latter will be discussed in Section 1.4.4.

Table 1.1 Worked example for the residence time in a rotary calciner with or without a dam. Units in bold highlight the difference in residence time between an empty rotary and a rotary with a dam.

		Equation number
Rotary diameter	0.91 m (3 ft)	
Slope	0.0104 (1/8 in./ft)	
Flow factor	1.00	
Dam height	0.0762 m (3 in.)	
Mass rate	136 kg/h (300 lb./h)	
Density	961 kg/m^3 (60 lb./ft^3)	
Angle of repose	35°	
Rotation rate	3 rpm	
L_{hz}	2.44 m (8 ft)	Via thermocouple measurement
R_t	**15.6 min**	1.2
h_e	0.053 m (2.1 in.)	1.7
$(H_d + h_e)/s$	12.5 m (41 ft)	1.8
V	0.12 m^3 (4.2 ft^3)	1.10
R_d	**50.5 min**	1.11

As an aside, when the slope is steep, there is another method that can be used, but one must make sure that the constraints are satisfied (see the italicized statement below). Allow the feed rate to halt and let the tube run out entirely. What is left is the material prior to the dam that is not able to make it over the dam. Call the volume of this remainder material V_v. With the help of Eq. (1.9) (set h_e equal to zero), the remainder volume is calculated from Eq. (1.12):

$$V_v = \frac{0.51D^3}{s}\left(\frac{H_d}{D}\right)^{2.5} \tag{1.12}$$

Equation 1.12 also provides an easy means of confirming the slope of the rotary calciner when the volume of the material is measured by vacuuming it out.

If the vacuumed material can be considered inside of the lined-out hot zone, then an approximation of the residence time can be calculated from Eq. (1.13):

$$R_d = R_t + V_v/Q \tag{1.13}$$

where R_t stands for the residence time for the empty tube and R_d stands for the residence time with a dam.

1.4.4 Residence Time Distribution in a Rotary Calciner

It is important to have a good idea of the mean residence time of the catalyst material as it travels through the lined-out hot zone of a rotary calciner. However, it is also important to know the residence time distribution. A wide residence time distribution means that some material remains in the hot zone for much less time than the mean residence time, and thus may be inadequately treated. Some material also stays in the rotary calciner for a much longer time than the mean residence time. Thus, undesired secondary reactions in the forming of the catalyst may render the catalyst less active or selective.

It is possible to experimentally measure residence time distributions with a tracer injection. Commercial rotary calciners require the preparation of 100% contaminant-free catalyst, so conducting this experimental measurement requires a sacrificial catalyst or, more often, a sacrificial support material. Capturing a good tracer signal when running at commercial rates also requires a substantial amount of tracer.

Opportunities to conduct the required testing in commercial units are rare. Pilot plant equipment affords more opportunities for tracer studies due to its smaller scale. Danckwerts [40] developed a pulse test method that measures the residence time distribution of a rotary calciner. Kohav et al. [41] performed stochastic particle trajectory simulations in a rotary calciner in order to discern factors that influence the axial dispersion. Gao et al. [23] applied the same method as Danckwerts in order to measure the residence time distribution function and followed the experimental technique practiced by Sudah et al. [42].

A catalyst batch is prepared in order to establish lined-out conditions in a pilot-scale rotary calciner. Equation (1.14) obtains experimentally the mean residence time R_t of the catalyst in an empty rotary calciner:

$$R_t = \frac{M}{F} \tag{1.14}$$

where M is the mass inventory in the rotary measured from the place where the catalyst falls from the feeding device into the rotary, while F is the lined-out rate of mass of catalyst exiting the rotary. Equation (1.15) calculates the axial velocity of the catalyst as:

$$v = \frac{L_r}{R_t} \tag{1.15}$$

where L_r is the distance from the feed point to the outlet of the rotary calciner.

The mass M must be corrected for the short section in the rotary between the back plate and the feed point.

Using a small tracer sample of the same material but impregnated and dried with a colorant provides a clear distinction between the catalyst and the tracer material. Once the non-colored catalyst flow is established, a small tracer sample is dropped on top of the moving bed toward the middle of the rotary. The location of the drop-off point represents a compromise between establishing a long sample path in the rotary calciner and the ability to deliver the sample there physically. Samples obtained at the exit of the rotary calciner are inspected for their colorant concentration.

The distance from the exit to the sample drop-off point is referred to as L_S. Equation (1.16) calculates the mean residence time R_s of the tracer sample:

$$R_S = R_t \frac{L_S}{L_r} \tag{1.16}$$

The samples are representative of the entire material outflow at each time a sample is taken. Equation (1.17) then calculates the residence time distribution function $E(L_S, t)$ at the exit of the rotary:

$$E(L_S, t) = C(t) / \int_0^\infty C(t)dt \tag{1.17}$$

where $C(t)$ is the concentration of the colorant sample at the exit of the rotary calciner at time t measured from when the sample is dropped. In addition, per Eq. (1.18):

$$\int_0^\infty E(L_S, t)dt = 1 \tag{1.18}$$

The residence time distribution function yields the normalized amount of sample $E(L_S, t)dt$ in the rotary calciner from the time the sample is dropped to the time interval $(t, t + dt)$ over which the sample particles exit the rotary.

At the calciner exit, Eq. (1.19) is obtained from the Taylor dispersion model (Levenspiel [43], Liu [44]):

$$E(\mathrm{L}_S, t) = \frac{v}{\sqrt{4\pi t D_a}} e^{-(L_s - vt)^2/4tD_a} = C(t))/\int_0^\infty C(t)dt \tag{1.19}$$

where D_a is the axial dispersion coefficient and is obtained via parameter estimation during the curve fitting of the experimental data using Eq. (1.19). The parameter estimation methodology of Eq. (1.19) obtains information on the amount of time a catalyst particle stays in the rotary calciner and also on whether this is a wide or a narrow distribution.

The residence time distribution function is calculated from the sample drop-off point to the exit and it is not the residence time distribution function of interest through the hot zone. With the measured distribution function at hand in

Table 1.2 Catalyst residence time distribution in the hot zone.

Paredes (2017) : v = 9cm/ min , L_{hz} = 50cm, D_a = 10cm²/min	
t/R_{hz}	$100\int_0^t E(L_{hz},u)du$ (the percentage of material no longer in the hot zone at a reduced time t/R_{hz})
0.8	5.8
0.9	21.7
1.0	47.1
1.1	71.4
1.2	87.6

Eq. (1.19), however, it is easy to calculate and obtain the residence time distribution through the hot zone. Equation (1.20) calculates the residence time distribution at the exit of the hot zone from the point where it enters the hot zone:

$$E(L_{hz}, t) = \frac{v}{\sqrt{4\pi t D_a}} e^{-(L_{hz} - vt)^2/4tD_a} \tag{1.20}$$

It is now easy to determine how much the actual residence time differs from the mean residence time. Table 1.2 gives an example for the calciner used in the work of Paredes et al. [24], together with the parameters used. In Table 1.2, it is shown that 5.8% of the catalyst exits early on from the hot zone of the rotary calciner with a residence time 20% less than the mean residence time, while there is a tail of 12.4% (87.6–100%) that remains 20% longer than the mean residence time. Only 49.7% (21.7–71.4%) or approximately half of the catalyst experienced a residence time in the rotary calciner within 10% of the mean residence time in the hot zone! Thus, there can be a substantial spread of heat treatment times experienced by the catalyst in a rotary calciner.

Gao et al. [23] found that the flow-ability of the catalyst was better for spherical shapes than for cylinders and quadrulobes. The better flow-ability of spheres over extrudates leads to a lower bed build-up and hence a lower residence time for spheres than for extrudates, as expected. In addition, for the axial dispersion coefficient, Gao et al. [23] found that:

1) The axial dispersion coefficient decreases when lowering both the rotation rate and the incline of the rotary calciner.
2) A high feed rate and a large angle of repose of the materials also reduce axial dispersion.

1.5 From the Laboratory to a Commercial Plant

1.5.1 Scale-up Technology

Properly scaling recipes and transferring technology to manufacturing plants requires the right equipment. The type of equipment evolves over years of trial and error and may require revisiting if a new plant is brought into the circuit.

Because mixing and mulling are very important parts of forming catalyst particles, correctly scaling up to commercial equipment is very important. The mill wheels weigh several hundred pounds and induce a lot of shear in the mixture. In small laboratory units, this same shear is not available and may need to be intensified through unit modification when performing experiments that aim to mimic a commercial unit. Studded pillar tubs have a number of smaller laboratory units available and can often scale up readily without unit modification.

1.5.2 Scale-down Technology

As much as scaling up is challenging, so is scaling down. Currently, companies that employ high-throughput experimentation often need equipment that allows the production of catalyst quantities on the scale of only 1–10 g, starting with mixing and mulling and moving on to extrusion.

I have a patent [45] on such a "machientje" that enables 10–20 catalyst compositions to be mixed and mulled simultaneously and readied for ram extrusion. The invention keeps the mixture totally enclosed, preventing a loss of moisture and enabling mixing and mulling in a similar way to the "studded pillar tub" technology. The properties and mechanical strengths of such extruded catalysts are on a par with pilot plant and commercial catalysts.

Nomenclature

A_s Area of a segment of a circle (m^2)
C_s Steam concentration in the rotary (vol.%)
C Concentration of tracer (arbitrary units)
D Diameter of a rotary calciner (m)
D_a Axial dispersion coefficient (m^2/s)
E Residence time distribution function (1/s)
f Flow factor for a rotary calciner ($f \geq 1$; *empty* = 1)
F Lined-out product rate (kg/s)
h_s Height of a segment of a circle (m)
h_e Maximum bed depth in an empty rotary (m)
H_d Height of the dam in a rotary calciner (m)

l	Length of the uphill of a poured sample (m)
l_d	Length of the material buildup for a rotary calciner with a dam (m)
L_r	Length of the rotary (m)
L_{hz}	Length of the lined-out hot zone of a rotary (m)
L_s	Distance of the tracer sample drop-off point from the exit (m)
M	Inventory of catalyst in an empty rotary (kg)
N	Rate of rotation (revolutions/s)
Q	Volumetric feed rate (m^3/s)
R	Radius of a circle (m)
R_d	Residence time in a rotary with a dam (s)
R_t	Residence time in a rotary without a dam (s)
R_s	Residence time of the tracer sample (s)
R_{hz}	Residence time in the lined-out hot zone of a rotary (s)
s	Fractional slope of a rotary (–)
t	Time (s)
v	Axial velocity of the catalyst (m/s)
w	Diameter of the footprint of the poured sample (m)
V	Volume of catalyst in a rotary calciner prior to the dam (m^3)
V_{nc}	Volume of non-condensable gas (m^3)
V_v	Inventory of catalyst before the dam that requires vacuuming (m^3)
W_s	Condensate sample weight (kg)
φ	Slope of the rotary (°)
θ	Dynamic angle of repose (°)

References

1 Stiles, A.B. (1983). *Catalyst Manufacture, Laboratory and Commercial Preparations*. Marcel Dekker.

2 Le Page, J.F. (1987). *Applied Heterogeneous Catalysis*. Paris, France: Institut Français du Pétrole Publications, Éditions Technip.

3 Neimark, A.V., Kheifets, I.J., and Fenelonov, V.B. (1981). Theory of preparation of supported catalysts. *Industrial & Engineering Chemistry Product Research and Development* 20: 439.

4 Chester, A.W. and Derouane, E.G. (2010). *Zeolite Characterization and Catalysis*. SpringerLink.

5 Marceau, E., Carrier, X., and Michelle, C. (2009). *Synthesis of Solid Catalysts*. Wiley-VCH Verlag GmbH & Co. KGaA. ISBN: 978-3-527-32040-0.

6 Lekhal, A., Glasser, B.J., and Khinast, J.G. (2004). Influence of pH and ionic strength on the metal profile of impregnation catalysts. *Chemical Engineering Science* 59: 1063–1077.
7 Lekhal, A., Glasser, B.J., and Khinast, J.G. (2007). Drying of supported catalysts. In: *Catalyst Preparation* (ed. J. Regalbuto), 375–404. CRC Press.
8 de Jong, K. (2009). *Synthesis of Solid Catalysts*. Wiley-VCH Verlag GmbH & Co. KGaA. ISBN: 978-3-527-32040-0.
9 Chester, A.W., Kowalski, J.A., Coles, M.E. et al. (1999). Mixing dynamics in catalyst impregnation in double-cone blenders. *Powder Technology* 102: 85–94.
10 Chester, A.W. and Muzzio, F.J. (2004). Development of catalyst impregnation processes. *American Chemical Society, Division of Petroleum Chemistry, Preprints* 49: 18–20.
11 Liu, X., Khinast, J.G., and Glasser, B.J. (2008). A parametric investigation of impregnation and drying of supported catalysts. *Chemical Engineering Science* 63: 4517–4530.
12 Shen, Y., Borghard, W.G., and Tomassone, M.S. (2017). Discrete element method simulations and experiments of dry catalyst impregnation for spherical and cylindrical particles in a double cone blender. *Powder Technology* 318: 23–32.
13 Romanski, F.S., Dubey, A., Chester, A.W. et al. (2010). Optimization of dry catalyst impregnation in a double cone blender: an experimental and computational approach. 244th National Fall Meeting of the American-Chemical-Society (ACS), Philadelphia.
14 Romanski, F.S., Dubey, A., Chester, A.W., and Tomassone, M.S. (2012a). Dry catalyst impregnation in a double cone blender: a computational and experimental analysis. *Powder Technology* 221: 57–69.
15 Romanski, F.S., Dubey, A., Shen, Y. et al. (2012b). Improved mixing in catalyst impregnation using a double cone incorporated with baffles. 244th National Fall Meeting of the American-Chemical-Society (ACS), Philadelphia.
16 Koynov, S., Wang, Y., Redere, A. et al. (2016). Measurement of the axial dispersion coefficient of powders in a rotating cylinder: dependence on bulk flow properties. *Powder Technology* 292: 298–306.
17 Kresge, C.T., Chester, A.W., and Oleck, S.M. (1992). Control of metal radial profiles in alumina supports by carbon dioxide. In: *Applied Catalysis A: General*, vol. 81, 215–226. Amsterdam, Netherlands: Elsevier Science Publishers B.V.
18 Liu, X., Khinast, J.G., and Glasser, B.J. (2010). Drying of supported catalysts: a comparison of model predictions and experimental measurements of metal profiles. *Industrial & Engineering Chemistry Research* 49: 2649–2657.
19 Liu, X., Khinast, J.G., and Glasser, B.J. (2012). Drying of supported catalysts for low melting point precursors: impact of metal loading and drying methods on the metal distribution. *Chemical Engineering Science* 79: 187–199.

20 Liu, X., Khinast, J.G., and Glasser, B.J. (2014). Drying of Ni/alumina catalysts: control of the metal distribution using surfactants and the melt infiltration method. *Industrial & Engineering Chemistry Research* 53: 5792–5800.

21 Chaudhuri, B., Muzzio, F.J., and Tomassone, M.S. (2006). Modeling of heat transfer in granular flow in rotary vessels. *Chemical Engineering Science* 61: 6348–6360.

22 Chaudhuri, B., Muzzio, F.J., and Tomassone, M.S. (2010). Experimentally validated computations of heat transfer in granular materials in rotary calciners. *Powder Technology* 198: 6–15.

23 Gao, Y.J., Glasser, B.J., Lerapetritou, M.G. et al. (2013). Measurement of residence time distribution in a rotary calciner. *AIChE Journal* 59: 4068–4076.

24 Paredes, I.J., Yohannes, B., Glasser, B.J. et al. (2017). The effect of operating conditions on the residence time distribution and axial dispersion coefficient of a cohesive powder in a rotary kiln. *Chemical Engineering Science* 158: 50–57.

25 Emady, H.N., Anderson, K.V., Borghard, W.G. et al. (2016). Prediction of conductive heating time scales of particles in a rotary drum. *Chemical Engineering Science* 152: 45–54.

26 Yohannes, B., Emady, H.N., Anderson, K. et al. (2016). Scaling of heat transfer and temperature distribution of granular flows in rotating drums. *Physical Review E* 94 (042902): 1–5.

27 Yohannes, B., Emady, H., Anderson, K. et al. (2017). Evolution of the temperature distribution of granular material in a horizontal rotating cylinder. *Powders & Grains* 140: 03012.

28 Davis, B.H. and Hettinger, W.P. Jr. (1982). *Heterogeneous Catalysis – Selected American Histories*. Washington, DC: American Chemical Society.

29 Thomas, J.M. and Thomas, W.J. (1967). *Introduction to the Principles of Heterogeneous Catalysis*. London, New York: Academic Press.

30 Masters, K. (1985). *Spray Drying Handbook*, 4e. New York: Wiley. ISBN: 0-470-20151-7.

31 Cybulski, A. and Moulijn, J. (2005). *Structured Catalysts and Reactors*. CRC Press. ISBN: 9780824723439.

32 Satterfield, C.N. (1980). *Heterogeneous Catalysis in Practice*. New York: McGraw-Hill Book Co.

33 Stiles, A.B. (1987). *Catalyst Supports and Supported Catalysts – Theoretical and Applied Concepts*. Stoneham, Massachusetts: Butterworth Publishers. ISBN: 0-409-95148-X.

34 Pereira, C.J., Cheng, W.-C., Beeckman, J.W., and Suarez, W. (1988). Performance of the minilith – a shaped hydrodemetallation catalyst. *Applied Catalysis* 42: 47–60.

35 Kubsh, J.E., Rieck, J.S., and Spencer, N.D. (1990). Ceria oxide stabilization: physical properties and three-way activity considerations. In: *Catalysis and Automotive Pollution Control II* (ed. A. Crucq) Studies in Surface Science and Catalysis, 125–138. Amsterdam, The Netherlands: Elsevier Science Publishers.

36 Beeckman, J.W. and Hegedus, L.L. (1991). Design of monolith catalysts for power plant NOx emission control. *Industrial & Engineering Chemistry Research* 30: 968–978.

37 Bai, C. (2018). Hydrogenation catalyst, its method of preparation and use. US Patent 9, 861,960, filed September 2, 2014, and issued January 9, 2018.

38 Sullivan, J.D., Macer, C.G., and Ralston, O.C. (1927). Passage of Solid Particles through Rotary Cylindrical Kilns. U.S. Bureau of Mines, Technical Paper, 384.

39 Saeman, W.C. (1951). Passage of solids through rotary kilns: factors affecting time of passage. *Chemical Engineering Progress* 47: 508–514.

40 Danckwerts, P.V. (1953). Continuous flow systems: distribution of residence times. *Chemical Engineering Science* 2 (1): 1–13.

41 Kohav, T., Richardson, J.T., and Luss, D. (1995). Axial dispersion in a continuous rotary kiln. *AIChE Journal* 41: p2475.

42 Sudah, O.S., Chester, A.W., Kowalski, J.A. et al. (2002). Quantitative characterization of mixing processes in rotary calciners. *Powder Technology* 126 (2): 166–173.

43 Levenspiel, O. and Smith, W.K. (1957). Notes on the diffusion type model for the longitudinal mixing of fluids in flow. *Chemical Engineering Science* 6 (4–5): 227–235.

44 Liu, X.Y. and Specht, E. (2006). Mean residence time and hold-up of solids in rotary kilns. *Chemical Engineering Science* 61 (15): 5176–5181.

45 Beeckman, J.W. (2019). Apparatus and method for mixing and/or mulling a sample. US Patent 10, 307,751, filed February 2, 2016, and issued June 4, 2019.

2

Extrusion Technology

2.1 Background

An economical way of manufacturing catalysts is by the process of extrusion. Catalysts are extruded through one of two methods: either auger extrusion or piston extrusion. Auger extrusion, also known as screw extrusion, is a continuous process and yields high production rates of extrudates that have a relatively simple shape. Ram extrusion, also known as piston extrusion, operates batch-wise and the extrusion rate is lower than for auger extrusion. Ram extrusion is favorable when the shape of the extruded body is large and complex and when a steady die pressure is required.

The pressure that develops at the extruder die can be very substantial. For safety reasons, always install the necessary pressure or torque cutouts in the equipment to avoid reaching unsafe pressure or torque levels. Always follow the implemented safety procedures and practices and never stand in front of the die when the extruder is operating.

In auger extrusion, the paste is carried forward inside a lined barrel. At the end of the barrel, the auger pushes the extrusion paste against a die plate, and under sufficient pressure, the paste exits through the die channels as long strands that are then conveyed to a drier. The strands can be cut to size at the die itself by means of a cutter blade or they can be sized later on in the manufacturing process. The extrudates are typically 1–3 mm in diameter and can have various cross-sectional shapes.

In ram extrusion, the extrusion paste is loaded batch-wise into a cylindrical barrel that is equipped with a die plate on the discharge end and a pusher piston on the other end. At a sufficient pressure, the paste will exit the die as large, shaped bodies, such as multichannel monoliths of an overall size of 5–50 cm in diameter.

Catalyst Engineering Technology: Fundamentals and Applications,
First Edition. Jean W. L. Beeckman.

Extrusion is the movement of a body (the extrusion paste) under stress against surfaces (the auger, the barrel, and the die plate channels). In addition to the resistance to flow through the die channels, in this chapter I will also consider the ability of the extrusion paste to become malleable and to deform, and how to accomplish malleability (*kneedbaarheid*, *malléabilité*, *formbarheit*) as the paste goes through the entire extrusion process.

The forced movement of the extrusion paste against surfaces and the deformation forces in play are of high importance. Forces acting in a body under stress fall into two classes: forces that act perpendicular to an arbitrarily chosen infinitesimal surface element and forces that act parallel to that surface element. Forces that act perpendicular to the surface element are called normal forces and for the intensity (stress) thereof, defined as the force per unit area, I will use the symbol σ. Forces that act parallel to the surface element are called shearing forces and for the intensity thereof I will use the symbol τ. Any arbitrarily directed force acting on an infinitesimal surface element can always be decomposed into a normal force and one or two shearing forces, depending on whether you are working in a two-dimensional or three-dimensional frame.

Think of normal forces as compressive or tensile. Shearing forces, depending on the material properties, may result in layers of the material moving parallel to each other at different speeds. A metal bar with a shearing force on a surface element will remain as a metal bar of near-identical geometry. A catalyst extrusion paste is a large agglomerate of individual tiny particulates, and a shearing force on a surface element can cause substantial movement of parallel layers. Substantial deformation of a catalyst paste can therefore come about and cause abrasion of the particulates in the moving layers.

Before going into further detail, let us consider some of the terms and properties related to extrusion.

2.2 Rheology

2.2.1 Shear Stress, Wall Shear Stress, and Shear Rate

For surfaces moving parallel to each other at different speeds, Eq. (2.1) defines the shear stress, τ, in pascal (Pa), working on a surface element of a deforming body as:

$$\tau = \eta_b \left(\frac{\partial v}{\partial y} \right) \tag{2.1}$$

where η_b is the dynamic viscosity of the body and $\left(\frac{\partial v}{\partial y} \right)$ is the velocity gradient of the body perpendicular to the chosen surface element.

The term $\left(\frac{\partial v}{\partial y}\right)$ is the difference in velocity between the two surfaces divided by the normal distance between them, and it represents the intensity of the shearing of the two layers. The velocity gradient is also known as the shear rate and is commonly reported in units of inverse seconds. Depending on the application and the properties of the body, the shear rate spans many orders of magnitude. Importantly, multiplying the shear stress by the area of the surface element yields the force acting on that surface element.

The size scale of the individual entities (particles) that make up an extrusion batch differs strongly from plastics to ceramics and catalyst powders. The size scale of the high-molecular-weight molecules in the extrusion of plastics is much smaller than that of an extrusion paste formed by powders. For an extrusion paste made from catalyst powders or ceramic powders, the shearing of layers parallel to each other occurs at the micron scale down to the nanoparticle scale. The friability of catalyst powders under shearing stress is important. Typical catalyst powders are much weaker than ceramic powders and are more susceptible to abrasion (also known as friability). Catalyst powders do have a terrific ability to absorb and adsorb liquids to a much greater extent than ceramic powders. Often, the powders will flow as a dry mass, even after adding a substantial amount of liquid, until a critical level is reached where the powders start to agglomerate.

When an interstitial layer of fluid is present between the common surface of a wall and a moving body, the picture of forces changes. The shearing stress acting on a surface element of the wall in contact with the interstitial fluid is called the wall shear stress. Equation (2.2) defines the wall shear stress as:

$$\tau_w = \eta_l \left(\frac{\partial v}{\partial y}\right)_{y=0} \tag{2.2}$$

where η_l is the dynamic viscosity of the fluid and $\left(\frac{\partial v}{\partial y}\right)_{y=0}$ is the velocity gradient of the layer of fluid at the wall, measured perpendicular to the wall.

Multiplying the wall shear stress by the area of the wall surface element yields the force acting on that surface element. The wall shear stress increases when both the viscosity of the fluid and the shear rate increase. As defined, the wall shear stress is independent of pressure as long as the viscosity of the liquid is not appreciably influenced, which is the case for every pressure range considered in this book. However, the thickness of the layer of fluid may be influenced by the pressure since an extruder is not a closed system. For an extrusion paste, some sources use Eq. (2.3) to represent the wall shear stress as:

$$\tau_w = \tau_0 + \beta v \tag{2.3}$$

where τ_0 is the wall shear stress extrapolated to zero velocity and v is the relative velocity of the surface of the extrusion paste versus the surface of the wall.

Only those values of τ_w where v is of practical value are important during actual extrusion. The parameter τ_0 is an approximation of what happens at near-zero velocities, and Eq. (2.3) is a linear model for τ_w with two parameters.

When the two surfaces remain at the same distance from each other with a fluid between them and when their velocity relative to each other is given, then you know the gradient. For a flat geometry, the velocity across the liquid changes linearly from zero (at the surface considered at rest) to the velocity of the moving surface. The shear rate in Eq. (2.2) can then be expressed as per Eq. (2.4):

$$\left(\frac{\partial v}{\partial y}\right)_{y=0} = \frac{v}{\delta} \tag{2.4}$$

where v then represents the velocity of the bulk paste relative to the wall and δ is the thickness of the liquid film.

Taking the malleability of a catalyst extrusion paste into consideration, it is neither certain nor clear when a liquid layer is present. If a liquid layer is present, then the pressure profile in the extruder likely plays a role in the thickness of the liquid layer, since the paste will try to squeeze the liquid out due to pressure. Hence, there may be a gradient of the thickness of the liquid layer along the length of the extruder auger. A thinner liquid layer in the high-pressure region of the auger with the same bulk velocity of extruder paste will yield a higher gradient and thus higher wall shear stress for the extruder paste at that location.

Friction occurs when there is no liquid film present and the extrusion paste drags along a surface in direct contact with the paste. The physics of the movement and the malleability of an extrusion paste may range from shear stress behavior to frictional behavior, or somewhere in between. I have seen both kinds in commercial extruders of catalyst paste. The literature uses both shear stress behavior and friction behavior to discuss extrusion, and the subject warrants more research. It is a fascinating area indeed.

2.2.2 Friction

Consider two bodies in uniform contact over an area and pressed together with a certain force normal to that area. The literature shows that a minimum force, called the static friction force, applied parallel to the surface is required in order to initiate the movement of the two bodies relative to each other. The static friction force is independent of the area of the contacting surface. The two phenomena – the minimum static force and the independence of the area of the contact surface – are called Amontons' laws (Guillaume Amontons, 1663–1705). The friction force points in the direction opposite to the movement. The ratio of the static friction force to the force normal to the surface is called the static friction coefficient μ_s and it is a dimensionless quantity. Knowing both the normal force and the friction coefficient automatically yields the static friction force no matter the area of the surface.

Consider two bodies pressed together with a certain force normal to the area of contact that are moving at a uniform speed relative to each other. The literature shows that a force parallel to the surface and opposite of what is called the kinetic friction force is required in order to keep the bodies moving. The ratio of the parallel force to the normal force is called the kinetic friction coefficient μ_k. The coefficient μ_k does not usually depend on the relative velocity of the two surfaces, nor on the size of the common area of the two bodies. The kinetic friction coefficient μ_k is different from the static friction coefficient μ_s and has a lower value. In the absence of lubricants, kinetic friction is called dry friction. It is ideally independent of the speed of motion, and this independence is called Coulomb's law.

Hence, one needs a certain minimum force – the static friction force – to initiate movement of a body along a surface. Once movement is initiated, further movement is easier. Anybody who has been assigned the unfortunate task of moving furniture around likely can testify to this.

The frictional contact area of two solid materials with one another is neither homogenous nor geometrically flat. The contact surface consists of point-like or line-like contacts that are called asperities. Asperities are irregularities in the surface at the micron, submicron, and nanoscale levels, and they belong in the multidisciplinary science of tribology. The multidimensional scale of asperities and their sizes, ranging over many orders of magnitude, have been modeled via fractal geometry.

A body of mass, m, at rest that makes contact with a surface over an area onto which a normal force, N, is applied by the body will only glide along the surface when this force is above a particular minimal value, T_s, as expressed by Eq. (2.5):

$$T_s = \mu_s(N + mg) \tag{2.5}$$

If the applied force T is less than the minimal force T_s, then no relationship exists between T and N from a friction perspective.

Once gliding is initiated, μ_k will determine the force necessary to sustain gliding. I use the term "glide" to describe the motion of a body over a surface in the presence of kinetic friction. Equation (2.6) calculates the force T_k required to sustain gliding:

$$T_k = \mu_k(N + mg) \tag{2.6}$$

If T is less than T_k, then the body comes to rest. Only when there is movement of the one surface with respect to another is there a relationship of the kinetic friction force with respect to the normal force.

Once gliding is initiated, and if T is greater than T_k, then the body moves along the surface and glides according to Eq. (2.7):

$$ma = m\frac{dv}{dt} = T - T_k = T - \mu_k(N + mg) \tag{2.7}$$

Many literature sources concerning this topic will show μ_k to display a linear proportionality to the velocity above a certain minimum value, as in Eq. (2.8):

$$\mu_k = \mu_{k,0} + \alpha v \text{ for } T > T_k \tag{2.8}$$

and Eq. (2.9):

$$\mu_k = \mu_{k,0} \text{ for } T = T_k \tag{2.9}$$

For gliding forces where T is greater than T_k, the body then accelerates according to Eq. (2.10):

$$m\frac{dv}{dt} = T - \left(\mu_{k,0} + \alpha v\right)(N + mg) \tag{2.10}$$

The body accelerates until a constant terminal velocity v_∞ is reached $\left(\frac{dv}{dt} = 0\right)$, as obtained from Eq. (2.11):

$$v_\infty = \frac{1}{\alpha}\left(\frac{T}{N + mg} - \mu_{k,0}\right) \tag{2.11}$$

It would be interesting to see, when applying different normal forces N, which parts of the surface of a body will glide and which will remain stationary against frictional surfaces. Consider the body of the extrusion paste along an auger and the barrel in an extruder. The body of extrusion paste is indeed fully enveloped by surfaces, but it is not entirely clear which parts of the paste surface will glide and which will remain stationary relative to a rotating auger and stationary barrel. Will the paste glide along the auger's surface, along the barrel's surface, or both?

Let us try an easy case of a very shallow cylindrical body that sits on top of a first surface with a friction coefficient $\mu_{s,1}$. Add a second surface that sits on top of the cylinder with a different friction coefficient $\mu_{s,2}$. A normal force N is applied as shown in Figure 2.1. When applying a gradually increasing force T parallel to

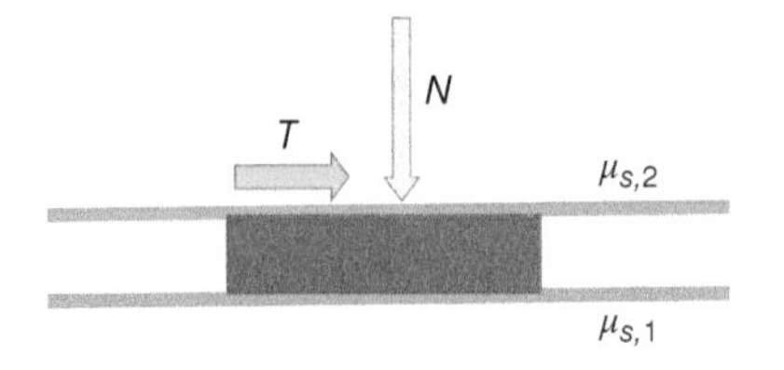

Figure 2.1 A shallow cylinder between two parallel plates.

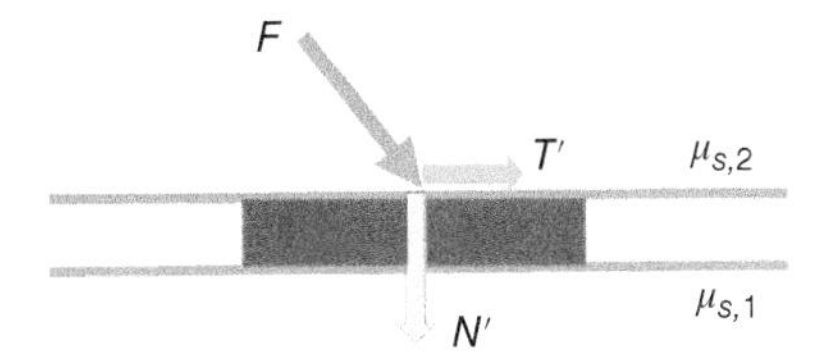

Figure 2.2 A shallow cylinder between two parallel plates with an angled force.

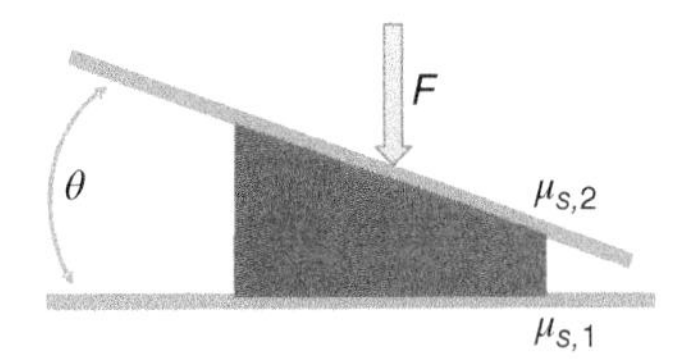

Figure 2.3 A shallow cylinder between two plates at an angle θ.

the surface, it becomes intuitively clear that the surface with the highest static friction coefficient will remain stationary, while the other will glide (like the pads that you can buy to place under furniture legs).

Now, let us make this a bit more interesting (see Figure 2.2) and apply a force F not parallel along the surface, but rather at an angle. You can now decompose this force F into a component T' along the surface and a component N' normal to the surface; hence, I did not really make any significant changes to the case in Figure 2.1.

But what would happen if one were to change the body in red as in Figure 2.3 and use a truncated cylinder instead of a cylinder? Take the case where there is only a force F normal to the first surface, hence with a zero parallel force component to the first surface. Clearly, no matter how small $\mu_{S,1}$ is, there will be no movement of the body of paste against the first surface. Depending on the angle θ that determines the ratio of the force components of F, the second surface may or may not move. Equation (2.12) establishes the condition that must exist for the second surface not to move:

$$\tan\theta < \mu_{s,2} \tag{2.12}$$

2.2.3 Rheometer Data

A good handbook for rheology is the one by Mezger [1] on rotational and oscillatory rheometers. To better understand and experience the rheology of catalysts for extrusion, a TA Instruments rheometer, Model TA Discovery HR-2 Hybrid Rheometer, was purchased.

The instrument enables you to place a properly kneaded and mulled mixture, ready for extrusion, between two parallel plates and to apply a constant normal

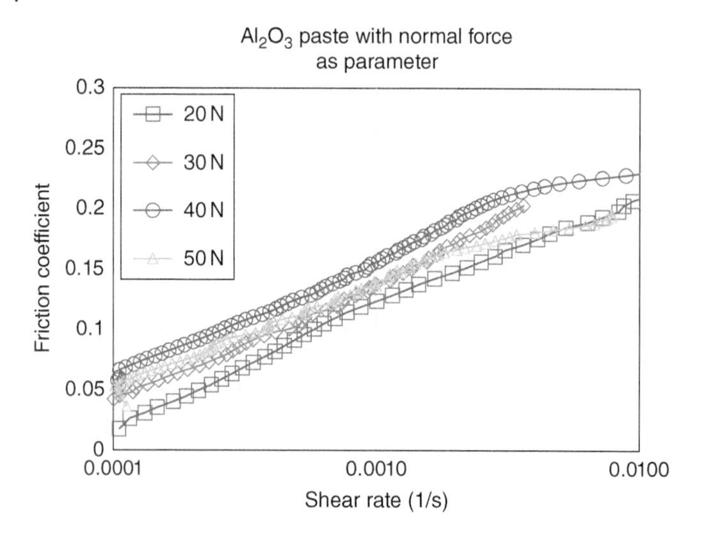

Figure 2.4 Friction coefficient versus shear rate for alumina.

force N onto the plates. The top plate is put in rotation over a range of increasing velocities and the instrument measures the torque required to rotate as the speed of rotation increases over a few orders of magnitude. The instrument takes the distance between the plates into consideration and experimentally I tried to keep this distance relatively constant. The shear stress is then determined as a function of the disk velocity. For pseudo-boehmite alumina that is properly mulled and kneaded at 40% solids, the shear stress is a function of the disk velocity. I found that the shear stress varies linearly with the normal force N applied and hence became indicative in my mind of friction behavior.

TA Instruments TRIOS Version IV software enables one to assess the friction coefficient, and Figure 2.4 shows the friction coefficient as a function of the shear rate. All of the curves for different normal forces come together into almost a single response curve reasonably well. This is the behavior expected for friction, and I think it is an appropriate means by which to study the extrusion of pseudo-boehmite alumina. The friction factor in Figure 2.4 can be written as per Eq. (2.13) for values of v greater than a small reference velocity v_0:

$$\mu(v) = \mu_0 + \kappa \ln (v/v_0) \tag{2.13}$$

with Eq. (2.14):

$$\mu(v_0) = \mu_0 \tag{2.14}$$

The reference velocity v_0 is very small in value, but for practical purposes, as far as the rate goes, it can essentially be considered equal to zero. For rate values below v_0, the friction factor is considered equal to μ_0.

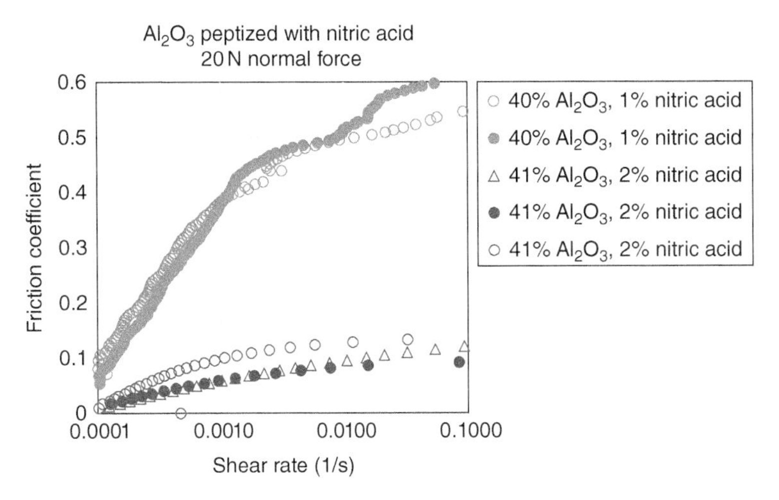

Figure 2.5 Friction coefficient versus shear rate for peptized alumina.

Later, peptizing the same alumina grade with 1 wt.% nitric acid (at 40% solids) and 2 wt.% nitric acid (at 41% solids) was then evaluated on the rheometer. Figure 2.5 shows a much higher friction coefficient for the lower-nitric-acid alumina than for the higher-nitric-acid alumina (even at the higher solids levels of the higher-nitric-acid alumina case). The acid enhances the peptizing of the alumina and makes the paste smoother for extrusion, which is a well-known phenomenon in the academic and commercial arenas. It is thus possible to observe a substantial difference in the properties of alumina pastes by varying the acid content. The variation of the measured friction factor with acid content (i.e. the extrusion recipe) makes the TA rheometer instrument a good candidate for conducting further research on various materials in the field of extrusion.

2.2.4 Comparing Friction and Wall Shear Stress

In Table 2.1, I provide a comparison of models used in the literature for the resistance to flow of extrusion pastes for plastics, ceramic powders, and catalyst powders. The extrusion of plastics is very well studied. There are few published studies on ceramic extrusion pastes, and even fewer on catalyst extrusion pastes. The references given in Table 2.1 are not intended to be comprehensive, but rather to give a flavor of some of the models used in the industry for the resistance of paste flow.

Ceramic paste extrusion uses pre-hardened raw material such as alpha-alumina powder. Required amounts of water, binder, and extrusion aid are added to the powder in order to extrude the "green" body. In the context of extrudates, the descriptor "green" means that the extrudate has not been fired yet or has

Table 2.1 Friction models and shear models used in the literature for rheology and extrusion.

Material	Smallest die-hole diameter (mm)	Friction	Wall shear stress	Comment	Reference
Ceramic, alpha-alumina	3.2	NA	$\tau_w = \tau_0 + \beta v^n$	Linear pressure profile	Benbow and Bridgwater [2]
Ceramic, alpha-alumina	3.0	NA	$\tau_w = \tau_0 + \beta v^n$	Linear pressure profile	Burbidge and Bridgwater [3]
Clay-like ceramics	4.8	$F_f = \mu F_N$	NA	Exponential pressure profile, die flow: $Q = a'P - b'$	Parks and Hill [4]
Solids compaction and biomass solids	NA	$F_f = \mu F_N$	NA	Exponential pressure profile	Orisaleye and Ojolo [5]
Ceramics	NA	$F_f = \mu F_N$	NA	Coulomb friction coefficient	Laenger [6]
Alumina	~1.25	$F_f = \mu F_N$	NA	$\mu_0 = 0.2$ with a small increase with velocity, $\mu(v) = \mu_0 + \kappa \ln(v/v_0)$	Beeckman (this book)
Silica/zeolite	~1.25	$F_f = \mu F_N$	NA	$\mu(v) = \mu_0 + \kappa \ln(v/v_0)$	Beeckman (this book)
Alumina/ zeolite	~1.25	$F_f = \mu F_N$	NA	$\mu(v) = \mu_0 + \kappa \ln(v/v_0)$	Beeckman (this book)
Alumina	NA	NA	Shear stress	NA	Winstone [7]
γ-Alumina/ boehmite pastes	NA	NA	Shear stress	NA	Mills and Blackburn [8]

not been exposed to its final heat treatment. With ceramics, minimizing the final pore volume and interstitial particle volume helps one to obtain a truly solid body of minimal porosity. Raw materials may have been previously fired in order to minimize pore volume and any shrinkage upon final firing, which may lead to microscopic defects. Because of the strength of ceramic pastes, much less abrasion and attrition will occur during shearing in the extrusion process.

Catalyst pastes, on the other hand, are delicate, leading the catalyst manufacturing industry to protect its processes through trade secrets or patents. It is important

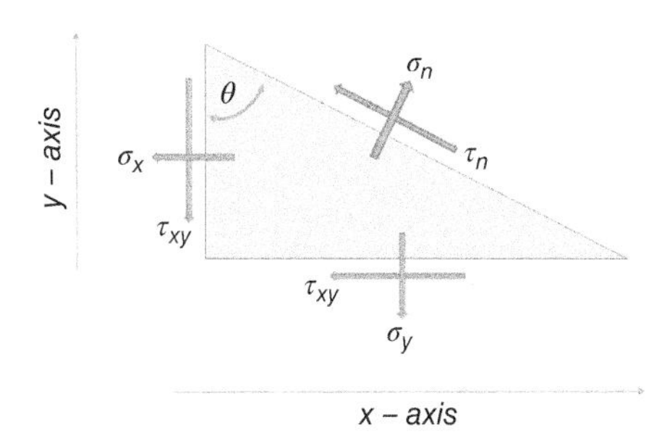

Figure 2.6 Force diagram for a material under stress.

for catalyst powders to keep their pore volumes and surface areas intact during the shearing experienced while mixing, mulling, kneading, and extrusion. The catalyst manufacturer's task is to keep the particulate structure intact while making the catalyst paste extrudable and delivering an extrudate of sufficient strength. During manufacture, catalyst strength has to be appropriate for keeping the extrudates intact to a practical degree during manufacturing, as well as during the final use of the catalyst in chemical reactors. The subjects of strength and catalyst integrity will be discussed in detail in Chapter 3.

2.2.5 A Paste under Stress

At each internal point of a paste in equilibrium under stress as shown in Figure 2.6, the forces acting on a triangle (using two dimensions for ease of representation) can be decomposed in the (x, y) frame, indicated by two normal components, σ_x and σ_y, and a single shear component, τ_{xy}. When rotating the (x, y) frame such that the positive x′-axis is in the direction of the angle for σ_n and the positive y′-axis is in the direction of the angle for τ_{xy}, then Eqs. (2.15) and (2.16) express the normal forces σ_n and shear forces τ_n working on a surface making an arbitrary angle θ:

$$\sigma_n = \frac{\sigma_x + \sigma_y}{2} + \frac{\sigma_x - \sigma_y}{2} \cos 2\theta + \tau_{xy} \sin 2\theta \tag{2.15}$$

$$\tau_n = -\frac{\sigma_x - \sigma_y}{2} \sin 2\theta + \tau_{xy} \cos 2\theta \tag{2.16}$$

When using an alternative orthogonal frame where σ_n is shown on the x-axis and τ_n is shown on the y-axis, according to Otto Mohr (1835–1918), all of the stress values are the coordinates of points that lie on the Mohr circle, with a radius calculated from Eq. (2.17):

$$R_M = \sqrt{\left(\frac{\sigma_x - \sigma_y}{2}\right)^2 + {\tau_{xy}}^2} \tag{2.17}$$

The origin of the Mohr circle is located on the x-axis, with Eq. (2.18) giving the coordinate as:

$$O_M = \frac{\sigma_x + \sigma_y}{2} \tag{2.18}$$

The angle between the radial from the origin of the circle to the point of interest on the circle's circumference and the x-axis is equal to 2θ. For a material paste with a particular internal friction factor μ_0 (which is the ratio of the shear stress to the normal stress), at no point in the paste should the Mohr circle be crossed by the internal friction factor line because otherwise the material for θ would shear. Hence, the stresses for a stationary paste must be less than those allowed by the tangent to the Mohr circle, as given by Eq. (2.19):

$$\tau_n \leq \mu_s \sigma_n \tag{2.19}$$

A paste that is just barely moving means that the Mohr circle is at a tangent with the internal friction line.

2.2.6 The Yield Strength of a Paste

If one holds a clump of extrusion paste in one's hand, nothing seems to happen. Clearly, the gravitational force acting on the clump against the surface of your hand is not strong enough to cause deformation, or it is too small to be noticeable. At first, squeezing the clump will accomplish some level of densification by expulsing air from within the clump, but once the air has been removed, the clump has gained some structural strength and requires more force to cause further permanent deformation. Experienced operators use this "consolidation" and "more force" phenomenon to judge whether a paste is ready for extrusion or whether it needs more "work".

Upon entering the extruder, a paste that is properly prepared will undergo more shear – it is impossible to avoid this. Because a paste has a certain structural strength, one has to apply a specific minimal pressure to make it deform. For example, if you place a paste under a stress of 2 MPa (~300 psi) in a cylinder, it may not flow even when an opening is created in the cylinder wall. However, it may flow (or yield) under the influence of an extra 0.2 MPa (~30 psi) of pressure through an opening in the cylinder wall.

This minimal 2.2-MPa of pressure is called the yield strength or yield stress of the paste. The yield strength of the paste plays a large role in estimating the amount of work required to change the shape of a paste or in calculating the work required to have a paste enter an extrusion channel. For instance, Benbow and Bridgwater [2] calculate the stress required to deform a paste so that it can move from a channel of diameter D_1 into a channel of diameter D_2, as expressed by Eq. (2.20):

$$\Delta P = 2\sigma_u \ln\left(\frac{D_1}{D_2}\right) \tag{2.20}$$

where σ_u is called the uniaxial yield stress. Typical values for the yield stress of a catalyst paste range from 0.5 to 10.0 MPa.

2.2.7 Paste Density

It is relatively easy to determine the paste density. The paste strands have a diameter equal to the diameter D_0 of the extrusion channel as they exit the die. Once exited from the die, the strands can either shrink due to loss of moisture during drying and calcination or, in some cases, there may also be swelling depending on the extrusion aids used. Assume that the calcined strands have a measured diameter D_{cal} and a measured density ρ_c. Define the solids content s of the extruder batch as the weight percent solids that remains after the heat treatment of the catalyst sample at a user-defined reference temperature. The reference temperature is a choice, but it is usually high enough to remove any water from the batch.

In order to obtain the consolidated paste density ρ_p, one needs to adjust the density of the strands after drying and calcination according to any change in diameter and length, as well as the solids content of the extrusion recipe by using Eq. (2.21):

$$\rho_p = \rho_c\left(\frac{D_{cal}}{D_0}\right)^3 \frac{100}{s} \tag{2.21}$$

The third power of the ratio of the diameters in Eq. (2.21) is needed in order to properly account for the dimensional changes in the three spatial directions. The easiest way I visualize dimensional changes is to consider that a body is made up of many tiny identical cubes that are neatly stacked and as such mimic the overall shape of the body. Identical dimensional changes in the (x, y, and z) directions for each tiny cube then translate to the total three-dimensional change of the body.

Based on the density of the paste, it is now straightforward to relate the mass rate of the extruder to the volumetric rate and hence the extrusion velocity in the die channels.

2.3 Extrusion

2.3.1 Ram Extrusion

Benbow and Bridgwater [2] wrote a very noteworthy textbook on extrusion, while Burbidge and Bridgwater [3] published a good paper on the single-screw extrusion of pastes. Benbow and Bridgwater analyze a ram extruder yielding a single strand or multiple strands. The overall pressure drop,ΔP, for a ram extruder according to Benbow and Bridgwater is given by Eq. (2.22):

$$\Delta P = 2(\sigma_0 + \alpha v)\ln\left(\frac{D_r}{D_0}\right) + 4(\tau_0 + \beta v)\left(\frac{L_0}{D_0}\right) \tag{2.22}$$

where ΔP is the sum of (i) the pressure drop through the barrel, (ii) the pressure drop caused by the paste entering the narrow extrusion channel, and (iii) the pressure drop through the channel. D_r is the diameter of the ram piston, D_0 is the diameter of the extrusion channel, and L_0 is the length of the extrusion channel.

The die extrusion channel is what Benbow and Bridgwater [2] call the "die land." The velocity of the paste in the extrusion channel is called v. The yield stress and the wall shear stress extrapolated to zero velocity are σ_0 and τ_0, respectively. The coefficients α and β indicate the effect that the paste velocity in the channel has on uniaxial yield stress and wall shear stress, respectively. In total, the model in Eq. (2.22) has four parameters (σ_0, α, τ_0, β).

Benbow and Bridgwater [2] also suggest Eq. (2.23) for more refined studies:

$$\Delta P = 2(\sigma_0 + \alpha v^{n_1})\ln\left(\frac{D_r}{D_0}\right) + 4(\tau_0 + \beta v^{n_2})\left(\frac{L_0}{D_0}\right) \tag{2.23}$$

Equation (2.23) increases the number of parameters by two (n_1, n_2) to a total of six.

The analysis is very elegant and shows how the channel design (L_0, D_0), the rheological properties (σ, τ), and the channel extrusion velocity (v) together determine the overall extrusion pressure for a ram extruder.

Burbidge and Bridgwater [3] extend this analysis for use with single-screw designs in flooded mode. Flooded mode means that the entire screw is filled with extrusion paste that has been fully de-aerated, and therefore the work is strictly for a screw with a known fill length.

With pilot extruders or even commercial extruders, it is often not possible to determine how much of the screw is flooded. I think it may be advantageous to resolve the situation by directly using the pressure at the die face. Unfortunately, the die pressure is not always available. However, I believe it is very worthwhile to obtain the pressure at the die face by means of a pressure transducer with only modest labor and at a reasonable cost.

The pressure at the die face is a process variable and a convenient tool with which to link the extrusion rate to the die design. For typical extrusion pressures of 5–10 MPa, Eq. (2.24) relates the die pressure to the extrusion rate through a single channel:

$$P_d = 4(\tau_0 + \beta v)\left(\frac{L_0}{D_0}\right) \tag{2.24}$$

Equation (2.24) represents a die pressure model with two parameters (τ_0, β). Measurements of the die pressure as a function of the extrusion rate would yield the parameters τ_0 and β. You could also determine τ_0 and β from pilot plant experiments or even from single-channel dies.

When using dies that are of the same design and material as commercial ones (except for the total number of channels), it may be possible to use Eq. (2.24) to make predictions about commercial extruders – more on this in Section 2.3.2.1. Equation (2.24) relates the extrusion pressure to the extrusion rate for a single channel. Pilot extruders have several tens of channels, while commercial extruders have several hundreds to thousands of channels. Combining the extrusion rate in a single channel with the total mass extrusion rate, R_e for a die having k channels, leads to Eq. (2.25) for the velocity in a channel:

$$v = \frac{R_e}{k\rho_p\Omega} \tag{2.25}$$

where Ω is the cross-sectional area of the channel.

2.3.2 Auger Extrusion

Figure 2.7 shows a typical auger and its cross section. Typical parameters are the outer diameter of the auger D_s, the diameter of the core D_C, the pitch of the auger p, and the thickness of the lip T_l. The open space in the auger close to the die is completely filled with extrusion paste, with all of the air having been squeezed out in an earlier part of the auger. As long as air is present, steady pressure cannot build up; the only thing that would be happening in this case is that the auger would still be getting packed.

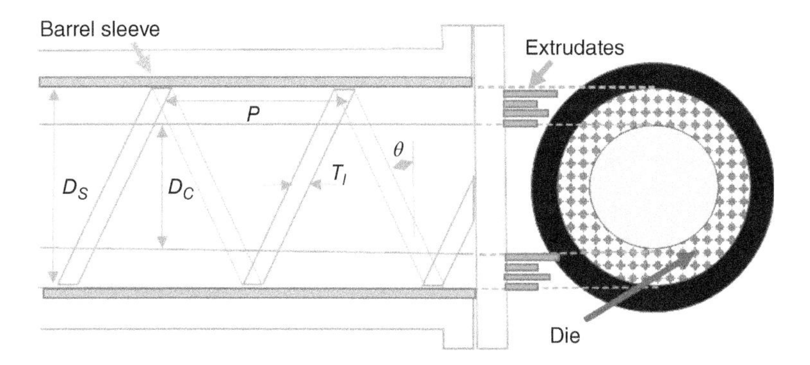

Figure 2.7 Typical extruder auger and cross section.

The packing of the barrel is often done with die plates that have larger channel openings to facilitate the compaction of the paste because it is easier to establish solid paste flow through the larger channels. A good paste flow is important in order to avoid prolonged "working" and "over-peptization" of paste in the extruder. The excessive "working" and "over-peptization" of a paste in the extruder can cause free liquid slurry to form at the barrel surface, consequently causing slip of the paste in the barrel. Once the barrel is properly packed, one can take the packing die off and put the die of interest in place; typically, extrusion begins and lines out, but it is not guaranteed. A continuous steady extrusion rate is particularly important, as is a uniform flow of the strands from the die.

"Uniform" does not mean that the material exits the die at every point at the same rate along the annulus at every instant of time. "Uniform" here means that the paste exits at the same rate along the annulus primarily when the end of the lip of the auger passes the extrusion channels in the die. For an auger with a single lip, this means every 360° turn of the auger. Some augers have a dual lip, which means that the material exits primarily upon every 180° rotation of the auger.

2.3.2.1 Die Equation or Die Characteristic

The die equation or die characteristic expresses the relationship between the extrusion rate and the pressure drop through the die. Benbow and Bridgwater [2] provided a detailed and comprehensive model for this relationship, reporting that the first term on the right-hand side of Eq. (2.22) is much larger than the second term; thus, a pressure tap located at the die may pick up this term in full or only partially.

The typically small gap between the die and the screw makes it difficult to judge where exactly to place the tap, since the pressure sensor area is often larger than the gap. Because pilot and commercial dies have a large number of holes, it makes sense to approximate the full expression in Eq. (2.22) as per Eq. (2.26):

$$\Delta P = \sigma_0^+ + \beta^+ R_e/\rho_p \tag{2.26}$$

where σ_0^+ and β^+ are now two parameters that have to be determined experimentally.

Both σ_0^+ and β^+ are functions of the setup: the length of the die land and the diameter of the extrusion channel. Since there are only a few die parameters (length, diameter, and shape) in pilot and commercial use and they do not change often, Eq. (2.26) can be of more practical use. ΔP, measured by the total pressure at the pressure tap at the die, requires a minimum value of σ_0^+ before achieving even a very small extrusion rate. As ΔP increases, the rate of extrusion rises linearly according to the second term on the right-hand side of Eq. (2.26). Parks and Hill [4], whose approach is built on friction, adopted the same relationship as in Eq. (2.26).

For Newtonian plastics, the literature shows a simple relationship between the pressure drop and the extrusion rate for Hagen–Poiseuille flow (laminar flow) in die channels per Eq. (2.27):

$$\Delta P = \frac{\eta_l L_0}{K} R_e / \rho_p \tag{2.27}$$

where for cylindrical channels of diameter D_0, the value of K is obtained per Eq. (2.28):

$$K = \frac{\pi {D_0}^4}{128} \tag{2.28}$$

while for rectangular channels of a width W and height H, Eq. (2.29) yields:

$$K = \frac{WH^3}{12} \tag{2.29}$$

2.3.2.2 Model 1: The Extruder Characteristic Equation

Birley et al. [9] provide an excellent write-up on the extruder characteristic equation approach that balances two rate terms. The first is the net flow of extrusion paste through the extruder and the second is the paste flow through the die.

The net paste flow through the extruder consists of the difference between two terms: (i) a constant positive geometric paste flow through the extruder due to the constant rotation of the auger reduced by (ii) an increasing flow of paste snaking backward along the auger with an increasing pressure at the die. The difference of the two terms leads to the net paste flow of extrusion paste versus the pressure drop ΔP over the die and is a monotonically downward operating line.

The paste flow through the die as a function of the pressure drop over the die is a monotonically upward operating line, as shown in Eq. (2.26).

The operating line for the net paste flow through the extruder and the operating line for the paste flow through the die cross at the operating point of the extruder. Birley et al. [9] use this approach for the extrusion of plastics, while Parks and Hill [4] use it for the extrusion of clay-like materials.

2.3.2.2.1 Geometric Paste Flow

The velocity v_b at the barrel surface is obtained from the screw rotational speed according to Eq. (2.30):

$$v_b = \pi D_b N_s \tag{2.30}$$

where D_b is the barrel diameter and N_s is the number of revolutions per unit time. This velocity translates to a velocity along the channel of the auger but still at the barrel surface according to Eq. (2.31):

$$v_c = \pi D_b N_s \cos\theta \tag{2.31}$$

where θ is the angle of the flight shown in Figure 2.7.

For plastics with Newtonian characteristics, it is possible to approximate the average velocity in the auger as half the velocity at the barrel (v_c at the barrel and zero at the auger core). Thus, it is possible to write the mass flow rate through a single channel of the auger as Eq. (2.32):

$$R_g = \frac{1}{2}\pi WHD_bN_s\rho_p cos\theta \tag{2.32}$$

where W is the width at right angles between two consecutive flights and H is the height of the auger groove.

For non-Newtonian plastics or ceramic or catalyst pastes, one will have to exercise care and judgment in order to obtain the correct physical representation for use in Eq. (2.32). For instance, for a catalyst paste, my opinion is that the paste moves as a plug through the screw channel. Hence, the one-half factor in Eq. (2.32) should either be dropped or adjusted. More research is required in order to shed further light on this.

2.3.2.2.2 Backflow Caused by Restriction at the Die

Birley et al. [9] reported that it is possible to represent the paste backflow in a single-screw extruder by Eq. (2.33):

$$R_b = \frac{WH^3\rho_p}{12\eta_l L_h}\Delta P \tag{2.33}$$

where η_l is the dynamic viscosity and L_h is the helical length of the screw. The authors assume the flow to be Poiseuille-like with zero velocity at both the barrel side and the screw side and a parabolic profile in between. This assumption perhaps needs further consideration for catalyst paste extrusion since it is not consistent with the assumption used for the geometric flow rate in Section 2.3.2.2.1. Here, as well, further work is required in order to achieve clarity in this matter. Equation (2.34) then expresses the net flow R_e exiting the die as:

$$R_e = R_g - R_b = \left(\frac{1}{2}\right)\pi WHD_bN_s\rho_p \cos(\theta) - \frac{WH^3\rho_p\Delta P}{12\eta_l L_h} \tag{2.34}$$

2.3.2.2.3 Extruder Operating Point

As a function of the pressure drop over the auger, Eq. (2.34) represents a linear relationship with a downward trend. Combined with Eq. (2.26), the crossover yields the operating point of the extruder. Equation (2.34) applies to plastics and needs adjusting for a catalyst extrusion paste, likely by dropping or adjusting the one-half factor from the expression for the geometric rate while carefully considering the backflow expression.

The extruder characteristic equation approach is graphically very attractive and clearly showcases the different parameters at play. It is similar to the treatment of the operating point of a centrifugal pump in a pipeline resistance network.

2.3.2.3 Model 2: Pressure Profile along the Auger and the Die

Burbidge and Bridgwater [3] apply a less graphic but perhaps more accurate approach. The authors numerically integrate the pressure profile through the auger, starting from atmospheric pressure and an assumed paste velocity through the auger channel. At the end of the auger, the calculated pressure at the die will yield an extrusion rate through the die from Eq. (2.26). The extrusion rate then has to match up with the assumed paste velocity at the start of the calculation. If the values do not match, then one will have to adjust the parameters until they do. Burbidge and Bridgwater [3] apply this approach, which yields an accurate calculation of extruder performance.

One caveat is that this approach assumes that the auger is flooded and that the effective auger length is known. That is not always possible for commercial extruders of catalyst paste, nor for pilot extruders. There may be a scenario in which the auger is only partially filled with paste and the length of the flooded part of the auger is a variable. According to the feed rate to the extruder, the auger fill length would need to come to a steady state. A different extruder feed rate would yield a different auger fill length. A pressure reading at the die would greatly help this effort and yield valuable information for extruder operation.

According to Benbow and Bridgwater [2], the materials they extrude are ceramics with little pore volume, and these are likely abrasive in nature. The ability of these materials to adsorb water is limited. Applying pressure may release some of the water and allow the water to form a thin film at the interface between the paste and the screw or at the barrel surface, as they mention.

In contrast, the materials typically used in catalyst research and development and in catalyst manufacturing are softer in nature than ceramic materials and have higher pore volumes and surface areas. It is necessary to monitor shear during mixing, mulling, and kneading, as well as further shear in the extruder itself during extrusion, in order to avoid over-mulling and creating shear-thinning conditions at the barrel side that make it difficult for the barrel to get a grip on the paste. A good grip is essential to creating movement through the barrel and pressure at the die face. A further distinction is that the typical diameters of the extrudates used in the catalyst industry are much smaller than those used in the work by Benbow and Bridgwater [2].

2.3.2.4 Model 3: Friction-based Models

2.3.2.4.1 General Background

The barrel, the auger, and the paste interact and drag the paste down the extruder. The interaction builds pressure as the paste finally pushes down on the die and exits through the die channels.

The analyses shown in Model 1 and Model 2 for single-screw extrusion are interesting and good methodologies for describing the paste flow in an extruder.

Some sources take into account the increased wall friction due to the increased pressure; other sources do not. The effect of wall friction and the increase of wall friction with pressure are common in analyses when storing granular materials in silos.

In regard to auger and paste movement, a principle may have been overlooked in the extrusion literature, one that is related to a marine capstan. A capstan is a cylinder that allows for a rope to be wrapped around it. This rope is connected to and intended to hold a load on the other end. Use of a capstan is widespread in mooring ships and tying down sails in marine applications. The principle is known as the capstan equation or belt friction equation. It was introduced and described in the 1700s and is quantified by the Euler–Eytelwein equation. The capstan equation relates the ratio of the load force to the hold force at the gliding point as a function of the friction factor of the rope against the capstan and the total angle of rotation around the capstan. The radius of the capstan has no influence on the ratio of the tensile forces; it is only the friction factor and the total angular rotation that play a role.

Extrusion paste spirals around the auger and goes from a nearly nonexistent feed-end load to a very appreciable load at the location of the die. As the paste moves forward along the auger (i.e. the rope around the capstan), it is under compressive stress rather than tensile stress, as with a capstan. The force differential brought about by the angular rotation of the auger, however, is the same as it is for a capstan, and that is the principle that may have been overlooked.

2.3.2.4.2 The Euler–Eytelwein Equation or Capstan Equation

The capstan equation yields the tensile forces acting at both end points of a rope that is wrapped around a capstan (a cylinder) over a certain fractional number of turns (expressed as angular rotations in radians). The equation is derived for a condition under which the forces along the rope and the capstan are in equilibrium due to friction between the rope and the capstan just before the rope slips.

The capstan equation in Eq. (2.35) expresses the relationship between the forces at both ends and is exponential in nature; thus, it is possible to hold a large tensile stress at one end (a mooring ship) with a much smaller force at the other end. Equation (2.35) expresses this relationship:

$$F_0 = F_L e^{-\mu\theta} \tag{2.35}$$

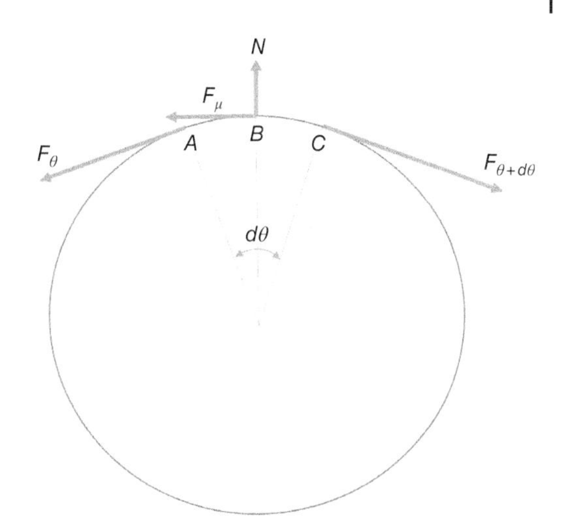

Figure 2.8 The capstan arrangement of forces.

where F_0 is the force at the mooring point (the capstan), F_L is the force at the load end (the ship side), μ is the friction coefficient between the capstan and the rope, and θ is the total angle of rotation in radians subtended by the rope.

The capstan equation is derived via a force balance over a differential angle $d\theta$ of the capstan, as shown in Figure 2.8. Here, $F_{\theta+d\theta}$ is the tensile stress on the rope at the load side, F_θ is the tensile stress on the capstan side, F_μ is the friction force at the point of slipping, and N is the normal force that the capstan exerts on the differential rope arc (ABC) subtending the two tensile stress forces. The force balance leads to Eq. (2.36) for a capstan:

$$\frac{dF}{d\theta} = \mu F \tag{2.36}$$

Integration of Eq. (2.36) then leads to the well-known capstan equation and yields the ratio of the forces in equilibrium at the gliding point.

Indeed, a baby with a proper capstan can keep a mooring ship from drifting away. That does not mean, however, that a baby could pull in a mooring ship with little force, because in that case friction is reversed. As can be seen in Figure 2.9, one would need the strength of Hercules to pull in a mooring toy ship via a capstan.

In addition, for conditions where the rope has a constant velocity v, the friction turns into a heat rate, W_h, expressed in Eq. (2.37):

$$W_h = v{*}F_L{*}\left(1 - e^{-\mu\theta}\right) \tag{2.37}$$

This mechanical picture of forces is very interesting and is potentially applicable to extrusion. Tensile forces for the rope would be replaced by compressive forces

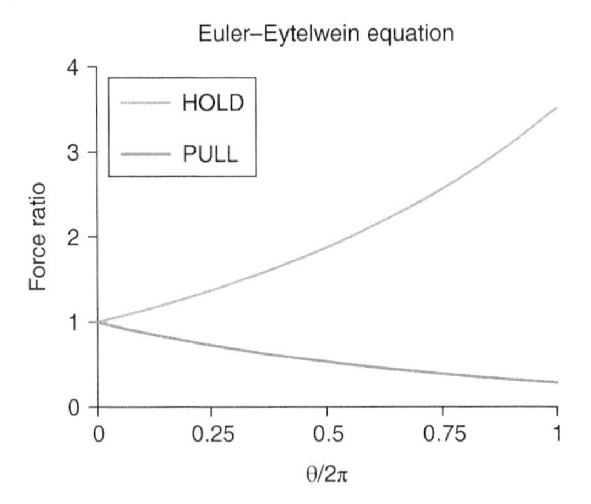

Figure 2.9 The capstan and the mooring ship.

for the paste. Friction forces along the auger are counteracted by evenly distributed frictional forces along the barrel surface instead of point forces at the rope ends. This type of behavior I have at times observed when the extrusion of a particular mixture comes to a grinding halt because it is either too dry or has not been worked enough.

2.3.2.5 Pictorial for Paste Movement against a Blind Die

Sometimes, the die becomes plugged and operates as if there were no extrusion channels. The rotating auger simply presses the paste against the "blind" die and there is evidently no net forward movement of the paste. The friction forces of the auger and the barrel determine the die pressure. Relatively speaking, the paste is forced to continuously snake back in a spiral motion along the screw surface flights and the core, as shown in Figure 2.10, and along the lines addressed in Section 2.3.2.2.2 on backflow due to pressure at the die.

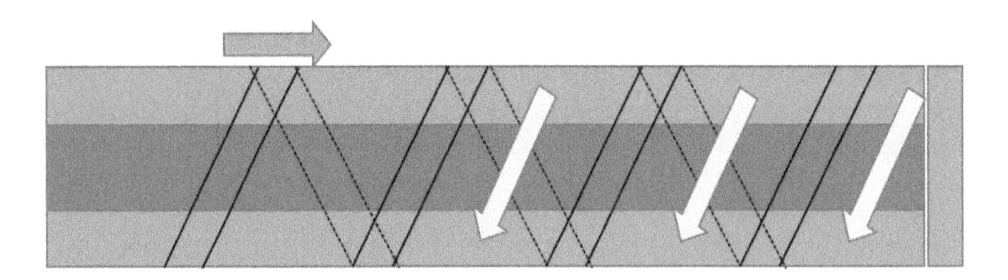

Figure 2.10 Paste snaking back in a screw auger.

During catalyst extrusion, the die may also "act" as if it has become plugged when the paste has been over-sheared. Over-shearing creates a layer of liquid (catalyst slurry) at the barrel surface and prohibits the development of a good pressure profile and sufficient pressure for the paste to extrude through the die. The paste turns along with the screw and slips at the barrel surface due to the layer of slurry.

The pressure obtained at the die can be calculated in a similar way as with the capstan equation in Eq. (2.35). Equation (2.38) is the differential equation that determines the compressive forces on the spiral snake ends:

$$\frac{dP}{d\theta} = \mu_c P + \mu_f P = \mu P \tag{2.38}$$

where μ_c is the effective friction coefficient against the auger core and μ_f is the effective friction coefficient against the auger flight.

Upon integration of Eq. (2.38), one gets:

$$P_D = P_0 e^{\mu\theta} \tag{2.39}$$

A paste snake would have a certain angle of rotation, θ. Only a few rotation angles would be necessary to create a very substantial pressure, P_D, on a clogged die, starting from an ambient pressure, P_0. Hence the need for safety cutouts. Some catalyst pastes extrude only with great difficulty and the safety device will cut out the extruder; in such cases, the recipes need to be adjusted.

2.3.2.6 Pictorial for Paste Movement against an Open Die

Paste motion with an open die is perhaps easier to visualize when viewing the barrel housing as rotating around a stationary screw. Burbidge and Bridgwater [3] and others also favor this view. The die rotates with the barrel and the paste snakes forward as a spiral around the auger. The friction force that the barrel exerts on the paste is the driving force, and the paste glides with friction along the screw.

The movement is easily envisioned when the angle of the screw flight is relatively small. If the angle were to increase, it would become increasingly difficult for the paste to glide against the screw. For even larger angles, the movement of the paste would no longer be of practical use.

Nomenclature

a	Acceleration (m^2/s)
a'	Coefficient ($Pa^{-1}\ m^3/s$)
b'	Coefficient (m^3/s)
D_b	Extruder barrel diameter (m)
D_{cal}	Diameter of a calcined strand of catalyst (m)

D_C Diameter of the core of a screw (m)
D_0 Diameter of an extrusion channel (m)
D_1 Diameter of a channel (m)
D_2 Diameter of a channel (m)
D_r Diameter of a ram extruder (m)
D_s Outer diameter of an auger (m)
F Force (N; also Pa.m^2)
F_f Force parallel to the surface (Pa.m^2)
F_N Force normal to the surface (Pa.m^2)
F_0 Force at the capstan (Pa.m^2)
F_L Force at the load end (Pa.m^2)
g Gravitational acceleration (m^2/s)
H Height of a rectangular channel (m)
k Number of extrusion channels in a die (–)
K Factor depending on channel shape (m^4)
L_h Helical length of an auger (m)
L_0 Length of an extrusion channel (m)
m Mass of a body (kg)
n Exponent (–)
n_1 Exponent (–)
n_2 Exponent (–)
N Force normal to a surface (Pa.m^2)
N_s Rate of rotation of a screw in revolutions per unit time (s^{-1})
O_M Location of the Mohr circle on the x-axis (Pa)
p Pitch of an auger (m)
P Pressure (Pa)
P_D Pressure on a clogged die (Pa)
P_0 Ambient pressure (Pa)
P_d Pressure at the paste side of a die (Pa)
ΔP Pressure drop (Pa)
Q Volumetric rate (m^3/s)
R_M Radius of the Mohr circle (Pa)
R_b Backflow mass rate in an auger (kg/s)
R_e Mass rate of extrusion (kg/s)
R_g Geometric mass extrusion rate (kg/s)
s Solids content of an extrusion paste (%)
t Time (s)
T Force applied parallel to a surface (Pa.m^2)
T_s Minimum force of static friction (Pa.m^2)
T_k Minimum force of kinetic friction (Pa.m^2)
T_l Thickness of the lip of an auger (m)

v Velocity (m/s)
v_b Velocity at the barrel surface (m/s)
v_c Velocity in an auger channel at the barrel surface (m/s)
v_∞ Terminal velocity (m/s)
W Width of a rectangular channel (m)
W_h Heat rate (J/s)
x Coordinate (m)
y Coordinate (m)
β^+ Parameter for the lumped shear/yield stress expression (Pa.s.m^{-3})
δ Thickness of a liquid layer (m)
η_b Dynamic viscosity (Pa.s)
η_l Dynamic viscosity of fluid (Pa.s)
μ Friction coefficient (–)
κ Parameter (–)
μ_0 Friction coefficient at zero velocity (–)
μ_c Friction coefficient against the auger core (–)
μ_f Friction coefficient against the auger flight (–)
μ_k Kinetic friction coefficient (–)
μ_s Static friction coefficient (–)
$\mu_{k,0}$ Kinetic friction coefficient at near-zero velocity (–)
ρ_p Density of a strand of catalyst (kg/m^3)
ρ_c Density of a calcined strand of catalyst (kg/m^3)
θ Angle (radians)
σ Normal stress (Pa)
σ_0^+ Lumped yield/shear stress for an extrusion channel (Pa)
σ_u Uniaxial yield stress (Pa)
σ_0 Uniaxial yield stress at zero velocity (Pa)
τ Shear stress (Pa)
τ_0 Wall shear stress at zero velocity (Pa)
τ_w Wall shear stress (Pa)
Ω Cross-sectional area (m^2)

References

1 Mezger, T.S. (2014). *The Rheology Handbook – For Users of Rotational and Oscillatory Rheometers*, 4e. Hanover, Germany: Vincentz Network. ISBN: 978-3-86630-842-8.

2 Benbow, J. and Bridgwater, J. (1993). *Paste Flow and Extrusion*. Oxford, UK: Oxford University Press.

3 Burbidge, A.S. and Bridgwater, J. (1995). The single screw extrusion of pastes. *Chemical Engineering Science* 50 (16): 2531–2543.

4 Parks, J.R. and Hill, M.J. (1959). Design of extrusion augers and the characteristic equation of ceramic extrusion machines. *Journal of the American Ceramic Society* 42 (1): 1–6.

5 Orisaleye, J.I. and Ojolo, S.J. (2017). Parametric analysis and design of straight screw extruder for solids compaction. *Journal of King Saud University – Engineering Sciences* https://doi.org/10.1016/j.jksues.2017.03.004.

6 Laenger, F. (2009). Rheology of ceramic bodies. In: *Extrusion in Ceramics* (ed. F. Händle), 141–159. Heidelberg, Germany: Springer-Verlag.

7 Winstone, G. (2011). Production of catalyst supports by twin screw extrusion of pastes. Doctorate thesis. University of Birmingham.

8 Mills, H. and Blackburn, S. (2002). Rheological behavior of γ alumina/boehmite pastes. *Chemical Engineering Research and Design* 80 (5): 464–470. https://doi.org/10.1205/026387602320224049.

9 Birley, A.W., Haworth, B., and Batchelor, J. (1992). Single-Screw Extrusion – The Extruder Characteristic. In: *Physics of Plastics: Processing, Properties and Materials Engineering*. Cincinnati, OH: Hanser.

3

The Aspect Ratio of an Extruded Catalyst

An In-depth Study

3.1 General

The aspect ratio of an extruded catalyst, also sometimes understood as the length-to-diameter ratio of an extruded catalyst, is an important property in commercial catalyst manufacturing. A short aspect ratio often leads to an excessive pressure drop in the reactor. Long aspect ratios tend to create a low-weight fill due to an excessively high void fraction. The shape and size of a catalyst is determined by the application and the conditions the catalyst will endure.

In this book, the aspect ratio of an extruded catalyst batch stands for the average length-to-diameter ratio of a representative sample of the catalyst batch. The aspect ratio is defined as the arithmetic average of the individual length-to-diameter ratios of each extrudate in the sample of interest.

The catalyst manufacturing plant's architecture and the available equipment determine the possible catalyst manufacturing methods. During handling, a catalyst can become damaged by breakage; this is called "natural breakage." It is an objective for a catalyst manufacturing plant to minimize this natural breakage because it hampers the production rate and is a source of waste. During loading in a reactor, the catalyst can become damaged as well. Careful loading procedures either via the use of a catalyst-oriented packing method or a sock-loading technique are often applied. At times, a catalyst can be very resilient to natural breakage, but it still needs to be sized appropriately. This sizing is called "forced breakage."

The mechanical strength of a catalyst and the severity of its handling govern both natural breakage and forced breakage.

The mechanical strength of a catalyst is an important subject of this book. As such, it goes hand in hand with the application of different methods for measuring strength in the general arena of natural and forced breakage of catalyst extrudates. This chapter investigates the breakage of extrudates caused by collision with a

Catalyst Engineering Technology: Fundamentals and Applications,
First Edition. Jean W. L. Beeckman.

surface and the breakage of extrudates under the influence of stress in a fixed bed. I then further apply what is learned to commercial plants and show how this methodology can quantify the severity of an operation and also the severity of an entire plant.

The bending strength of extrudates, especially the force at rupture in bending mode, is a major strength property of an extrudate. The bending strength can be used to mathematically model breakage caused by collision and stress.

Although I have already published some of my basic research, I am compiling all of my research and the experimental work of my coauthors in these publications into this chapter for ease of use. I also add applications for commercial manufacturing plants.

One objective in this chapter is to find a way to measure both catalyst breakage and mechanical strength in a well-controlled manner, and then to combine the two phenomena in order to obtain a basic model. A second objective is to better predict catalyst extrudate breakage in commercial plants and perhaps relate it to the architecture and equipment of a plant.

These objectives apply to the catalyst industry, although others, such as the food or pharmaceutical industries, may also find this approach interesting or worthy of further pursuit.

I consider two phenomena of catalyst breakage: first, catalyst breakage caused by an impact against or collision with a surface. This mode of breakage is highly complex because of the varied number of ways in which an extrudate can hit a surface. In addition, the condition of the surface – hard or soft – plays a major role. The surface may also be a bed of extrudates that may be stationary or in motion. Nevertheless, I will demonstrate that a reasonable agreement is obtained between well-controlled experimental collisions and a rather general model for the prediction of the reduction of the aspect ratio of an extruded catalyst.

The second phenomenon of breakage is when the catalyst is in a fixed-bed arrangement and when stress is applied to the bed. Again, the physics is highly complex and varied, yet a rather simple mathematical model is developed that allows one to predict the reduction of the aspect ratio along with the formation of fines.

In this book, I will only consider the average length-to-diameter ratio of a sample and not the distribution of length-to-diameter ratios. All of the research and direct applications in this book deal with the arithmetic mean of the distribution. It is important to determine this average from a representative sample of the material. The material can be a small 20-g sample, but it can also be a 500-kg catalyst bin or a 100 000-kg catalyst fill of a reactor vessel.

Taking a representative sample is a very important task. I suggest consulting the appropriate sources or contacting vendors who can assist with sampling equipment.

3.2 Introduction to Catalyst Strength and Catalyst Breakage

For an excellent overview of current thinking regarding the measurement of catalyst strength, see Le Page [1], Woodcock and Mason [2], and Bertolacini [3]. Some tests have been around for decades; thus, a very large but often difficult-to-consult database exists within some companies and published works.

With the exception of the American Society for Testing and Materials (ASTM), test operating procedures are sometimes described rather nonspecifically, which makes it difficult to compare results obtained. Some companies, institutions, and specialized analytical laboratories will conduct a round-robin test that then allows to compare the experimental test values obtained at different sites.

Wu et al. [4] and Li [5] have shown the difficulties of catalyst strength measurement and the need for standardized methodologies. A standardized methodology is probably one of the most important facets of catalyst strength measurements.

The crush strength of a catalyst is often measured by the force that crushes a catalyst extrudate that is contacted over a certain distance called the width of the anvil of the instrument. The measurement is repeated several times and the arithmetic average is reported as the crush strength of the extrudate. The individual measurements of the crush strength of extrudates have a wide distribution, which is typical when measuring catalysts and catalyst supports. I will discuss this more in Section 3.4.3.

The ASTM D7084-04 test method describes another method known as the bulk crush strength, which entails the following steps:

- Fill a shallow cylinder with catalyst.
- Apply a force on top of the bed up to a specific value.
- After unloading the catalyst, sieve the catalyst sample.
- Report the fines obtained due to breakage for the force applied.
- Repeat the test with a fresh catalyst sample at a different force level and determine the amount of fines as a function of the force applied.

It is the amount of fines that is chosen as a measure of the strength of the extruded catalyst.

Recently, the bending strength of a catalyst has also come to the forefront, but it is perhaps not yet widely applied. I refer to Li et al. [6] and Staub et al. [7] for more details on this matter in the literature. The methodology is presented, but measurements are done on fairly large-diameter extrudates and hence the studies are beyond the scope of this book. The results and discussions are more of an analytical nature rather than having an immediate application to pilot plants or commercial plants. In Beeckman et al. [8], I measured and applied the bending strength

Figure 3.1 Leonhard Euler.

method on commercial catalysts in the range of their typical sizes and shapes and found it to be a practical tool for describing a catalyst's strength and its resistance to breakage in pilot units and commercial plants.

Around 1750, Leonhard Euler (see Figure 3.1) and Daniel Bernoulli (see Figure 3.2) developed a mathematical theory to accurately describe how beams bend under specific force loads. This leap in understanding led to the concept of the Euler–Bernoulli modulus of rupture. In catalyst terms, I applied the Euler–Bernoulli theory to catalyst extrudates of typical commercial sizes and shapes. I used a three-point bending test with two support points and with the rupture force applied in the middle. The modulus of rupture is the tensile stress on the extreme underside of the extrudate in bending mode that causes breakage, as shown in Figure 3.3. The symbol for the modulus of rupture (or the tensile strength) is σ, and it is typically reported in units of pascal (Pa).

The maximum bending stress is located at the side of the extrudate opposite to where the force is applied; that is, the side where the extrudate fails in bending mode. The bending strength is the arithmetic average of several measurements on an adequate number of extrudates. There is a wide distribution of values for the bending force at the breaking point, but that just happens to be a characteristic of extrudates.

Figure 3.2 Daniel Bernoulli.

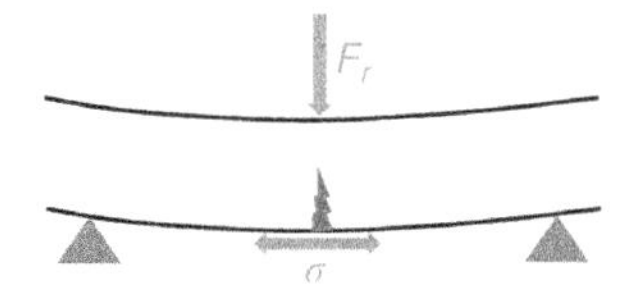

Figure 3.3 Three-point bending of an extrudate.

It is crucial to calibrate instruments frequently and prepare extrudates before taking measurements. In addition, moisture in the ambient atmosphere plays a role in the strength of extrudates and requires control through proper engineering design.

The strength of an extruded body is determined by the bonds among the particles making up the catalyst extrudate body. The bonds may not be as frequent, many, or homogeneous as one might expect. For instance, the bending stress at the breaking point is located in a very small area of the catalyst opposite to where the force is applied. The structure of the catalyst in that locale can hardly be called homogeneous "on a molecular scale," and hence that may help explain the broad range of strengths typically measured on individual catalyst extrudates of the same batch.

Beyond the strength of catalysts, there is the breakage of catalysts. In his thesis, Papadopoulos [9] describes the collision of particulate solids against a surface and describes the attrition and the comminution of the material. Since the particulates have an approximately spherical shape, however, similar results for extrudates are unlikely.

Salman et al. [10] and Subero-Couroyer et al. [11] describe a similar collision between granules and a flat surface, but again, this is not directly applicable to extrudates.

Data on extrudates with an aspect ratio typical of a commercial catalyst is scarce, perhaps because of the tediousness of collecting manual aspect ratio measurements. But today, the procedure is nearly fully automated such that the measurements are no longer a challenge.

Bridgwater et al. [12, 13] conducted a comprehensive study on particle breakage and looked at granular materials, beads, and extrudates. Unfortunately, the aspect ratio of the extrudates was very small and beyond the range of typical commercial catalysts.

Li [14] applied a statistical analysis to breakage and developed a probabilistic model to predict the breakage of particles with an extrudate shape. The aspect ratio here was also very small since the catalyst was obtained by tableting, a technique that is well studied and practiced in the pharmaceutical industry.

The discrete element method (DEM) applied to particle breakage provides knowledge at a very fundamental level. Heinrich [15] investigated the breakage and attrition of granules with a multiscale strategy concerning bending, compression, and shear using DEM calculations. Other researchers such as Wassgren [16], Potapov [17], Hosseininia and Mirgashemi [18], Potyondy [19], and Carson [20] applied the DEM to large-grain ensembles subjected to forces and showed how those forces distribute from grain to grain. Such a detailed approach enables those studying breakage in individual particles to understand the mechanism of particle breakage when stressed.

In Beeckman et al. [8, 21], I modeled catalyst breakage in a collision test and established a finite-difference model to describe the breakage with two physical parameters:

- The first parameter is the asymptotic aspect ratio obtained after a large number of impacts of the same severity; for example, the aspect ratio after a large number of drops from the same height of a sample of extrudates against a surface, measuring the aspect ratio after every drop.
- The second parameter is the asymptotic aspect ratio obtained after a single drop of a sample of extrudates that are progressively longer before the drop.

The model I use is heuristic and is a first-order Padé approximant as well as a Riccati equation. I have applied the model in simple applications of severity conditioning and severity sequencing.

Later, in Beeckman et al. [22], I was able to express the two parameters in a fundamental correlation using Newton's second law and the Euler–Bernoulli modulus of rupture, which led to a dimensionless group that is at the core of the breakage phenomenon.

I also applied the bending strength to breakage of extrudates in a fixed-bed setting under the influence of stress. I modeled the reduction of the aspect ratio using the bending strength and was able to describe the aspect ratio reduction as a function of the applied stress. Just as with collision, it is possible to define a dimensionless group that describes breakage caused by stress in a fixed bed. I was also able to model the formation of fines in such a fixed bed under stress.

Wu et al. [23] have shown the importance of breakage in a fixed bed, particularly with respect to the increased pressure drop due to a smaller aspect ratio and a larger fines content of the catalyst bed from breakage. More research is needed in the area of particle breakage due to the inherent variety of the mechanical strength and physical properties of catalysts. Studies and knowledge tend to be ad hoc, and few rules of thumb exist. For catalysts with an aspect ratio above two to three, sources are scarce when describing the breakage of typical catalysts from collision or in a fixed bed under stress.

3.3 Mechanical Strength of Catalysts

Let us elaborate on the knowledge gained over the past few decades concerning catalyst strength, how to use that knowledge to compare materials, and how to judge optimal material size and shape for catalyst manufacturing.

A body under strain (a body under deformation) caused by external forces will show tensile stresses throughout at various levels. It is the area of the body under maximum tensile stress that, when compared to σ (the tensile strength at rupture), determines material failure or material breakage. The value of σ is a property of the material and is theoretically not a function of the extrudate shape (quadrulobe or cylinder), size (diameter), or form (extrudate versus bead).

Since the method used to make catalysts can influence the many strength and rupture intricacies experienced during production, it is wise to measure the strength as a property of interest for the application at hand. The formal relationships between strength measurements such as side crush strength and modulus of rupture do apply to extruded materials and also to catalysts in beaded form. These relationships all derive from the theory of elasticity and can be used as a guide to compare catalyst strength.

The tensile strength of a catalyst – defined as the maximum stress that can be applied perpendicular to an internal surface without breakage (rupture) – determines its mechanical strength. This tensile strength is not necessarily a single value, but depends on the stage of the catalyst manufacturing process. For instance, a catalyst that is extruded and dried is often less resistant to breakage than that same catalyst after calcination.

Furthermore, a calcined catalyst often loses strength when impregnated with metal solutions. Humidity and seasonal changes can also play a role in the breakage of catalyst extrudates.

In addition to tensile stress, there is also shear stress for a body in deformation. The breakage of catalyst bodies by collision or stress in a fixed bed is essentially caused by bending; research shows that tensile strength governs breakage.

The industry also uses other units of measurement of catalyst strength in addition to tensile strength. The average side crush strength of single extrudates is expressed as a newton/mm (N/mm). Another is the bulk crush strength when stress is applied to a bed of extrudates, and this is measured in pascal. For spherical catalysts, the average force in newton (N) that crushes the catalyst is often used as a measure of strength. Though all of these strength variants appear different at first sight, they all can be brought together under the same roof and are, in the first approximation, exhibits of the tensile strength.

Let us explore in more detail how all of these measures of mechanical strength are determined and how they are related. It is necessary to exercise care and caution when predicting various catalyst strength measures based on tensile strength. The shape and size of a catalyst can influence the mechanical strength properties during the actual forming of the catalyst. Therefore, if it is possible to measure a specific strength property, then it is worth taking this route.

This book documents formal relationships between the modulus of rupture for extrudates, the side crush strength of extrudates, the crush strength of spherical beads, and the bulk crush strength of extrudates. These formal relationships have been documented in the literature before and are based on elasticity theory. They serve as a guide to compare and predict particle strengths, but they should not replace individual strength measurements because of the many intricacies in catalyst preparation. Predictive capabilities do lead to a more fundamental understanding and insight with respect to catalyst mechanical strength properties.

3.3.1 Bending Strength of Extrudates

The bending strength of materials is a well-known property in metals and other construction materials such as wood and concrete. The theory of elasticity is very well developed and mature, as represented in Timoshenko and Goodier's [24] excellent textbook, for instance.

The modulus of rupture relates the force required to break a catalyst extrudate to the intrinsic strength of a catalyst in a three-point bending test mode. Figure 3.4 is a schematic of a catalyst extrudate exposed to an external bending force perpendicular to its length axis. A catalyst is a brittle material that bends when a load is applied up to a point. When the load is released, the body returns to its original

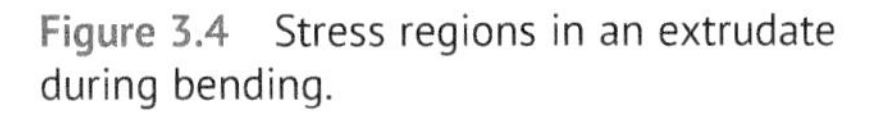
Figure 3.4 Stress regions in an extrudate during bending.

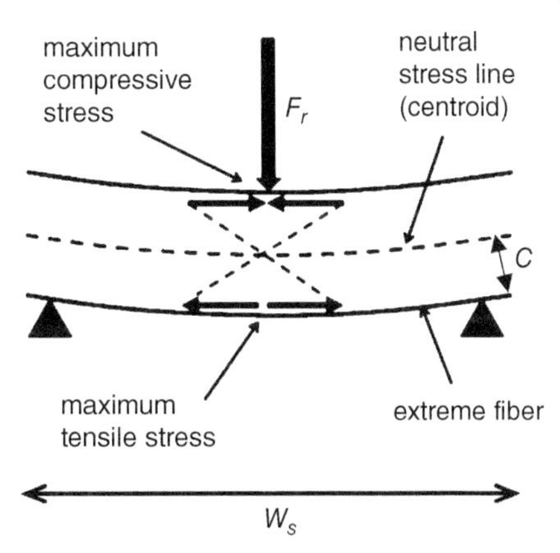

shape. The deformation is very small and cannot be observed with the naked eye unless the extrudates are very long. This behavior defines the elastic regime. When the load is increased further, the body bends more until a breaking point is reached, and the rupture force applied at that point defines the modulus of rupture.

There are two regions in a catalyst extrudate where different material strength properties are at work due to bending. The force for bending is exerted on the top of the extrudate. At the top, the stress along the axis of the extrudate is compressive (the area where arrows are pointing toward each other in Figure 3.4) and it acts against the compressive strength of the extrudate – think of folding the Yellow Pages in half.

The force exerted at the bottom of the extrudate is the area of maximum tensile stress by bending – which means the maximum stress is in what is called the extreme fiber (located where arrows are pointing away from each other in Figure 3.4). The bending force acts against the bending strength of the material or tensile strength of the extreme fiber.

The forces along the cross section of the extrudate go from compressive at the top of the extrudate to tensile at the bottom of the extrudate and are in opposing directions; hence, there is a line of zero stress (called the neutral stress line) where the extrudate has neither tensile nor compressive force applied to it.

Recently, I applied the concept of the modulus of rupture to commercially sized and shaped catalysts in order to quantitatively predict how catalysts break – and to what extent – during collision and compression in a fixed bed.

The modulus of rupture is determined in a bending experiment. The catalyst extrudate is laid to rest on two supports at a distance W_S apart. A downward force is applied in the middle between the two supports and slowly increased until rupture occurs at a rupture force F_r. Equation (3.1) calculates the modulus of rupture (MOR) for a cylindrical extrudate of diameter D as:

$$MOR = sF_r W_s / D^3 \tag{3.1}$$

where s is an extrudate shape factor equal to $8/\pi$ for a cylinder. The modulus of rupture is equal to the tensile strength expressed in Eq. (3.2):

$$MOR = \sigma \tag{3.2}$$

Table 3.1 lists the shape factors for extrudates of several shapes in the industry. For two extrudates made of the same material and with the same quality, the force required to break a square extrudate is approximately 70% higher ($16/3\pi$) than a cylindrical extrudate. Many sources exist that calculate the various shape factors that come into play through mechanical properties such as the section modulus, the moment of inertia, and the location of the neutral axis in the extrudate. For quadrulobe- and trilobe-shaped catalysts, the information was not available in the literature and I performed the analysis numerically using Mathcad 15.

In the absence of any analytical instrument, it is possible to gather comparative information on the tensile strength of an extrudate, as shown in Figure 3.5. Simply collect a few specimens that are as straight as possible. Without creating a bending action, slowly pull the extrudate along its axis until rupture occurs. The force applied at rupture is a measure of the tensile strength.

3.3.2 Extrudate Side Crush Strength

To obtain the side crush strength of a catalyst, place an extrudate flat on a surface and measure the force, F_r, required to crush or rupture the extrudate, as shown in Figure 3.6. To achieve this result, an anvil of a specific width W_a oriented perpendicular to the axis of the extrudate is lowered at a slow rate. The ratio of the force required to crush the extrudate to the width of the anvil is called the catalyst side crush strength. The anvil's width is 1–5 mm and the side crush strength is expressed in Eq. (3.3) as the rupture force per linear distance:

$$SCS = F_r / W_a \tag{3.3}$$

It would appear at first sight that the extrudate would be in a compressive mode with no tensile forces involved, but this interpretation is incorrect. Figure 3.7 shows the location of maximum tensile stress at the center of the extrudate axis, perpendicular to both the applied force and the extrudate axis. To visualize this experiment, I find it most helpful to think of the catalyst extrudate as a cylindrical elastic material such as rubber. This material would deform into an ellipsoidal cross section under the applied force. The material located along the long axis of the ellipse is stretched by the material above it that pushes the body outward.

Table 3.1 Moment of inertia, neutral axis, and shape factors of the various cross sections considered.

centroid C	CYLINDER	TRILOBE	QUADRULOBE	HOLLOW CYLINDER	BEAM
	D	D, αD	D, αD	D, αD	D, αD
α	—	$\frac{1}{1+\sqrt{3}/2}$	$1/(1+\sqrt{2})$	$0<\alpha<1$	$0<\alpha<1$
φ	$\pi/4$	$\frac{\pi+4\sqrt{3}}{(2+\sqrt{3})^2}$	$\frac{16+\pi}{4(1+\sqrt{2})^2}$	$(1-\alpha^2)\pi/4$	α
c/D	1/2	$\frac{1+\sqrt{3}}{(3+2\sqrt{3})}$	1/2	1/2	1/2
I/D^4	$\pi/64$	0.0463	0.0567...	$(1-\alpha^4)\pi/64$	$\alpha/12$
S	$8/\pi$	2.282...	2.2046...	$(8/\pi)/(1-\alpha^4)$	$3/(2\alpha)$
ψ	2	1.65	1.81	$2/(1+\alpha^2)$	3/2

Source: Courtesy of Wiley, doi: 10.1002/aic.15231.

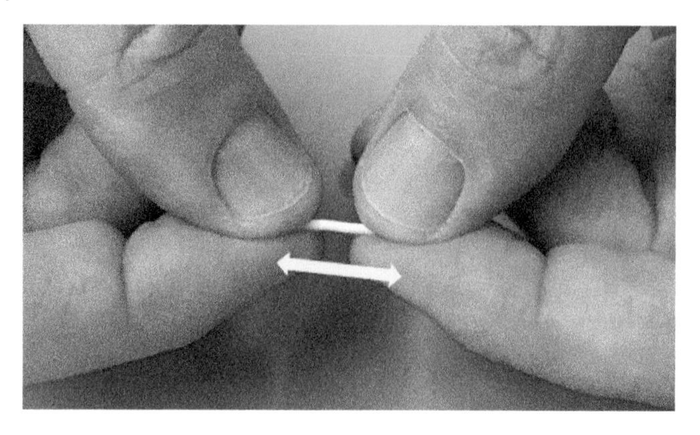

Figure 3.5 Manual tensile strength of an extrudate.

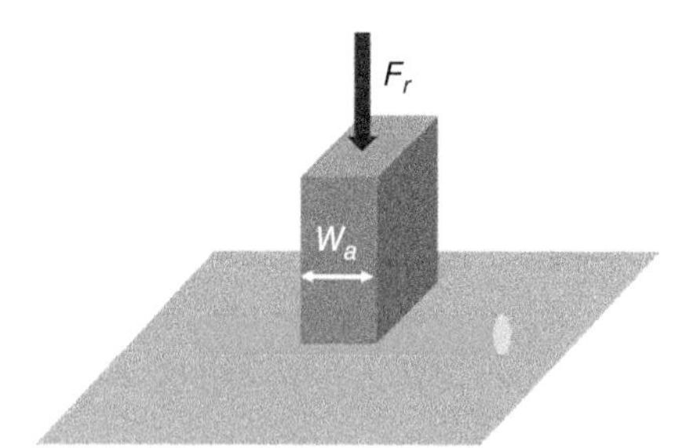

Figure 3.6 Side crush strength measurement of an extrudate.

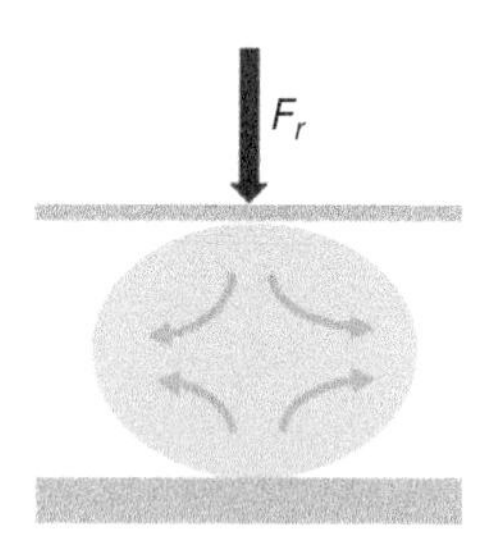

Figure 3.7 Tensile stress in a cylindrical extrudate due to an external force.

This action brings the material along the long axis under tensile stress [24]. According to the theory of elasticity, Eq. (3.4) expresses the relationship between the tensile strength and the rupture force for a cylindrical body:

$$SCS = \pi D\sigma/2 \tag{3.4}$$

Figure 3.8 Manual hand crush of an extrudate.

External references do not supply a theoretical value for a quadrulobe shape, but I will offer data that give an approximation, showing how it is comparable to a cylinder of the same diameter.

Especially at a plant demonstrating a new or improved recipe, it can be stressful (no pun intended) to wait and see how well extrudates are resisting a crushing force in the quality control laboratory. In the absence of analytical data, it is possible to perform a crush test by hand using a coin with a flat rim, as shown in Figure 3.8. Of course, this technique is highly subjective and yields no numerical value. On a comparative basis, however, if done by the same person, the process can ease anxiety among plant operators or engineers experienced with the technique.

3.3.3 Extrudate Bulk Crush Strength

Bulk crush strength measures the breakage of a catalyst loaded as a shallow fixed bed in a vertical cylindrical container when applying a compressive stress on top of the bed. ASTM D7084-04 describes this method in detail. In the industry, the amount of fines generated as a function of the applied pressure is taken as a measure of the catalyst's breakage and strength. Personally, I also consider the drop in the aspect ratio Φ of the extrudates as another good measure of the breakage of catalyst in the bed.

Equation (3.5) expresses the dependence of the aspect ratio on the applied stress and tensile strength of the extrudate [25]:

$$\Phi = \Psi\,(\sigma/sP)^{1/3} \tag{3.5}$$

where s is an extrudate shape factor and Ψ is a factor equal to $6^{1/6}$ or approximately 1.35. Equation (3.5) then also allows one to calculate the tensile strength of the catalyst from the aspect ratio and the applied stress P.

3.3.4 Crush Strength of a Sphere

In Figure 3.9, when applying a crushing force F_r to a spherical catalyst (also known as a bead catalyst), the maximum tensile stress is located in the center of the bead. Le Page [1] provides an excellent write-up on the subject. When the stress exerted on the bead has reached the tensile strength of the bead, the bead ruptures. The crush strength of a sphere or a bead, called CSB, is the force that crushes the sphere between two hard parallel surfaces. Equation (3.6) expresses its relationship to tensile strength:

$$CSB = F_r = \pi D^2 \sigma / 2.8 \tag{3.6}$$

This relationship demonstrates that, at least theoretically, the crush strength of a spherical catalyst is proportional to the square of its diameter.

3.3.5 Young's Modulus of Elasticity

Young's modulus of elasticity, also known simply as Young's modulus, measures the stiffness of an extrudate. This property is also expressed in units of pascal. The modulus of elasticity is the slope of the stress versus strain curve. The stress is expressed in units of pascal, while the strain is the fractional deformation of the body.

Equation (3.7) calculates the modulus of elasticity for an arbitrary but constant cross section beam in a three-point bending test:

$$E = \frac{Fw^3}{48I\delta} \tag{3.7}$$

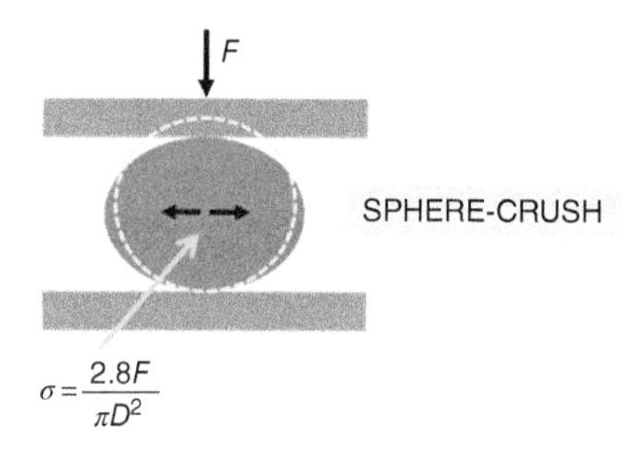

Figure 3.9 Crush strength of a sphere.

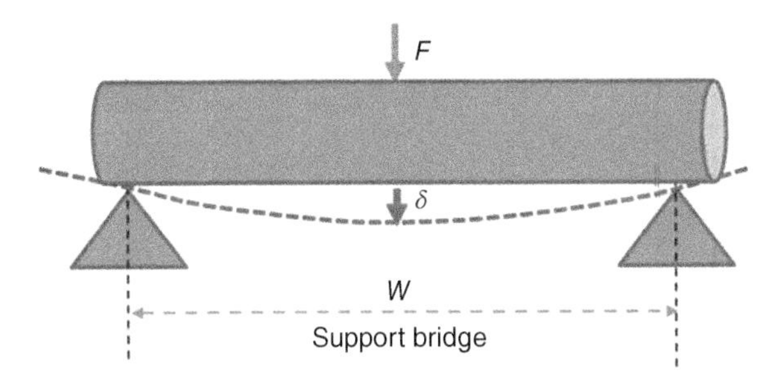

Figure 3.10 Young's modulus of elasticity measured in a three-point bending test.

where F is the actual force acting in the center between two support points a distance w apart from each other, I is the moment of inertia of the extrudate given in Table 3.1, and the deflection δ is the actual measured deflection at the center point, as shown in Figure 3.10.

Instron software has only two hard-coded settings for the shape of the cross section – a beam with a rectangular cross section or a beam with a circular cross section – and calculates Young's modulus of elasticity according to these settings. Because a variety of shapes exist in the catalyst industry, I suggest always using the circular (cylinder) shape as input to Instron. Use the correction factors in Table 3.2 for shapes other than cylinders.

Table 3.2 Correction factors for E from Instron values.

Extrudate Shape	Multiplicative conversion factor for Instron E to catalyst E	Comment
Cylinder	1.0	
Hollow cylinder	$1/(1-\alpha^4)$	α is the ratio of the inner over the outer diameter
Quadrulobe	0.86	
Trilobe	1.06	
Rectangular beam	$0.59/\alpha$	α is the ratio of the width over the height of the beam (the force exerts over the width of the beam)
Hexagon	0.82	

3.4 Experimental Measurement of Mechanical Strength

3.4.1 Bending Strength, or the Modulus of Rupture Instrument

I used the Instron Model 5942 single-column tabletop system shown in Figures 3.11 and 3.12, in conjunction with Instron's Bluehill 3 software, to collect and process the experimental data presented here. The 5942 system has the benefit of being on the smaller side, and it fits nicely on a lab bench. It is taller than it is wide. The system includes a load frame with a controller, a load cell, a fixture (in this case an anvil and three-point bending fixture shown in Figure 3.13), and the software.

The system has two 2530-series load cells: one with 10-N force capacity and one with 50-N force capacity. According to Instron's website, the linearity of both load cells is equal to or better than ±0.25% for loads ranging from 100% down to 1% of the load cell capacity.

Most testing was performed using the 10-N load cell. Figure 3.14 shows the system's controller. The controller is responsible for communication between the transducers and the computer. The load frame itself contains a frame

Figure 3.11 Instron Model 5942 single-column tabletop system.

Figure 3.12 Instron Model 5942 general view.

Figure 3.13 Anvil and micro three-point bend fixture.

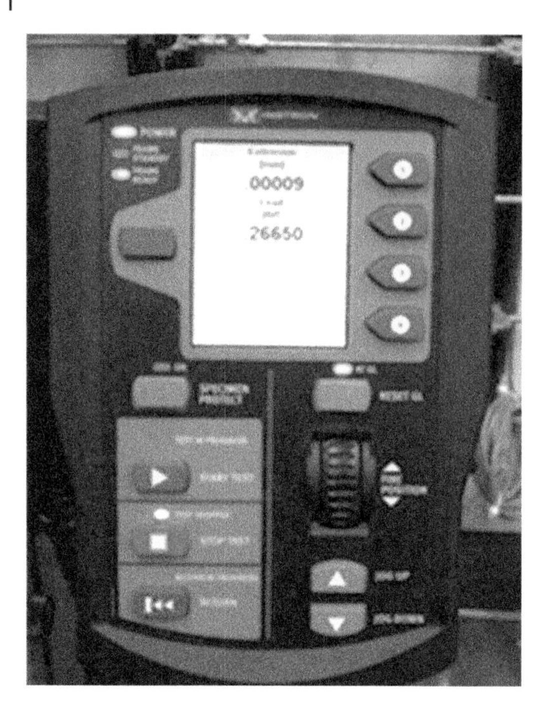

Figure 3.14 Instron Model 5942 controller.

interface board that links the electrical components of the frame and enables communication with the controller. Bluehill 3 software handles the selection of test parameters, system operation, and data collection. This software is used to specify the method for each user containing parameters tailored to differences in samples such as shape and diameter.

The creation of the specifications also includes the choice of anvil speed, length of the support span, rate of data collection, choice of measurements and calculations, and determination of the end of test for a given specimen. Operators can choose live displays and report options. During sample testing, the operator places the extrudate across the three-point bending fixture and starts the test with a mouse click or by pressing the Start Test button on the control panel. The starting position of the anvil on the crosshead is easily controlled using the Jog Up/Down and Fine Position functions on the control panel and can be set as the repeated return point between individual specimens.

Upon sensing the operator-specified decrease in load as the extrudate begins to break, the crosshead returns to its starting position and the system stands by for the next sample specimen. Safety features of the Instron Model 5942 system include an emergency stop button and manually set limit stops to protect against any unforeseen crosshead movement.

3.4.1.1 Strain Rate Sensitivity

The measured rupture force I found is strain-rate sensitive at a high anvil speed, which means that the force at which an extrudate breaks depends on the speed of the anvil. At anvil speeds at or below 2.5 cm/min, the force can be considered constant. Above 2.5 cm/min, the rupture rate decreases when the anvil speed increases.

As an example, Figure 3.15 shows the rupture force as a function of the inverse of the anvil speed. For practical purposes, it is best to use 1.25 cm/min as the standard anvil speed. Some earlier work by my coauthors and I [21, 25] was performed on an older instrument at a higher anvil speed of approximately 25.4 cm/min. Comparing the modulus of rupture values on the Instron at low and high anvil speeds, the results show that the value of the modulus of rupture at a low anvil speed is approximately fivefold higher than the modulus of rupture observed at a high anvil speed. This fivefold factor appears to be similar for the majority of the catalysts sampled.

It is important to conduct the analysis with a consistent anvil speed, whether at a low speed or at a high speed, but a low speed is preferred. Quality control of the instrument is important, and frequent calibration and annual checkups will help facilitate the development of a consistent database.

3.4.1.2 Bridge Width Sensitivity

The width of the support bridge is the distance between the support points of the three-point bending test. One can set the distance at 3, 5, or 7 mm. I prefer the 5- and 7-mm settings, but I use 5 mm as the standard, as it does not create operational issues. In addition, at this setting, the ratio of the gap to the extrudate

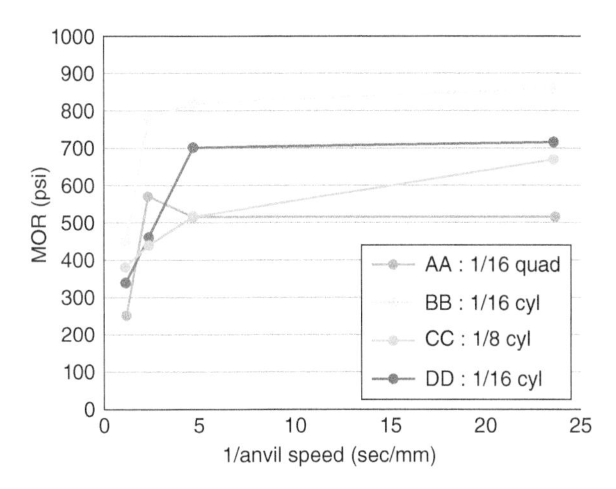

Figure 3.15 Rupture force as a function of the inverse of the anvil speed (5-mm bridge width) (catalyst diameter in fractions of an inch, 1 inch = 25.4mm).

diameter is 3.1 for a 1.6-mm diameter extrudate. For thicker extrudates, such as 3.2 mm in diameter, I prefer 7 mm with a gap-to-diameter ratio of 2.2.

Figures 3.16–3.18 show that the modulus of rupture is fairly constant and independent of the gap, as it should be per the Euler–Bernoulli equation. The rupture force itself is sensitive to the gap and has hyperbolic behavior with respect to the width of the gap.

3.4.1.3 Influence of the Length of the Extrudate

I also checked whether the actual aspect ratio of a sample had an influence on the modulus of rupture. It is certainly possible to make the casual point that the longer extrudates held up very well for their aspect ratio during the handling history of the sample, and that those long extrudates consequently would be stronger than the average strength of the sample. Similarly, it is possible to consider that shorter extrudates are weaker than the average from an aspect ratio perspective.

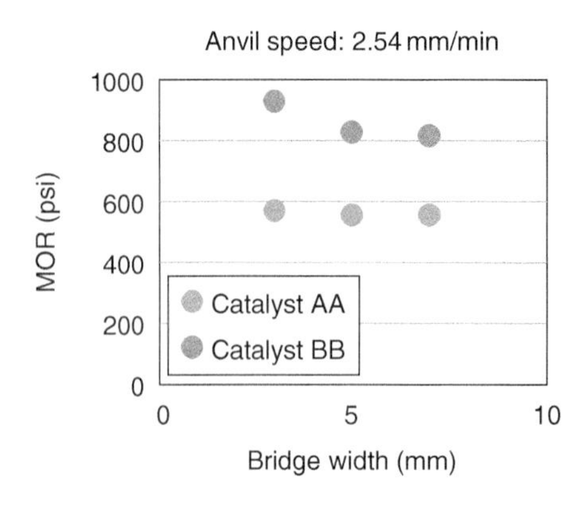

Figure 3.16 Impact of support span (bridge width) at a 2.54-mm/min anvil speed.

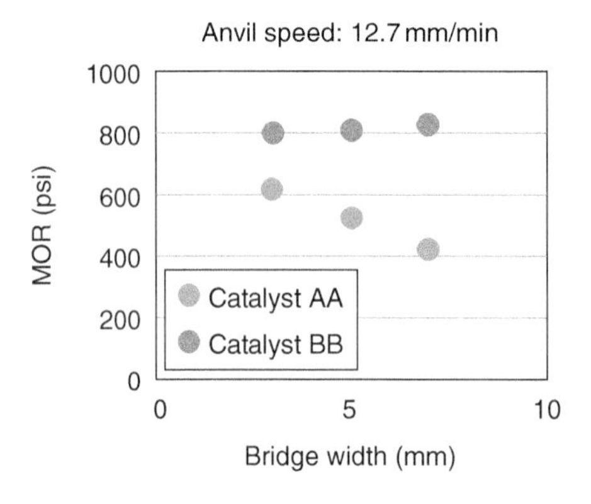

Figure 3.17 Impact of support span (bridge width) at a 12.7-mm/min anvil speed.

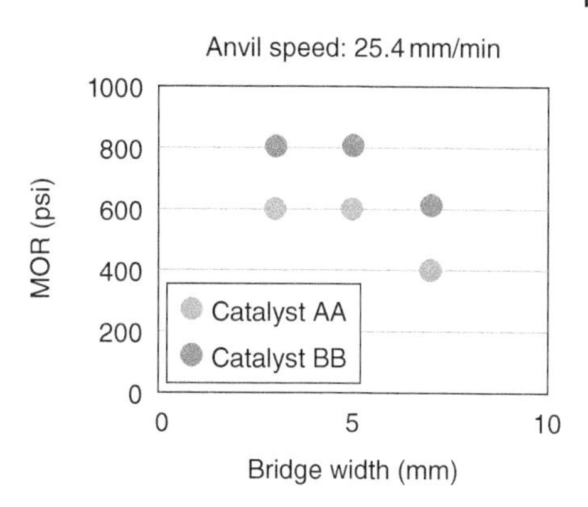

Figure 3.18 Impact of support span (bridge width) at a 25.4-mm/min anvil speed.

Table 3.3 Commercial cylinder catalyst sample (1.59-mm diameter).

Φ	Rupture force (N)	MOR (MPa)	E (MPa)
5.4	4.62	14.8	615
3	4.66	15.0	589

Overall, a question arises: Is the modulus of rupture a function of the sampled extrudate length? To answer this question, we selected a sample and sieved the extrudates into a long fraction and a short fraction – the short fraction falls through US Mesh No. 14 and the long fraction falls through US Mesh No. 12 and stays on No. 14 – and measured both for modulus of rupture, rupture force, and Young's modulus of elasticity. Table 3.3 shows the results. The length of the sampled extrudates does not influence the values measured for the strength properties, and the casual point mentioned earlier is generally not valid. Be aware that materials that have not been prepared homogeneously and show a wide distribution of strengths or catalysts that have many defects may induce the effects that are mentioned in the first paragraph of the current section.

3.4.1.4 Modulus of Rupture Reproducibility

Figure 3.19 shows data on the reproducibility of the test as a function of time. The repeatability is good: ±14% of the mean for two-thirds of the samples, as indicated in Table 3.4.

Measured extrudates always have a certain degree of curvature along the length axis. This curvature may be pronounced or may be hardly visible. Depending on the width of the bridge and the length of the extrudate, once the extrudate is placed on the bridge, it will naturally tend to rest in its most stable state. Thus, cylindrical extrudates that are much longer than the support bridge will tend to end up resting

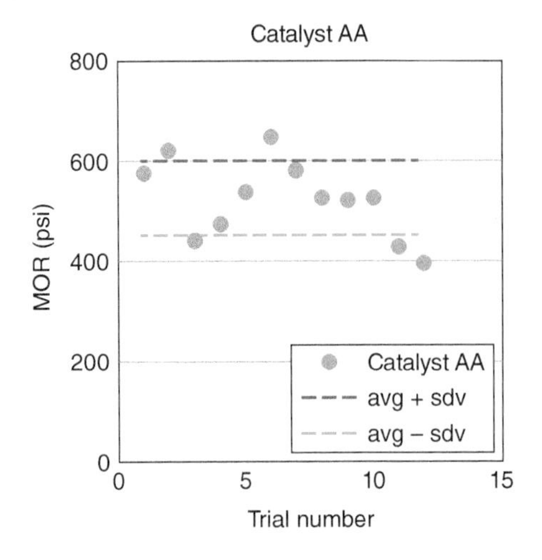

Figure 3.19 MOR repeatability.

Table 3.4 Results of the repeatability study.

Quadrulobe extrudate (1.59-mm diameter)					
	Catalyst ID: AA				
Measure #	**Speed (cm/min)**	**Span (mm)**	**Maximum load (N)**	**MOR (MPa)**	**Load cell (N)**
1	2.54	5	1.44	3.98	10
2	2.54	5	1.55	4.27	10
3	2.54	5	1.11	3.05	10
4	2.54	5	1.19	3.27	10
5	0.254	3	2.26	3.73	10
6	1.27	3	2.70	4.46	10
7	2.54	3	2.43	4.01	10
8	0.254	5	1.32	3.64	10
9	1.27	5	1.31	3.62	10
10	0.254	7	0.94	3.64	10
11	1.27	7	0.77	2.98	10
12	2.54	7	0.71	2.73	10
			avg	3.62	
			st. dev.	0.50	

bowed upward. Cylindrical extrudates that are long enough to span the bridge – but not much longer – tend to end up resting bowed downward. Extrudates resting bowed upward will likely display a greater rupture force than those resting bowed downward.

Such natural states are unavoidable when measuring the bending strength unless one were to cherry-pick extrudates, and that is not preferred if one wishes to obtain a true representation of bending strength variability.

3.4.1.5 Wet Modulus of Rupture

I tried wetting down the extrudates before testing to determine whether this had any effect on the modulus of rupture. I anticipated that this experiment might be important with respect to the aspect ratio during ion exchange and impregnation with liquids at commercial plants.

To verify my hypothesis, I compared "dry" versus "wet" moduli of rupture according to this procedure:

- Riffle the sample and calcine two splits of it at the same time in air with a ramp of 3 °F/min up to 1000 °F, then hold at that temperature for one hour.
- After calcination, run the dry sample on the Instron immediately upon unloading under a N2 blanket.
- Place the sample to be wetted in a desiccator containing deionized water overnight.
- Remove the humidified sample from the desiccator and place it in a container containing a moist piece of soft supporting material.
- Drip deionized water onto the sample until it is wet. Seal the lid onto the container with tape and leave the sample for a just over an hour.

During testing, the lid remained on the container, with no N2 blanket.

Table 3.5 shows the results of this experiment. The wetting mode indeed lowers the modulus of rupture for this specific catalyst by approximately 20%. It may be

Table 3.5 "Wet" versus "dry" modulus of rupture.

Catalyst AA			
Anvil speed: 0.2 mm/s			
Bridge width: 5 mm			
	Force (N)	MOR (MPa)	E (MPa)
Dry	1.21	3.34	353
Wet	1.01	2.78	306

more or it may be less for other catalysts, but it would be prudent to measure the wet modulus of rupture before commercial exchange or impregnation. Be aware that the effect may be greater with exchanges and impregnations of actual impregnation liquids because those can have a non-neutral pH and may weaken the binder or the active phase even more.

3.4.1.6 Modulus of Rupture Report

Figure 3.20 is a typical report showing the force that breaks the extrudate (the Maximum Load column), the calculated modulus of rupture (the Modulus of Rupture column), and the Young's modulus of elasticity (the Young's Modulus column) for an extrudate with a quadrulobe shape. Values are reported in non-metric units (1 in. = 25.4 mm, 1 lbf = 4.45 N, 145 psi = 1 MPa). The graph shows the measured extension, which is quite small, at about 2/1000th of 1 in. It is striking that the Young's modulus of elasticity, which is proportional to the slope of the ascending force versus the strain curve, is fairly constant among the different samples.

In performing a back-of-the-envelope calculation, consider specimen no. 12, with a reported Instron modulus of rupture of 1008 psi and a Young's modulus of elasticity of 63 000 psi in Figure 3.20. For Young's modulus of elasticity, the software assumes a cylinder.

For the modulus of rupture, the numbers are:

- Rupture force (F) = 0.5686 pound force (lbf).
- Shape factor s (quad) = 2.20 (see Table 3.1).
- Bridge width (L) = 5 mm = 0.197 in.
- Diameter (D) = 1/16 in.
- The modulus of rupture is then $s \times F \times L/D^3 \approx 1009$ psi (agreeing with the value shown in Figure 3.20).

For Young's modulus of elasticity, the numbers are:

- Slope of the ascending part in the graph: slope = (0.58 lbf)/(0.0443–0.0422) in. = 276 lbf per in.
- Bridge width (L) = 5 mm = (0.5/2.54) in. = 0.197 in.
- Extrudate radius (R) = (1/16 in.)/2 = 0.031 in.
- Young's modulus of elasticity = $\text{slope} \times L^3/(12 \times 3.14 \times R^4)$ = 60 700 psi (agreeing with the value shown in Figure 3.20 for sample 12 of 62 867 psi, considering a rough estimate of the slope). The value for the true E would then be a factor of 0.86 lower because the extrudate was a quadrulobe (see Table 3.2).

	Maximum Load (lbf)	Modulus of Rupture (psi)	Young's Modulus (psi)
1	0.2224	394.4709	52317.78733
2	0.3759	666.7147	49114.35792
3	0.1322	234.5332	20799.61631
4	0.4023	713.6626	57337.99941
5	0.4019	712.9948	55858.49030
6	0.2104	373.1438	49375.89648
7	0.2450	434.5604	55939.56052
8	0.1180	209.2925	43022.92650
9	0.4904	869.9615	64782.03840
10	0.3836	680.3971	48830.66682
11	0.3408	604.5298	70396.28880
12	0.5686	1008.6680	62867.88301
13	0.2967	526.2983	50745.07853
14	0.2456	435.6625	58136.57411
15	0.0952	168.8623	25243.86399
16	0.1687	299.1885	38454.71466
17	0.3062	543.1682	63003.93149
18	0.1299	230.4667	39138.11505
19	0.2429	430.9302	50777.73460
20	0.1299	230.4583	29487.60534
21	0.1637	290.3108	47206.76494
22	0.5452	967.1774	62016.28157
23	0.1509	267.6368	64423.55003
24	0.1308	231.9489	33393.83650
25	0.3080	546.3337	65800.33401
Mean	0.2722	482.8549	49938.87603
Standard Deviation	0.1378	244.4382	12992.38389

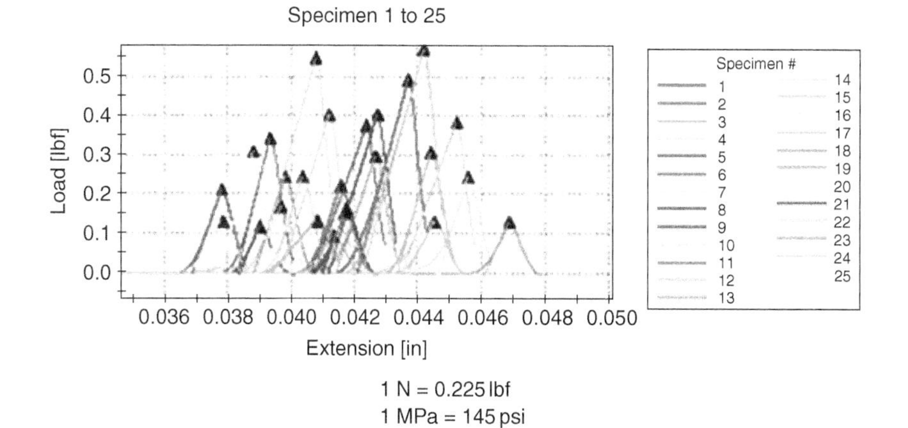

1 N = 0.225 lbf
1 MPa = 145 psi

Figure 3.20 Instron MOR report.

3.4.2 Side Crush Strength, Bead Crush Strength, and Bulk Crush Strength

Side crush strength, bead crush strength and bulk crush strength have been studied abundantly, and ASTM procedures are available. It is interesting to see, however, how the strength properties such as the side crush strength, the bead crush strength, and the bulk crush strength line up with the bending strength and eventually the tensile strength of the catalyst. Figure 3.21 compares side crush strength as a function of the modulus of rupture for quadrulobes, while Figure 3.22 offers the same comparison for cylinders.

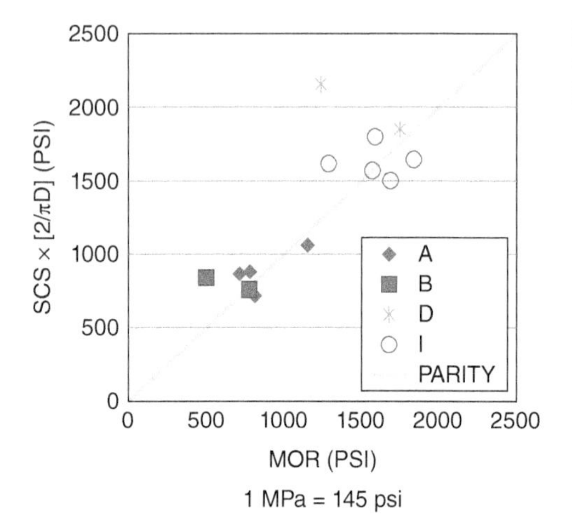

Figure 3.21 Side crush strength versus MOR for quadrulobes (1 MPa = 145 psi).

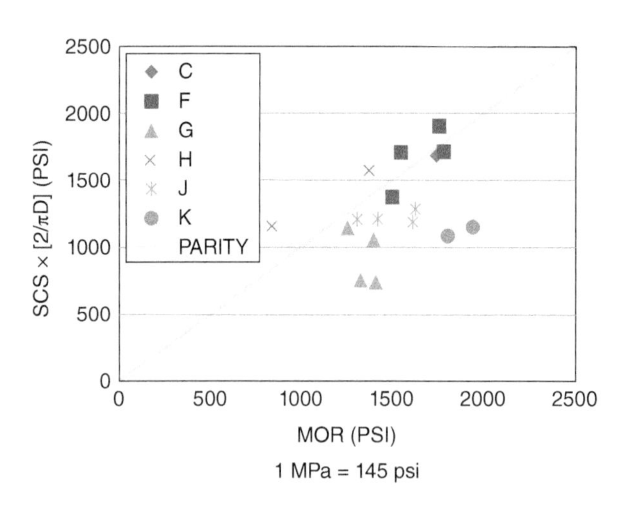

Figure 3.22 Side crush strength versus MOR for cylinders (1 MPa = 145 psi).

Table 3.6 Comparison of catalyst strength as extrudate or bead for different moduli of rupture.

	Side crush strength (N/mm)	Crush strength of a bead (N)
MOR (MPa)	1.59-mm cylinder	1.59-mm sphere
3.45	8.6	9.8
6.90	17.1	19.5
10.34	25.7	29.3

As mentioned earlier, given the absence of external references, I used the same correlation for quadrulobes as I did for cylinders of the same diameter. The estimate turns out to be quite accurate. Most grades look excellent in the prediction, but some are off (either higher or lower), and without measurement, it is not possible to make a prediction. Table 3.6 compares the modulus of rupture, the side crush strength, and the bead crush strength for extrudates or beads with a diameter of 1.59 mm.

This comparison is calculated from the correlations given earlier in Eqs. (3.4) and (3.6). For beads, no direct data is available to establish a comparison, but the crushing force is well in range of typical values.

The modulus of rupture yields the bending strength of the extrudate at the breaking point. The modulus of rupture is a physical/mechanical property of the material regardless of extrudate shape and size. It is a dimensional property that is measured in units of pressure or stress. The modulus of rupture is sensitive to all aspects of catalyst preparation that influence extrudate bending strength. Such aspects include chemical makeup, mulling recipe, extruder conditions, and extrudate quality, as well as any post-treatments such as drying and calcination. In that respect, the modulus of rupture is an ideal tool to help with troubleshooting in commercial plants or to scale up catalyst production from a laboratory and pilot plant to commercial manufacturing plants from an aspect ratio perspective.

3.4.3 A Speculation on the Variability of Strength from Extrudate to Extrudate

Extrudates made in quantities of many kilograms or hundreds of tons carefully follow recipes and are extruded under tight control. The materials are homogeneously and uniformly made, and pilot plant and commercial plant operators take great pride in the quality of their work.

Yet these same catalysts evaluated in the laboratory – even in a small sample – will show considerable variance from extrudate to extrudate. It is hard to believe that one extrudate is indeed so much weaker or so much stronger than the next. So, the question is why there is such a high variance in the rupture force from extrudate to extrudate. I speculate that in the modulus of rupture testing, the *position* of the anvil along the length of the extrudate significantly and abruptly affects the force required to break the extrudate. The reason for this is that the rupture force in a very specific location along the length of the extrudate is sampled from a strength distribution. By changing the anvil's position just a little bit, the instrument registers another strength value sampled from the distribution. The cover of this book shows this "white noise"-like variation for a uniform distribution that I use in Chapter 4 to describe amorphous materials. I speculate that ever so slightly moving the anvil to a nearby position, if that were possible, would change the rupture force and register the extrudate as either a weak, average, or strong extrudate.

In a single extrudate, I think there are many hundreds or thousands of measurable force values (or more) depending on the exact position of the anvil. What is constant is the distribution of strength values. Therefore, at the end of the day, I think that both the laboratory and the plant are correct. The laboratory measures the strength distribution and its variance, while the plant manufactures a catalyst with a consistent and constant strength distribution.

For the bending strength, measurement is destructive, and so it is not possible to test this speculation. However, the same variance is also often apparent in the side crush strength. Strictly from a research perspective, it would be possible to make measurement in this case nondestructive by having a support platform that could accurately move the contact point of the anvil over micron distances, while the anvil width could be changed from 3 mm to 1/100 mm and measurements made every one-tenth of a millimeter along the extrudate axis. It would be necessary to keep the anvil's penetration into the extrudate very small (e.g. 1/100 mm) and to record the applied force. This could constitute a very interesting experiment and lead to apparatus development, which in my opinion would attract great interest from research laboratories.

3.5 Breakage by Collision

3.5.1 Background

During catalyst manufacturing, a catalyst extrudate travels through the plant and meets the equipment as designed by the plant. This design is often many years old and has been optimized over the years. The journey of an extrudate from one piece of equipment to another exposes it to breakage in various ways – either through exposure to the equipment itself or through exposure to neighboring extrudates during the transfer – at different levels of severity.

For example, catalyst transport often requires dropping catalyst extrudates from a belt dryer onto conveyors or from conveyors into bucket elevators. From there, the catalysts may drop from one conveyor onto another. In addition, feeding catalysts via chutes and then dropping them onto or into another piece of equipment such as rotary calciners or fixed-bed calciners are common practices.

During transport, the catalyst breaks and shows a number of stepwise reductions in the aspect ratio. Sieving and sizing, which typically separate out the large extrudates called "overs" and those below specification as "dust, chips, and fines" from catalyst production, are other examples of handling operations with specific severities that can break the catalyst and reduce the aspect ratio.

When catalyst extrudates are particularly strong, it may be necessary to use a hammer mill to force them to break, thus reducing the aspect ratio per customer specifications.

Stepwise reductions in the aspect ratio can be deceptively small, but they do add up, and the final aspect ratio can disappointingly miss the specification. In such instances, the manufacturing plant will incur production losses, as well as cost and effort to resolve the issue.

The catalyst's aspect ratio exhibits transient behavior as it passes through equipment and transfers from one piece of equipment to another. If a catalyst passes a second time through this equipment line, it will likely break to an even lower aspect ratio, but the second breakage will be less than the breakage observed during the first pass.

There are also times when aspect ratio changes are minimal, being more of the kind encountered with catalyst attrition. This behavior is called asymptotic behavior because the aspect ratio asymptotes toward a limit value. It is only possible to correctly model asymptotic behavior based on very specific experiments that standardize the conditions of catalyst breakage. These experiments reveal the transient behavior of the aspect ratio throughout each step until it finally transitions into asymptotic behavior.

I have previously published work [8,21,22,26] linking the drop in aspect ratio to the bending strength of extrudates, focusing on the catalyst breakage caused by repeated impact against a surface or by stress in a fixed bed of catalyst.

The theory of elasticity applies to the deformation and breakage of catalysts. A natural result of the Euler–Bernoulli beam theory was the concept of breakage by bending to the point of rupture. This rupture point is of immense value in the design of bridges and buildings in order to help identify and quantify what loads a structure can withstand without failure, and although one can apply the same concept to the bending strength of catalyst extrudates, references are scarce and not part of typical curricula taught in the technical centers of manufacturing plants.

So far, side crush strength and bulk crush strength have been used as measures of strength. The database that tries to correlate the catalyst aspect ratio with these strength measurements is enormous, but it is also very ad hoc, very dispersed, and lacking general correlations.

I developed a two-parameter model for predicting the reduction in the aspect ratio of catalyst extrudates due to breakage by impulsive forces, as experienced in a collision test. It is possible to correlate both parameters as defined in the model with the strength of the extrudates and the severity of the drop test because both are asymptotic in nature.

For breakage caused by impulsive forces, the model reveals that the extrudate aspect ratio will drop according to a second-order or a pseudo-second-order break law.

Applying the model to cases of severity sequencing and severity conditioning reveals the very nonlinear behavior of aspect ratio predictions and yields results that are often nonintuitive and hard to obtain without the ability to conduct specific engineering analyses.

The aspect ratio of a catalyst is a critical value and typically has a minimum specification of about two to three, often set according to a plant's experience with a particular catalyst grade or by the application. At times, a plant invests precious time and raw materials in improving the aspect ratio, sometimes with mixed results.

Transferring technology from the laboratory to the plant requires on-site demonstrations based on data from laboratory or pilot extruders. These laboratory- or pilot-made materials are comprehensively tested and checked, but the commercial aspect ratio is a property that is difficult to determine up front. Commercial catalysts need to satisfy many specifications, and the aspect ratio is just one of them. If the aspect ratio does not meet these specifications during the demonstration, then delays and costs carried by the manufacturer are the result.

Often, there is no choice but to accept the side crush strength or the bulk crush strength as a guide for resilience against aspect ratio reduction, but this may be misleading. The aspect ratio of a catalyst for commercial demonstration so far can only be obtained during the demonstration. If the aspect ratio is above four, there are typically no issues in later commercial manufacturing operations. For aspect ratio values below four, there is concern that the final aspect ratio will not meet the specification during production campaigns because of a variety of factors such as:

- The specific manufacturing site.
- Possible double handling upon storage/calcination.
- Sizing and sieving equipment.
- Aspect ratio losses from bucket elevators and drops in chutes, on distributor cones, and in calciners.

- Ambient temperature and humidity.
- The human element: experienced crews and operators may have different opinions on how to handle aspect ratio issues.

All of these factors enhance the degree of difficulty, widen the variability of the product aspect ratio, and make it exceptionally difficult to reliably estimate the outcome from an aspect ratio perspective. Prediction of the aspect ratio of extruded catalysts has been a long-standing issue and so far has eluded many efforts toward its correlation with independently measurable catalyst properties.

Often, manufacturing plants have very specific and closely guarded trade secrets and technical know-how that has been gathered over the years to deal with the aspect ratio of the different catalyst grades produced. Therefore, it would be beneficial to identify a methodology and a tool that would enable professionals to quantify the structural strengths of catalysts from an aspect ratio perspective with the ability to be scaled up.

One could extend this technical know-how further and apply it to catalyst regeneration quality control where *ex situ*/*in situ* procedures can lead to catalyst breakage during regeneration and reloading into reactors.

This chapter also deals with the modeling efforts of a laboratory drop test. In a drop test, extrudates of various shapes and sizes are dropped from variable heights onto a hard empty surface, followed by an analysis of the aspect ratio of the extrudates after the drop. Modeling occurs through a discrete finite-difference method and employs a first-order Padé approximant with two parameters. This modeling effort is only a first-order approximation of the change of the aspect ratio with regard to breakage due to impulsive forces generated from repeated impacts against a surface.

3.5.2 Mathematical Modeling of Extrudate Breakage

3.5.2.1 Experimental

The aspect ratio of catalyst extrudates in commercial plants is an important variable, and much effort is often spent on getting this variable into an acceptable range. For extruded catalysts, the diameter and the shape of the cross section are essentially constant but still require close monitoring because extrusion dies do exhibit wear and the diameter and shape have specifications. Commercial dies do not always wear uniformly and need close attention from the operating crew or technical personnel.

One variable that is prone to a wide distribution is the length of the extrudate when it is finally delivered to the customer. Extrudate length is controllable by setting specifications on the average aspect ratio of the extrudates and by controlling the fines and overs in production. The overs control the maximum length of the extrudate and are returned to the process when and where it makes the most practical sense. The fines can be dust and chips or are extrudates that became too short. The fines are sieved out and may be reworked or labeled as waste and discarded. The waste stream is a loss, so it requires close monitoring in order to minimize its impact on production.

The aspect ratio of a catalyst is experimentally determined by riffling a sample of about 200 extrudates from the entire catalyst specimen and performing an optical scan. The scan is then transferred to the Alias analysis software that determines every individual length-to-diameter ratio and then calculates their arithmetic average or the "aspect ratio" of all of the extrudates. A video in the *Journal of Visualized Experiments* (*JoVE*) by Beeckman et al. [26] shows the procedure in full and can easily be accessed by a Google search for "Beeckman *JoVE*" or doi: 10.3791/57163.

After extrusion and drying, a catalyst experiences a number of drops of variable heights onto a surface or a bed. Each drop is a collision; the extrudate typically survives many or all of the drops, but some meet their match and break. Collision is typically caused by a gravitational fall and the impact surface can be quite varied. It can be:

- Hard, as on a metal plate.
- Semi-hard, such as falling onto a bed of catalysts.
- Semi-soft, such as falling onto a bed of vibrating extrudates, as in a vibratory drier.
- Soft, such as falling onto padding.

I wanted to construct a well-developed drop test where it was possible to control the height of the gravitational drop and standardize the impact surface. I will elaborate here on my original experimental procedure, but acknowledge the potential for many improvements.

Figure 3.23 shows the apparatus I developed to do a drop test. It consists of a plastic tube mounted vertically with a horizontal metal plate at the bottom. A gap between the plate and the bottom of the tube allows the extrudates to slide out of the way after impact. Using several tubes of different lengths, it is possible to vary the drop height.

The lowest drop height was 0.20 m and the highest drop height was 6.40 m. I obtained the most data when using a drop height of 1.83 m. Very low drop heights affected how the extrudate fell through the ambient air; the very short time duration of the

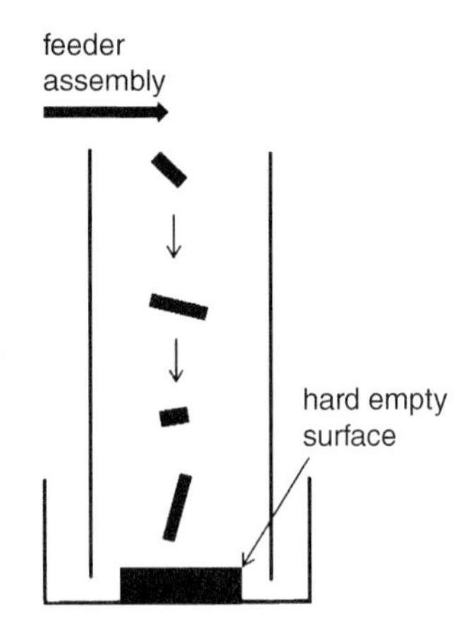

Figure 3.23 Drop test apparatus. Source: Courtesy of Wiley: doi: 10.1002/aic.15046.

drop may favor certain impact angles over others. While the prevalence of certain impact angles is not necessarily an issue, it is important to be aware of it.

The extrudates are released from a chute; the height difference between the chute and the plate is considered the drop height. The chute is horizontal and has a V-shaped cross section. My coauthors fed the extrudates from the feed vessel one at a time. After the extrudate impacted the collision plate, in order to minimize any secondary breakage that would affect the results, my coauthors installed appropriate padding around the bottom plate area. I did not further consider any secondary breakage part in the experimental procedure.

Again, following the procedure carefully and consistently should yield results that are themselves consistent from drop to drop. Perform the drop test not just once but several times consecutively, starting with a sample, dropping it, collecting it, measuring the aspect ratio, and then dropping it again and again. Using a small enough number of extrudates to allow the sample to be used in its entirety in the length-to-diameter ratio instrument is most convenient. Any error due to sampling of extrudates is then eliminated.

Eliminate any chips and broken pieces after performing the drop test, because the software used to interpret the scanned sample for the aspect ratio would very likely misinterpret the diameter or length in the presence of these chips and pieces, as well as any pieces that are no longer extrudates. The software does have a convenient way to discard these chips and pieces.

To achieve a good, repeatable test, keep the drop plate free of broken pieces that could change the impact of collisions. It helps to use a low feed rate and essentially to use a rate of one extrudate at a time. Keeping track of the amount of fines may also help the analysis because this may help the scale-up efforts when taking the catalyst to a commercial setting because of the upfront knowledge of the production of fines.

The feed rate may affect collisions on stationary vibrating beds. A high feed rate of extrudates places colliding extrudates in close proximity and will create a containment effect (i.e. the sheltering of catalyst extrudates by the extrudates surrounding it). I consider these effects as being lumped into a plant's severity factor. For now, the specific impact of the feed rate on the severity of a plant has not been taken into account.

Because my objective was to develop correlations and a methodology that would be applicable to extrudates of all kinds, with pretreatments typical in pilot and commercial plants, I used broad brush strokes when looking for a library of extrudates. Instead of focusing on a large sample of extrudates all of the same kind and varying the breakage tests, I used a broad set of samples representing different extrudates and pretreatments and looked at a few well-controlled tests on breakage in order to develop the model.

Table 3.7 Catalysts used in the drop test.

Catalyst	Zeolite	Binder	Shape	Diameter (mm)	Temperature (°C)
A	1	Silica	Cylinder	1.59	~120
B1	2	Silica	Cylinder	1.59	~120
B2	2	Silica	Cylinder	1.59	~120
B3	2′	Silica	Cylinder	1.59	~120
B4	2′	Silica	Cylinder	1.59	~120
B5	2′	Silica	Cylinder	1.59	~120
C1	2′	Silica	Cylinder	1.59	~540
C2	2′	Silica	Cylinder	1.59	~540
C3	2′	Silica	Cylinder	1.59	~540
C4	2′	Silica	Cylinder	1.59	~540
C5	2′	Silica	Cylinder	1.59	~540
D1	3	Alumina-1	Quadrulobe	1.59	~540
D2	3	Alumina-2	Quadrulobe	1.59	~540
D3	3	Alumina-3	Quadrulobe	1.59	~540
D4	3	Alumina-4	Quadrulobe	1.59	~540
D5	3	Alumina-5	Quadrulobe	1.59	~540

Source: Courtesy of Wiley, doi: 10.1002/aic.15046.

Table 3.7 shows the variety of extruded materials used and the variations in the nature of the zeolite, the kind of binder, the die designs, the mulling recipes, the extrudate shapes, and the heat treatments, ranging from drying to calcination. The same catalyst diameters were used because these catalysts are simply more abundantly available.

Table 3.8 shows a catalyst dried at about 120 °C with a starting aspect ratio of about 24. I changed the drop heights from 0.2 m all the way up to 6.4 m and repeated the drops several times in order to see the entire scope of extrudate collision responses. This was the first set of experiments where it became clear that, for a large number of drops, the extrudate aspect ratio reaches an asymptotic value and does not simply keep on breaking into smaller and smaller values. This interesting revelation was not evident before these experiments.

I also thought that it was necessary to test materials on a pilot-unit scale; Table 3.9 shows the results. I tested the extrudates in both dried-only and dried and calcined states.

Table 3.10 lists drop data where the extrudates are bound with alumina and extruded as 1.6-mm-diameter quadrulobes with varying mulling recipes. In all

Table 3.8 Aspect ratio drop test data on catalyst A.

# of drops	Height 0.20 m	Height 0.61 m	Height 1.83 m	Height 6.40 m
0	~24	~24	~24	~24
1	6.09	4.54	3.31	2.67
3	4.13	2.93	2.14	2.01
5	3.45	2.62	2.02	1.92
7	3.68	2.35	1.95	1.73
9	3.69	2.21	1.72	1.81
11	2.99	2.12	1.75	1.62
16	2.99			
21	3.28			
31	2.73			

Source: Courtesy of Wiley, doi: 10.1002/aic.15046.

Table 3.9 Aspect ratio drop test data (each drop from 1.83 m high) for catalysts B1–B5 and C1–C5.

Catalyst	Φ_O	1 drop	2 drops	3 drops	4 drops	5 drops
B1	7.92	3.73	2.92	2.70	2.53	2.40
B2	6.94	3.94	3.08	2.68	2.44	2.14
B3	6.52	2.65	2.38	2.18	1.95	1.90
B4	7.03	2.94	2.37	2.31	2.01	1.95
B5	3.96	2.06	1.90	1.65	1.63	1.46
C1	7.92	4.28	4.24	3.98	3.83	3.57
C2	6.94	5.30	4.33	4.19	3.83	3.57
C3	6.52	3.22	2.92	2.60	2.60	2.44
C4	7.03	3.22	2.92	2.77	2.45	2.63
C5	3.96	2.17	2.06	1.88	1.84	1.78

Source: Courtesy of Wiley, doi: 10.1002/aic.15046.

of the drop tests, I had to remove dust and chips before measuring the aspect ratio because they created unnecessary difficulties for the software. Removing chips and dust commercially is necessary to deliver a high-quality product and reduces any potential for an increased pressure drop.

Table 3.10 Aspect ratio drop test data (each drop from 1.83 m high) for catalysts D1–D5.

Catalyst	Φ_0	1 drop	2 drops	3 drops
D1	4.37	3.28	3.08	2.86
D2	4.01	2.78	2.47	2.31
D3	4.31	2.92	2.55	2.38
D4	4.46	3.54	3.16	3.00
D5	4.26	3.81	3.47	3.13

Source: Courtesy of Wiley, doi: 10.1002/aic.15046.

3.5.2.2 Modeling via a First-order Padé Approximation

3.5.2.2.1 Two Fundamental Parameters

In Section 3.5.2.1, I mentioned that as one repeats dropping extrudates over and over again, the aspect ratio lines out to an asymptotic value fairly quickly, which is essentially only a function of the drop height. Intuitively, this makes sense. Everyone at some point has dropped an object such as a coffee mug or a drinking glass and seen it break into pieces upon hitting the floor; many of the smaller pieces may bounce and land on the floor again but not break further.

One day, I found myself at the back end of a commercial belt drier that took extruded catalyst strands and dried them to specification. I wondered what would happen when dropping very long extrudates a single time, and how those in such a single-break event would fare. Would these extrudates simply break into two or three pieces, or would the number of breaks increase with the length of the extrudate? Certainly, it seemed obvious that the momentum of the extrudate would increase with the length, likely giving more credence to the second scenario.

I performed this experiment in the same controlled setting as I used for the drop tests; Figure 3.24 shows the results. After just a single drop, the aspect ratio of such long pieces gave rise to a second asymptotic value. This second asymptote was another property that I found necessary to incorporate into a break model.

As a catalyst is transported through a plant, the drops that it encounters – from a drier belt, from one conveyor to another, into a bucket elevator, onto distribution equipment, into vibratory sieving equipment, or into a sampling jar – all have a certain severity. Each of these drops will reduce the aspect ratio slightly, but as the catalyst continues to travel and as the impacts accumulate, the total reduction in the aspect ratio can be substantial. Some materials are exceptionally strong and actually require help in order to break down to a practical size; in-line equipment such as hammer mills can accomplish the forced breakage of strong extrudates.

The development of a model equation should have a starting aspect ratio as a necessary input or initial condition. I adopted Φ_0 as the symbol to represent the

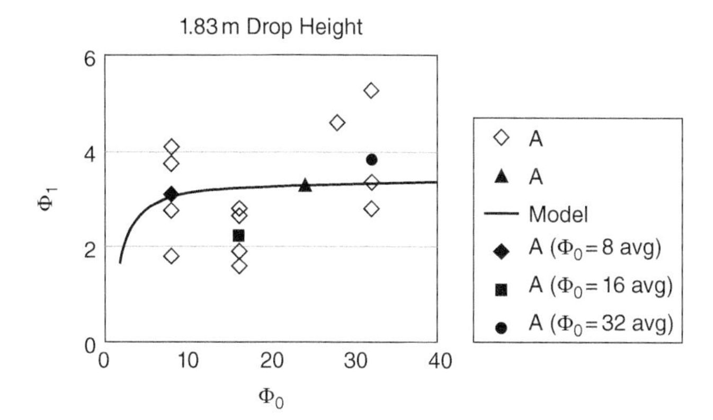

Figure 3.24 Length-to-diameter ratio after one drop of long extrudates. Source: Courtesy of Wiley: doi: 10.1002/aic.15046.

aspect ratio of the starting material. Φ_∞ represents the asymptotic aspect ratio after a large number of drops under the same conditions. Φ_α represents the asymptotic aspect ratio after a single drop of the extrudate when the starting aspect ratio is very large. This leads to a model with two parameters, Φ_α and Φ_∞, and one initial condition, Φ_0.

I selected a first-order Padé approximation in Beeckman et al. [8], and that publication includes the detailed derivation of the breakage model. The outcome of the modeling effort is shown here. Equation (3.8) expresses the aspect ratio Φ_j after j drops (0 included):

$$\Phi_j = (\gamma\Phi_0 + j\Phi_\infty)/(\gamma + j) \quad \text{for } j = 0, 1, 2, \ldots \tag{3.8}$$

with the factor γ expressed in Eq. (3.9) as:

$$\gamma = (\Phi_\alpha - \Phi_\infty)/(\Phi_0 - \Phi_\infty) \tag{3.9}$$

Equation (3.8) can also be represented as a finite-difference Riccati equation as shown in Eq. (3.10):

$$\Phi_{j+1} = \left(\Phi_\alpha\Phi_j - \Phi_\infty^2\right)/\left(\Phi_\alpha - 2\Phi_\infty + \Phi_j\right) \quad \text{for } j = 0, 1, 2, \ldots \tag{3.10}$$

It is easy to verify that:

$$\Phi_\alpha = \lim_{\Phi_0 \to \infty} (\Phi_1) \tag{3.11}$$

$$\Phi_\infty = \lim_{j \to \infty} (\Phi_j) \tag{3.12}$$

$$\Phi_0 = \lim_{j \to 0} (\Phi_j) \tag{3.13}$$

When the initial aspect ratio Φ_0 is sufficiently large, the general Eq. (3.8) reduces to an asymptotic solution given by Eq. (3.14):

$$\Phi_j = \Phi_\infty + (\Phi_\alpha - \Phi_\infty)/j \text{ for } j = 1, 2, \ldots \tag{3.14}$$

$$\gamma \cong 0 \tag{3.15}$$

$$\gamma\Phi_0 \cong \Phi_a - \Phi_\infty \tag{3.16}$$

It is possible to rearrange certain finite-difference equations to yield a linear relationship, which allows one to graph the data and sometimes to obtain the parameters from the slope and the intercept. The model equations in the general case can be rewritten as Eq. (3.17):

$$\Phi_0 - \Phi_j = j(\Phi_0 - \Phi_\infty)/(\gamma + j) \tag{3.17}$$

The change in aspect ratio in Eq. (3.17) is cumulative and thus has less experimental error than the point-to-point differences. Writing the inverse leads to Eq. (3.18):

$$1/\left(\Phi_0 - \Phi_j\right) = 1/(\Phi_0 - \Phi_\infty) + (\Phi_\alpha - \Phi_\infty)/\left[j(\Phi_0 - \Phi_\infty)^2\right] \tag{3.18}$$

So graphing the inverse of the cumulative change as a function of the inverse of the number of drops yields a straight line for the general model. The intercept and slope are functions of the two asymptotic parameters as well as the starting aspect ratio. Looking at the ratio, u, of the slope over the intercept leads to an expression (Eq. (3.19)) independent of Φ_0:

$$u = \Phi_\alpha - \Phi_\infty \tag{3.19}$$

For a given set of data, the model will force the aspect ratio of the starting material exactly through Φ_0, even though it is an experimentally determined initial condition and therefore has an error bar on it. Based on a set of experimentally determined aspect ratios of identical drops, it is possible to use standard parameter estimation methods to determine the asymptotic aspect ratios. A fairly large starting aspect ratio can obtain a good estimate for Φ_α. If the starting aspect ratio is small, then treat the estimate for Φ_α carefully because it could have a large error bar. In addition, if the initial aspect ratio is small, then it could well be that the severity of a given drop test (i.e. from a given height) does not lead to an adequate severity; thus, the drop may not reduce the aspect ratio by much and only cause some essential loss in aspect ratio by loss from chips and fines.

All of the aspect ratios Φ_j that are experimentally determined have an error bar, including Φ_0. When minimizing the sum of squares of errors, one can include this starting aspect ratio in the minimization of the residual sum or squares of errors and essentially treat Φ_0 as a dummy parameter.

I consider the reduction in aspect ratio during a drop test to be a two-parameter model with a severity dependent on the drop height and the nature of the impact plate.

3.5.2.2.2 General Two-parameter Model Fitting

When testing a new material with no upfront knowledge of its properties, use the two-parameter model to model the drop test. One may need to adjust the drop height to make the aspect ratio change in a reasonable and practical manner. Avoid conditions that produce many fines and chunks, unless the process application warrants the conditions.

I treated all of the data from Tables 3.8–3.10 as independent drop curves of individual materials and determined the parameters Φ_α and Φ_∞ by minimizing the nonlinear residual sum of least squares. Both parameters for each case are given in Table 3.11. The solid lines in Figures 3.25–3.28 show that the final fit of the data

Table 3.11 Parameter estimates.

Catalyst	Height (m)	Φ_α	Φ_∞
A	0.20	6.70	2.89
A	0.61	4.93	1.93
A	1.83	3.44	1.60
A	6.40	2.73	1.62
B1	1.83	4.46	1.94
B2	1.83	6.19	1.40
B3	1.83	2.97	1.72
B4	1.83	3.36	1.67
B5	1.83	2.45	1.34
C1	1.83	4.59	3.57
C2	1.83	9.46	2.57
C3	1.83	3.58	2.25
C4	1.83	3.46	2.36
C5	1.83	2.35	1.69
D1	1.83	3.86	2.53
D2	1.83	3.38	1.92
D3	1.83	3.62	1.92
D4	1.83	4.93	2.39
D5	1.83	36.17	0.00

Source: Courtesy of Wiley, doi: 10.1002/aic.15046.

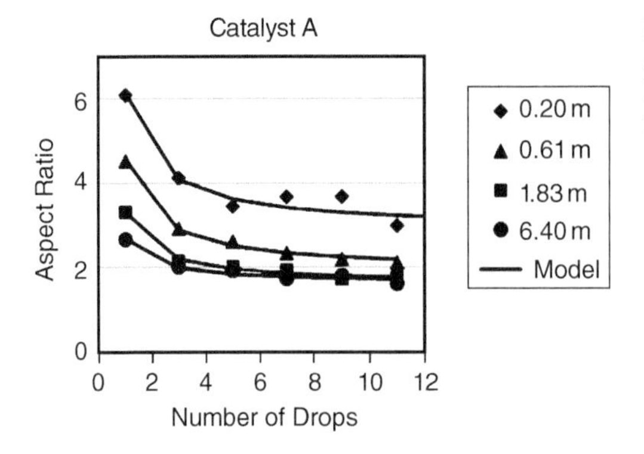

Figure 3.25 Data fit for catalyst A for different drop heights, $\Phi_0 = 24$. Source: Courtesy of Wiley: doi: 10.1002/aic.15046.

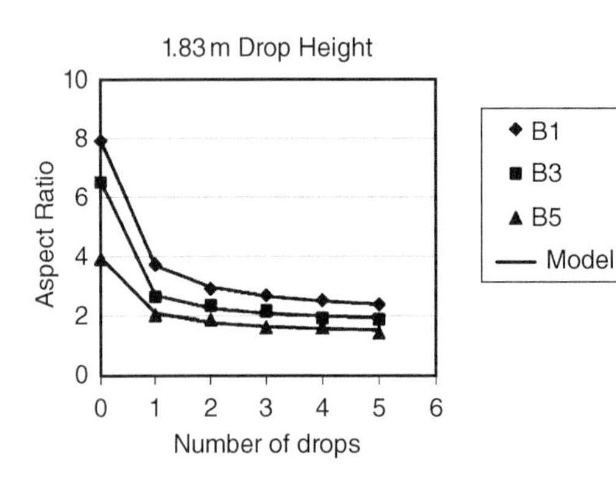

Figure 3.26 Data fit for catalysts B1, B3, and B5. Source: Courtesy of Wiley: doi: 10.1002/aic.15046.

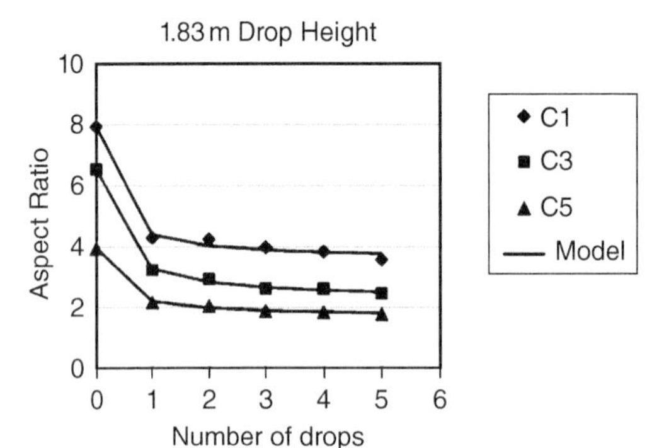

Figure 3.27 Data fit for catalysts C1, C3, and C5. Source: Courtesy of Wiley: doi: 10.1002/aic.15046.

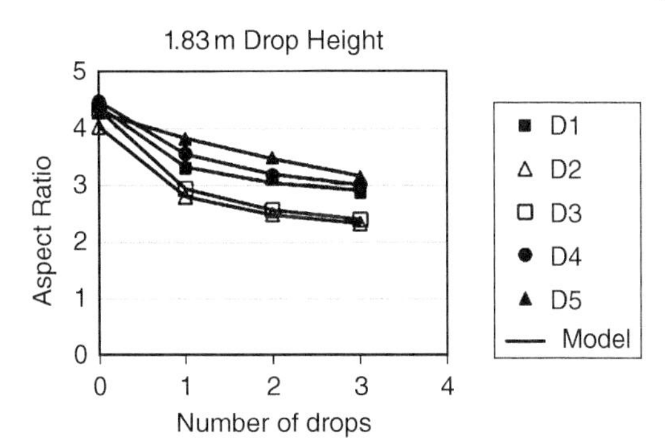

Figure 3.28 Data fit for catalysts D1–D5. Source: Courtesy of Wiley: doi: 10.1002/aic.15046.

is good. The pronounced bend in the onset of the drop curve is typically very sharp, and the two-parameter model seems to handle it well. The solid line is not smooth but rather represents the connection of individual points coming from the finite-difference equation. If one wanted more points for a smooth line, one could start from an arbitrary but appropriately chosen Φ_0 and apply the finite-difference Riccati equation with the obtained parameter values. The error bar on the aspect ratio is sometimes such that, with the data at hand, the parameter estimation becomes tricky. For instance, Figure 3.28 shows sample D5 to have a linear reduction in aspect ratio. One will need more data extended over more drops to get the lined-out value of the aspect ratio.

3.5.2.2.3 Ideal Materials

Parameters Φ_α and Φ_∞ are specific for a material and for the conditions of the drop test. They both entail breakage by collision. Thus, it is not surprising that one may want to look for a relationship between them. For all of the drop tests performed, including those from different heights, I graphed both parameters as a function of each other in Figure 3.29. One can see that there is a clear correlation between the

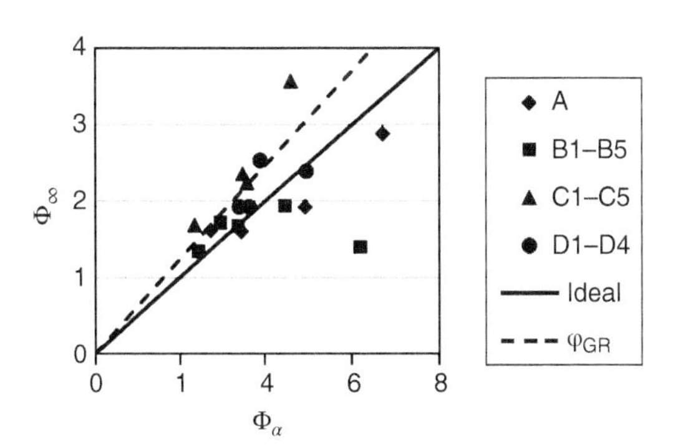

Figure 3.29 Correlation between parameters Φ_α and Φ_∞. Source: Courtesy of Wiley: doi: 10.1002/aic.15046.

parameters; for about 60% of the cases, the parameter Φ_α is about twice the value of Φ_∞. I call these materials/collisions ideal. Recall that I selected the materials for their broad differences in makeup and methods of preparation.

Given ideal materials, the model equations simplify to Eqs. (3.20)–(3.22):

$$\Phi_j = (\gamma\Phi_0 + j\Phi_\infty)/(\gamma + j) \text{ for } j = 0, 1, 2, \ldots \tag{3.20}$$

with

$$\gamma = \Phi_\infty/(\Phi_0 - \Phi_\infty) \tag{3.21}$$

or also as:

$$\Phi_{j+1} = 2\Phi_\infty - \Phi_\infty^2/\Phi_j \text{ for } j = 0, 1, 2, \ldots \tag{3.22}$$

When the initial aspect ratio for ideal materials is large, one can use Eqs. (3.23)–(3.25):

$$\Phi_j = \Phi_\infty + \Phi_\infty/j \text{ for } j = 1, 2, \ldots \tag{3.23}$$

$$\gamma \cong 0 \tag{3.24}$$

$$\gamma\Phi_0 \cong \Phi_\infty \tag{3.25}$$

The aspect ratio reductions for ideal materials are particularly simple; graphing Φ_{j+1} as a function of the inverse of Φ_j yields a straight line. Be aware, however, that the slope and the intercept of the line are both functions of the parameter Φ_∞; thus, they cannot be chosen independently. Once one chooses the intercept, one knows the accompanying slope, and vice versa. It is probably most practical simply to perform the parameter estimation for Φ_∞ and perhaps complete the graph as an illustration of fit.

3.5.2.2.4 Physical Arguments Supporting Ideal Materials

The existence of Φ_α and the observation that $\Phi_\alpha \cong 2\,\Phi_\infty$ for many materials perhaps invites one to make a physical argument.

An extrudate of a given aspect ratio increases its momentum as mass multiplied by velocity increases during its fall and reaches an impact momentum during collision. Without breakage, the change in the momentum is twice the impact momentum, and from Newton's second law it follows that the force exerted on the extrudate is the momentum change divided by the contact time. The longer the extrudate, the higher the aspect ratio and the higher the mass. In this simple argument, the impact velocity caused by gravitational acceleration is assumed to be the same for both large and small aspect ratios.

Further, I consider the contact time independent of the aspect ratio because of the lack of data (and it keeps the argument simple).

The strength of the extrudate for breakage drops the longer it gets from a torque perspective; a pencil breaks more easily than half a pencil. Thus, the force required to break an extrudate drops hyperbolically with the aspect ratio. This is also clear from the modulus of rupture of an extrudate. An extrudate with an aspect ratio equal to Φ_∞ is too short to break further upon collision. An extrudate with an aspect ratio equal to $2\Phi_\infty$ or Φ_α has the momentum to break and does so in two pieces; thereafter, neither of the two pieces has the required momentum to break further. An extrudate with an aspect ratio equal to $2\Phi_\alpha$ has twice the momentum of one with an aspect ratio of Φ_α and has enough momentum to cause two breaks, leading to three pieces and hence an aspect ratio equal to $2\Phi_\alpha/3$. Repeating the argument, an extrudate with an aspect ratio equal to $n\Phi_\alpha$ would lead to an aspect ratio of $n\Phi_\alpha/(n+1)$ after the collision, and in the limit for n, very large, very long extrudates would drop and be reduced to an aspect ratio of Φ_α, QED.

3.5.2.2.5 The Cumulative Break Function

Let us calculate the average number of breaks per extrudate required in order to reduce the aspect ratio measured after k drops to an aspect ratio observed after j drops and call it the break function, $H_{k,j}$. To calculate $H_{k,j}$, let N_k represent the number of extrudates observed after k drops and let N_j represent the number of extrudates observed after j drops. Neglecting the loss in fines, Eq. (3.26) expresses the balance for the total aspect ratio:

$$N_k\Phi_k = N_j\Phi_j \tag{3.26}$$

Each time an extrudate breaks in the set after k drops, it increases the number of extrudates by one. Therefore, the number of breaks in the sample after k drops compared to the number of breaks after j drops is $N_j - N_k$. Equation (3.27) then calculates the break function:

$$H_{k,j} = \left(N_j - N_k\right)/N_j \tag{3.27}$$

Substituting the aspect ratio balance equation results in Eq. (3.28):

$$H_{k,j} = \Phi_k/\Phi_j - 1 \tag{3.28}$$

Equation (3.29) is obtained by inserting the general difference Eq. (3.8) into Eq. (3.28) and rearranging to yield:

$$1/H_{0,j} = \Phi_\infty/(\Phi_0 - \Phi_\infty) + \Phi_0(\Phi_\alpha - \Phi_\infty)/\left[j(\Phi_0 - \Phi_\infty)^2\right] \tag{3.29}$$

where $H_{0,j}$ is the break function between sample j and the as-is sample. By the "as-is sample" I mean the starting sample or the sample that has not experienced any drops. Therefore, the inverse of the break function yields a linear relationship with respect to $1/j$.

3.5.2.2.6 Golden Ratio Materials

A tie-in may also exist with the well-known Golden Ratio (*Gulden Snede*), which is famous for the inherent aesthetic value it can impart in areas such as architecture and artwork. The Golden Ratio is present in many natural phenomena. The asymptotic parameters Φ_α and Φ_∞ have a physical meaning in that they are consequences of a physical process: here, a collision with a surface. These parameters are determined by the nature of the surface, the strength of the particle, the way in which the particle approaches the surface, and so on. I found it interesting to see if perhaps the Golden Ratio of a line segment could somehow be related to the collision of an extrudate with a surface.

The Golden Ratio is defined as a particular division of a line segment into a large section and a short section. The length-ratio of the long section to the short section is the same as the ratio of the entire section to the long section. Equation (3.30) expresses the Golden Ratio mathematically as:

$$\varphi_{GR} = \frac{1+\sqrt{5}}{2} = 1.618... \tag{3.30}$$

I found a link with the Golden Ratio for the case expressed as Eq. (3.31):

$$\Phi_\alpha = \varphi_{GR}\Phi_\infty \tag{3.31}$$

or Eq. (3.32):

$$\varphi_{GR}(\Phi_\alpha - \Phi_\infty) = \Phi_\infty \tag{3.32}$$

Thus, Φ_∞ is the large segment of the Golden Section of Φ_α. The dashed line in Figure 3.29 is an example of this relationship. The ratios of the asymptotic parameters for about half of the data points shown are close to the Golden Ratio. In Beeckman et al. [8], I explored some more possible relationships.

3.5.2.2.7 A Second-order Break Law for Collision

The finite-difference scheme found in Eqs. (3.8) and (3.10) describes the change in aspect ratio of extrudates due to collisions very well. It is a fun exercise to see if perhaps Eq. (3.8) allows one to deduce a break law that expresses the change more fundamentally. I found that by expressing changes in aspect ratios on a fractional basis, it is possible to derive the following equations.

Equation (3.33) defines χ_j as the forward fractional change from drop j to drop $j+1$:

$$\chi_j = (\Phi_j - \Phi_{j+1})/\Phi_j \tag{3.33}$$

and Eq. (3.34) defines Γ_j as the largest possible forward drop in aspect ratio as:

$$\Gamma_j = (\Phi_j - \Phi_\infty)/\Phi_j \tag{3.34}$$

Inserting the finite-difference Eq. (3.22) for ideal materials results in Eq. (3.35):

$$\chi_j = \Gamma_j^2 \tag{3.35}$$

Equation (3.35) demonstrates that the forward fractional change from one drop to the next for ideal materials follows a second-order break law. This second-order behavior also explains the pronounced strong bend in the drop curve. Initially, the difference in aspect ratio with the asymptotic ratio Φ_∞ is large; thus, the change squared is large as well. The quick drop in aspect ratio makes the difference, with the asymptotic ratio Φ_∞, much smaller. Hence, because it is squared, the so-called driving force for breakage becomes much smaller. It is remarkable that once the changes are expressed as forward fractional changes the break law has no adjustable parameters, not even Φ_∞. Equation (3.36) expresses the break law for the general case where both asymptotic parameters are independent as:

$$\chi_j = Z_j\Gamma_j^2 \tag{3.36}$$

where Z_j is defined as per Eq. (3.37):

$$Z_j = \Phi_j / \left(\Phi_j + \Phi_a - 2\Phi_\infty\right) \tag{3.37}$$

It is easy to verify that Z_j is equal to unity for ideal materials. In the general case, the forward fractional changes follow semi-second-order behavior. The factor Z_j as defined is not a constant but in fact changes with the aspect ratio.

3.5.2.3 Application to Operational Severity

It is now possible to apply the knowledge gained so far in the modeling of extrudate breakage to manufacturing plants in order to determine how plant architecture makes a difference in the aspect ratio of catalyst production. Let us look at two cases where the severity of the handling of the catalyst in a plant can perhaps be reduced. In the first case, I change the sequence of operational drop heights in going from one piece of equipment to the next. In the second case, I replace a high drop height with a series of low drop heights that span the same total drop height.

3.5.2.3.1 Optimal Sequencing of Drops with Different Levels of Severity

The objective in this case is to sequence two operations: one with low severity and on with high severity. Let us assign a 1.83-m drop as a high-severity operation and a 0.61-m drop as a low-severity operation. Assume that there is freedom to change the sequence during plant design.

Table 3.12 (Catalyst A) lists the calculations. The model used to compare both cases is the general model, with the values for the asymptotic parameters Φ_α and Φ_∞ shown in Table 3.11. The model shows that sequencing the high-severity drop

Table 3.12 Severity sequencing results.

	High severity → low severity	Low severity → high severity
Φ_0	24.00	24.00
Drop (m)	1.83	0.61
Φ_1	3.30	4.57
Drop (m)	0.61	1.83
Φ_2	2.87	2.74

Source: Courtesy of Wiley, doi: 10.1002/aic.15046.

before the low-severity drop is beneficial to obtaining the highest aspect ratio after both operations. The difference is modest at a 0.13 aspect ratio benefit, but just a few of these optimally sequenced drops can lead to a substantial retention of aspect ratio over lesser equipment arrangements.

Testing the optimal sequence in the laboratory gave a 0.1 aspect ratio advantage when sequencing the high-drop-height test before the low-drop-height test, confirming the model prediction. The retention sensitivity of the aspect ratio for drop heights can be communicated and discussed with equipment vendors and can play a role in equipment purchases. After all, equipment purchases require a substantial capital investment. Once selected and installed, equipment in a plant will be there for years.

3.5.2.3.2 Managing the Severity of a Drop

This case investigates whether changing a drop to a number of smaller drops but covering the same total drop height is advantageous. For example, a drop from a drier belt onto a conveyor can be substantial; a set of horizontal slats, such that the catalyst can fall from one slat to the next mounted just a little lower, would cover the drop height more gently. Let us compare a single drop of 1.83 m versus three drops of 0.61 m each, with a starting aspect ratio of the material on the drier belt of 24.

Using the model equations, Table 3.13 shows the results for three initial aspect ratios. Interestingly, an aspect ratio of 24 offers an advantage for a single high-severity drop when compared to three low-severity drops. Smaller initial aspect ratios show a much smaller advantage, while an initial aspect ratio of three offers an advantage for three horizontal slats.

Figure 3.25 shows that six slats a distance of 0.2 m apart would give an almost equal aspect ratio. The optimization becomes an interesting exercise because of the nonlinearity of breakage, but the model equations can help us to find the best

Table 3.13 Severity conditioning results.

	1 drop	3 drops	1 drop	3 drops	1 drop	3 drops
Φ_0	24.00	24.00	4.00	4.00	3.00	3.00
Drop (m)	1.83	0.61	1.83	0.61	1.834	0.61
Φ_1	3.30	4.57	2.64	3.15	2.40	2.72
Drop (m)		0.61		0.61		0.61
Φ_2		3.33		2.80		2.55
Drop (m)		0.61		0.61		0.61
Φ_2		2.88		2.60		2.44

Source: Courtesy of Wiley, doi: 10.1002/aic.15046.

solution. More gentle drops (more slats) may even bring us to a point where the material no longer breaks because it does not reach the necessary momentum. A "many gentle drops" design ("many slats" design) is likely more expensive and may require a larger equipment footprint, but it could help protect against catalyst breakage. Overall, this kind of engineering analysis can help draw more attention to equipment design and also to the overall architecture of the plant.

3.5.3 Fundamentals of Breakage by Collision

As stated, two parameters exist for extrudates related to collision: an asymptotic value obtained after many collisions and an asymptotic value obtained after just a single drop of very long extrudates. I have shown how these parameters relate to impact severity. The aspect ratio is the primary observable parameter in a sample linking to collision intensity.

As I have mentioned, collision intensity is very complex and many parameters exist, including:

- Impact velocity.
- The angular rotation rate from edge to edge and along the extrudate axis.
- The angle at the impact point.
- The impact force versus time diagram, which may have discontinuities.
- The roughness of the impact surface, including:
  - A bed of extrudates.
  - The presence of neighboring extrudates during the fall and the impact.

It is a momentous task to deal with any individual impact characteristic, and it could lead to an inability to see the forest for the trees. What I do is deal with a collision from the standpoint of an "average collision" and apply the physics observed.

A question that arises is why the extrudates would not keep breaking until they are finally reduced to an aspect ratio close to unity. The answer is that the shorter the extrudate, the greater the force required to break the extrudate. The momentum developed by an extrudate during its gravitational fall also becomes smaller as the aspect ratio becomes smaller. It becomes evident that there will be a crossover such that the force developed by a momentum change is no longer able to break the extrudate.

In a manufacturing plant, there will be many individual drops of varying severity. It is intuitively clear that high severity corresponds with high impact velocity, leading to a high momentum change. High severity also corresponds with a small contact time of the collision, which increases the force acting on the extrudate per Newton's second law.

It is also intuitively clear that very long extrudates break into a number of pieces and that there are more pieces after the drop as the aspect ratio increases. The dynamic modeling effort shown in Eqs. (3.8) and (3.10) establishes that it is possible to bring both parameters for extrudates related to collision into a single finite-difference equation and thus incorporate both parameters in the model. The starting aspect ratio is also in the model.

Applying the model to any impulsive breakage step is very helpful when applied in a manufacturing plant setting, but one still needs a mathematical model that can deal with many drops of different severities. Each drop in a plant occurs only once, but there are several of these individual drops as the catalyst travels through a manufacturing plant.

In a single drop, the aspect ratio will change according to Eq. (3.38):

$$\Phi_1 = (\gamma\Phi_0 + \Phi_\infty)/(\gamma + 1) \tag{3.38}$$

with

$$\gamma = (\Phi_a - \Phi_\infty)/(\Phi_0 - \Phi_\infty) \tag{3.39}$$

Both asymptotic aspect ratios have a physical-mechanical meaning, and the goal is to link them to the conditions of the impact.

Equation (3.40) writes Eq. (3.38) as a finite-difference Riccati equation:

$$\Phi_1 = \left(\Phi_a\Phi_0 - \Phi_\infty^2\right)/(\Phi_a - 2\Phi_\infty + \Phi_0) \tag{3.40}$$

where Φ_0 is the initial aspect ratio and Φ_1 is the aspect ratio after one drop.

I use the assumption for ideal materials that $\Phi_a = 2\Phi_\infty$, which simplifies to Eq. (3.41):

$$\Phi_1 = (\gamma\Phi_0 + \Phi_\infty)/(\gamma + 1) \tag{3.41}$$

with

$$\gamma = \Phi_{\infty}/(\Phi_0 - \Phi_{\infty}) \tag{3.42}$$

or Eq. (3.43), written as a finite-difference Riccati equation:

$$\Phi_1 = 2\Phi_{\infty} - \Phi_{\infty}^2/\Phi_0 \tag{3.43}$$

With Eqs. (3.41) or (3.43), an individual collision in a manufacturing plant is expressed in a single relationship with a single parameter that needs to be determined experimentally. Many of these individual steps exist in any catalyst manufacturing plant and none is the same. It is my goal to link this single parameter (here, the asymptotic aspect ratio) to the intensity of the collision and to define the severity of the collision based on fundamentals and first principles. It would also be very helpful to be able to decide and demonstrate which plant in the catalyst manufacturing circuit has the least severity based on existing commercial data.

3.5.3.1 Modulus of Rupture

Recall that the modulus of rupture is a measure of the inherent bending strength or flexural strength of a catalyst. It can be obtained from the rupture force required to break a catalyst extrudate in a bending test.

Equation (3.44) expresses the rupture force as a function of the properties of the catalyst:

$$F_r = \sigma D^3/sW_s \tag{3.44}$$

where F_r is the force required for rupture of the extrudate, σ is both the modulus of rupture and the tensile strength, D is the extrudate diameter, s is a shape factor, and W_s is the distance between the support points.

In this relationship, consider the distance between the support points to be equal to the length of the extrudate. The length of the extrudate can be rewritten as the product of the aspect ratio, Φ, times the diameter of the extrudate, as shown in Eq. (3.45):

$$F_r = \sigma D^2/s\Phi \tag{3.45}$$

The relationship between the rupture force and the strength of the extrudate in Eq. (3.45) is very simplified. For example, the location of the rupture force along the axial coordinate of the extrudate is assumed to be at the midway point of the extrudate. However, no data is available to shed light on the specific location of the rupture force during an actual collision. In fact, there may be a multitude of possible locations. I chose to apply the relationship expressed in Eq. (3.45), since at least directionally I believe the relationship yields the right trends.

3.5.3.2 Impact Force from Newton's Second Law

When an extrudate impacts against a surface, it experiences a force obtained from the change in linear momentum given by Newton's (Figure 3.30) second law shown in Eq. (3.46):

$$F_i = 2mv/\Delta\tau \tag{3.46}$$

where v is the impact velocity, $\Delta\tau$ is the contact time of the collision, m is the mass of the extrudate, and F_i is the force experienced during the impact.

The time of contact is likely not always singular (i.e. during a collision); there may be one or more times in which the catalyst makes contact with the surface. Thus, the time of contact is likely not contiguous and hence the force is likely jagged and spiked as a function of time. By "contiguous," I mean that the extrudate would make contiguous contact with the surface during the time interval.

I define the time interval as the time difference between first contact and last contact, even if the time interval is discontinuous. I assume that the collision is ideal and apply the change in linear momentum to obtain an estimate for the force and the time interval of contact from Newton's second law. I do not consider the change in angular momentum.

Figure 3.30 Sir Isaac Newton.

When extrudates freely fall and tumble through ambient air, one can calculate the impact velocity from literature correlations. Extrudates dropped from considerable height reach terminal velocity and thus no longer change their impact velocity. The way the average impact is defined here is by equating the rupture force to the force during impact.

3.5.3.2.1 Elastic Collision without Breakage

Not all impacts between an extrudate and a surface necessarily lead to breakage. For instance, the impact velocity can be very low or the extrudate can be very short. Even an extrudate that is long enough to break but hits at a certain angle can survive impact. For the reasoning in this book, assume that impacts are at right angles with the surface: what I mean is that the trajectory of the extrudate is perpendicular to the impact surface. The extrudate axis itself can still make a certain impact angle with the surface, but I will not consider the exact angle of the extrudate with the surface as a variable. In addition, when there is no breakage, the collision is termed "elastic." For an extrudate with a trajectory that hits a surface at right angles, without breakage and with conservation of kinetic energy, Eq. (3.47) applies:

$$F = 2mv/\Delta\tau \tag{3.47}$$

From this simple relationship, it follows that the impact force is linearly proportional to the impact speed, linearly proportional to the mass (and hence the aspect ratio), and inversely proportional to the contact time. Thus, a surface interaction that leads to a long contact time yields a lower impact force and therefore a "soft landing." A drop test on a hard plate is considered a "hard landing" because of the short contact time.

3.5.3.2.2 Collision with Breakage

As the velocity of an extrudate increases, there will evidently come a point at which the extrudate will break. Some of the kinetic energy of the extrudate will be needed to overcome the energy of the break, E_B. Therefore, the kinetic energy of the extrudates before and after the collision is no longer the same: hence the interaction is now termed "inelastic."

Figure 3.31 illustrates the bodies in motion before and after the collision. The plate with mass, M, is stationary before the collision and gets a small velocity component, v_P, after the collision. The impact velocity, v, of the extrudate before the collision becomes the lower release velocity, v', after the collision. See Eq. (3.48) for the linear momentum conservation:

$$mv = mv' + Mv_P \tag{3.48}$$

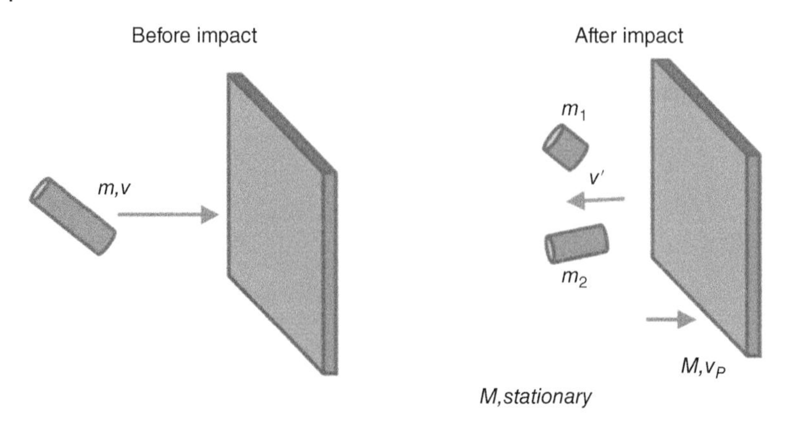

Figure 3.31 Simplified picture of the velocities before and after impact. Source: Courtesy of Wiley, doi: 10.1002/aic.15231.

The velocity is measured positively from left to right. The conservation of energy results in Eq. (3.49):

$$\frac{1}{2}mv^2 = \frac{1}{2}mv'^2 + \frac{1}{2}Mv_P^2 + E_B \tag{3.49}$$

Solving Eqs. (3.48) and (3.49) leads to Eq. (3.50) for the release velocity:

$$v' = \left\{ v - \frac{M}{m}\sqrt{v^2 - 2E_B\left(\frac{1}{M} + \frac{1}{m}\right)} \right\} / \left(1 + \frac{M}{m}\right) \tag{3.50}$$

Assuming that $M >> m$ and $\frac{1}{2}mv^2 > E_B$ results in Eq. (3.51):

$$v' = -\sqrt{v^2 - 2E_B/m} \tag{3.51}$$

or Eq. (3.52):

$$E_B = \frac{1}{2}m\left[v^2 - v'^2\right] \tag{3.52}$$

The relationship expressed in Eq. (3.52) for the break energy is simple, and it is possible to obtain the break energy by measuring the pre- and post-collision velocity with high-speed photography. Comparing extrudate positions frame by frame allows one to determine the distances the extrudates travel during the time elapsed between frames, and hence their velocity. The break energy acts as a constraint, and the kinetic energy of the extrudate before impact should be larger than the break energy for the break to be able to occur. Extrudates that impact at velocities and kinetic energies larger than the break energy will likely break, but this is not a certainty. In Section 3.5.3.3, I will discuss in more detail what finally determines whether an extrudate breaks or not and its relationship to strength.

3.5.3.2.3 Impact Velocity and Terminal Velocity

Obviously, a higher impact velocity makes it more likely that an extrudate will break when hitting a surface. Concerning a drop test, the impact velocity can be calculated well from existing sources. In Beeckman et al. [22], I detailed the entire procedure of the calculation and also compared the results with some experimental validations.

Extrudates that fall from a small height accelerate fast and encounter little drag from the ambient air because the velocity is still low. Thus, the impact velocity is determined by the drop height alone. For larger heights, one must take drag into account; when the height is substantial, the velocity reaches the terminal velocity and the impact velocity becomes a function of the drag characteristics. Equations (3.53)–(3.55) calculate the terminal velocity, v_t, the impact velocity, v, and the time to impact, t:

$$v_t = \sqrt{g/\kappa} \tag{3.53}$$

$$v = v_t\sqrt{1 - e^{-\left(\frac{2gh}{v_t^2}\right)}} \tag{3.54}$$

$$t = \frac{v_t}{g}\, a\cos h\left(e^{\frac{gh}{v_t^2}}\right) \tag{3.55}$$

$$\kappa = 0.565\left(\frac{\rho_g D}{\rho\Omega}\right) \tag{3.56}$$

where ρ is the density of the extrudate and ρ_g is the density of ambient air.

Figure 3.32 shows the impact time for a drop height of 10 m in ambient air. The aspect ratio, density, and shape were varied, as shown in Table 3.14. Under these conditions, neither the aspect ratio nor the initial orientation play a role in the time to impact. For smaller drop heights, I used the calculated values expressed in Eq. (3.55), in accordance with the literature, because there is too much error with short-duration stopwatch measurements.

3.5.3.2.4 Duration of the Contact Time in Collisions

In theoretical work, Hertz [27, 28] developed a theory for determining the contact time of two colliding spheres. Other investigators used his methodology for guidance on the measurement and treatment of the contact time of colliding bodies. Other references by Gugan [29], Leroy [30], and Bokor and Leventhall [31] confirmed the theory of Hertz with the experimental measurements shown in Table 3.15.

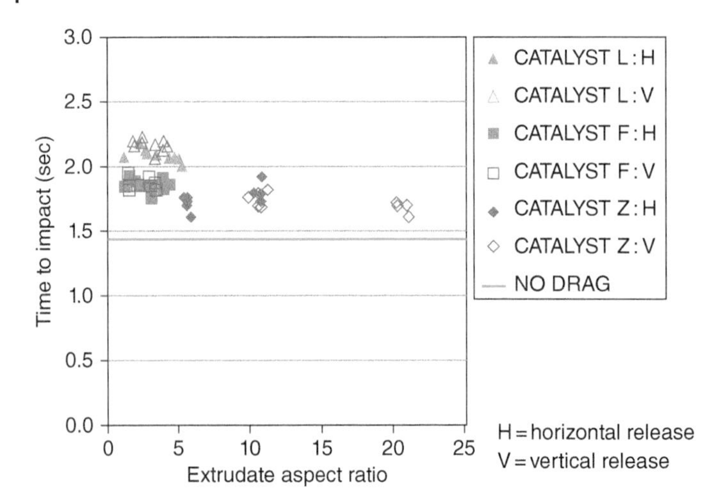

Figure 3.32 Experimental drop times at a 10-m drop height for extrudates L, F, and Z. Source: Courtesy of Wiley, doi: 10.1002/aic.15231.

Table 3.14 Average time to impact for various drop heights and various catalyst densities and shapes.

Catalyst ID	Shape	Density (kg/m^3)	Diameter (m)	Height (m)	Time (experimental) (s)	Time (calculated) (s)
A	Cylinder	1059	1.56E-03	10.1	2.62	2.63
Z	Cylinder	7210	1.90E-03	10.0	1.73	1.58
L	Trilobe	965	2.84E-03	10.0	2.12	2.25
F	Cylinder	2350	2.70E-03	10.0	1.85	1.76
F	Cylinder	2350	2.70E-03	3.2	0.87	0.87
Y	Cylinder	1870	8.30E-04	3.2	1.14	1.06
E1	Quadrulobe	1104	1.46E-03	10.0	2.53	2.59

Source: Courtesy of Wiley, doi: 10.1002/aic.15231.

The theoretical approach predicts that the contact time is inversely proportional to Young's modulus of elasticity to the power 0.4. The stiffer the body, the smaller the contact time becomes, which is intuitively very acceptable. The Hertz theory also predicts that the contact time is inversely proportional to the impact velocity to the power 0.2. A higher velocity leads to a smaller contact time, which is not very

Table 3.15 Contact times reported in the literature.

Author	Contact time (ms)
Gugan [29]	0.79–0.97
Leroy [30]	0.17
Bokor and Leventhall [31]	0.18

Source: Courtesy of Wiley, doi: 10.1002/aic.15231.

intuitive. One could argue that a higher-velocity impact leads to a higher degree of breakage of the body and thus a higher degree of deformation of the body, leading to a higher contact time.

But be careful, because the collision theory of Hertz is for colliding metallic spheres, whereas colliding extrudates are solid porous bodies of limited strength. The effect of the impact velocity is small because of the small value of the exponent. Plus, the complex behavior of actual extrudates impacting a surface makes contact time nearly intractable experimentally.

My hypothesis is that perhaps the actual real point of breakage of an extrudate against a surface happens upon full contact of the cylinder with the surface; that is, upon contact of the mantle of the cylinder along the axis thereof with the surface of impact. Think of a swimming pool; a belly flop hurts the most – enough said.

An extrudate will typically make contact at an angle with a surface; only very rarely is there full body contact. In many cases, however, the extrudate perhaps slides at the point of contact depending on the angle of friction, then establishes full body contact and breaks. If the angle is too large, the extrudate may bounce back from the surface unbroken. These effects are all speculative on my part, of course; only experiments using ultra-high-speed photography can determine and corroborate my hypothesis.

I did obtain some standard high-speed photography (up to 10 000 frames per second) that does indeed show the complexity of the impact. First, as already indicated in Section 3.5.3.2.2, it is possible to accurately determine the difference in speed of the approach of the extrudate before collision and the speed of the broken piece(s) leaving the surface by reviewing a frame-to-frame comparison of pictures taken from synchronized cameras installed at an angle to each other. Indeed, catalyst extrudates break upon impact and leave the impact point in a cone-like shape, and so at least two cameras are likely required to catch the exact trajectory. Perhaps a clever setup with mirrors or a parabolic mirror might be able to catch the event with a single camera, and so eliminate the need to synchronize the cameras.

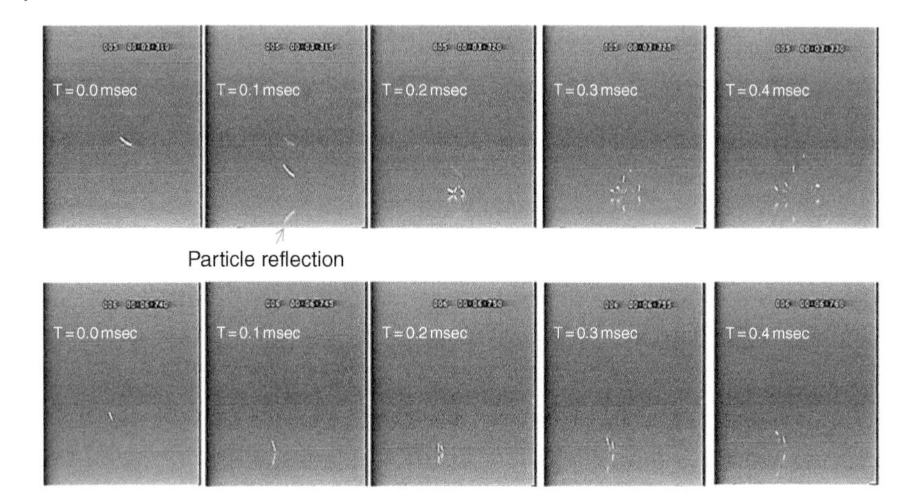

Figure 3.33 Impact of extrudates on an empty polycarbonate surface. Source: Courtesy of Wiley, doi: 10.1002/aic.15231.

The top row of Figure 3.33 shows the sequence of frames of an extrudate bouncing and breaking against a surface. The surface, which is made of Plexiglas, conveniently shows the reflection of the extrudate and helps define the exact impact point. Defining the contact time as the time difference between first contact and final contact in the series of photographs indicates a contact time on the order of 0.1 ms, or perhaps a fraction thereof. Certainly, the contact time will be a function of each particular situation, and it will likely be almost intractable to define an average, even based on multiple collision scene shots.

The average contact time will depend, of course, on the definition adopted for the contact time. From the sequence of photographs, the first time of contact is likely around 0.16 ms. At 0.2 ms, the leftmost end of the extrudate has not broken yet and still needs to impact between 0.2 and 0.3 ms in order to show a photograph of pieces at 0.3 ms. It is also clear that the actual break from the standpoint of a potential belly-flop contact is likely much shorter than 0.1 ms, perhaps due to the violent vibrational behavior of the extrudate against the surface.

3.5.3.3 Force Diagram at Twice the Asymptotic Aspect Ratio Φ_∞

It is now an opportune time to think about how to relate the impact force and the rupture force to the asymptotic aspect ratio, Φ_∞. The impact force can be calculated from Newton's second law, and in the simple form used here it is certainly not better than a first-order approximation of the impact phenomenon. Using the rupture force in the simple form as used in a three-point bending test is also at best a first-order approximation. However, both approximations lend themselves to practical measurement and application; thus, I will use them here to form a link to the asymptotic aspect ratio. As I explained in Beeckman et al. [22], upon reaching the asymptotic aspect ratio, the ratio of the impact force over the rupture force is a parameter called β, a collision interaction factor, shown in Eq. (3.57):

$$\beta = F_i/F_r \tag{3.57}$$

β is a factor with which one can link the impact force (as calculated from Newton's second law) to the rupture force (as measured in a three-point bending test), and it is likely a function of the severity of the test or the severity of the surface. For instance, a sizer like a hammer mill will have a certain β value, or a drop into a bucket elevator will have a certain β value.

Once measured, the parameter β can be tested against the data and validated, and this is exactly the methodology being tested in a drop test later on.

Equation (3.58) calculates the rupture force of an extrudate:

$$F_r = \frac{\sigma D^2}{s\Phi} \tag{3.58}$$

Equation (3.59) calculates the impact force:

$$F_i = 2mv/\Delta\tau \tag{3.59}$$

Figure 3.34 shows the representation of both Eqs. (3.58) and (3.59). The impact force increases proportionally with the aspect ratio while the rupture force drops proportionally with the aspect ratio; they are bound to cross. The question is: Where do they cross on the x-axis? In my opinion, they will cross at twice the asymptotic aspect ratio. My reasoning for this is that as experiments have shown, extrudates will break at twice the asymptotic length, and once they break (in half) they reach the asymptotic aspect ratio, after which they no longer break.

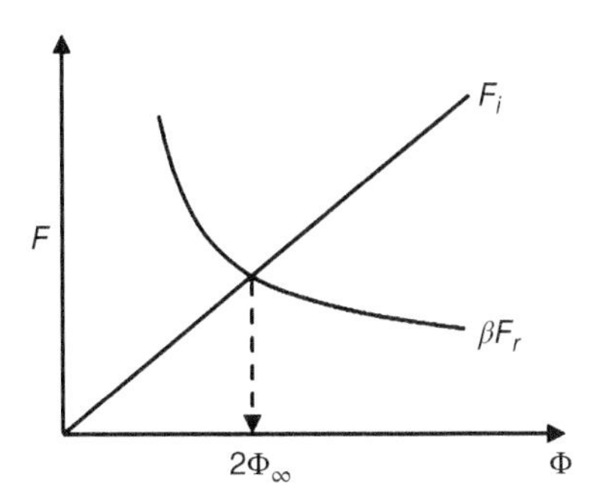

Figure 3.34 Balance of the impulsive force with the rupture strength. Source: Courtesy of Wiley, doi: 10.1002/aic.15231.

3.5.3.3.1 A Dimensionless Group Be for Breakage by Collision

Equation (3.60) relates the rupture force to the impact force at twice the asymptotic aspect ratio:

$$F_i = 2(D\Phi_\infty \Omega \rho a) = \beta F_r = \beta \frac{\sigma D^2}{2s\Phi_\infty} \tag{3.60}$$

where a is the deceleration of the extrudate during impact and Ω is the cross-sectional area. Equation (3.61) solves this very simple quadratic equation and leads to:

$$\Phi_\infty = \left(\sqrt{\beta}/2\right)\sqrt{Be} \tag{3.61}$$

where Be is a dimensionless group given by Eq. (3.62):

$$Be = \sigma/\psi \rho D a \tag{3.62}$$

while ψ stands for a shape factor defined in Eq. (3.63):

$$\psi = s\Omega/D^2 \tag{3.63}$$

3.5.3.3.2 Development of the Severity Functional of a Collision

The dimensionless group Be is clearly defined in Eq. (3.62), but it is not yet in a practical form. Application of Eq. (3.62) is difficult, since the deceleration during collision cannot be easily measured. An estimate leaves much to be desired. To circumvent this, I decided to normalize Be to the gravitational acceleration in Eq. (3.64):

$$Be = \sigma/\psi \rho D a = (g/a) \times \ \sigma/\psi \rho D g = (g/a) \times Be_g = 1/G \times Be_g \tag{3.64}$$

which leads to Eq. (3.65):

$$\Phi_\infty = \frac{1}{S_\infty} \times \sqrt{\sigma/\psi \rho D g} = \frac{1}{S_\infty} \times \sqrt{Be_g} \tag{3.65}$$

with Eq. (3.66):

$$S_\infty = 2\sqrt{G/\beta} \tag{3.66}$$

Besides the impact velocity, the factor S_∞ now contains the unmeasurable part of the collision. Inserting the dependence on velocity leads to Eq. (3.67):

$$S_\infty = \sqrt{8v/\beta g \Delta\tau} \tag{3.67}$$

The group S_∞ becomes larger when the impact velocity increases. It also becomes larger for a smaller contact time and leads to a greater impact severity. Hence, I call the dimensionless functional S_∞ simply the severity of the impact.

I used the index "∞" as subindex in the severity symbol to clearly indicate that the severity is based on the asymptotic aspect ratio Φ_∞. Equation (3.67) leads to a very practical correlation in order to express the asymptotic aspect ratio as a function of catalyst strength, catalyst properties, and severity of the impact, as expressed by Eq. (3.68):

$$\Phi_\infty = \frac{1}{S_\infty} \times \sqrt{\sigma/\psi\rho Dg} = \frac{1}{S_\infty} \times \sqrt{Be_g} \tag{3.68}$$

The term $\sqrt{\beta\Delta\tau}$ offers a welcome surprise. In first approximation, it works out to be a constant for the wide variety of catalysts tried, including a variety of drops from vastly different impact heights onto the same plate. The constancy of $\sqrt{\beta\Delta\tau}$ leads to an important simplification in my mind, in that the conditions of the impact determine the severity; it is equipment-specific, whereas the effect of impact velocity on severity has a simple square-root functionality.

Measuring Φ_∞ and Be_g in a given drop test enables one to calculate the severity, S_∞, which one can then use to predict the asymptotic aspect ratio for catalysts with different properties for that same experimental setup.

3.5.3.3.3 *Learnings from* Be

The dimensionless group *Be* does offer some basic guidance on how different parameters can affect the asymptotic aspect ratio of a catalyst:

- An increase in the square root of the modulus of rupture or the tensile strength of the extrudate yields a proportional increase in the aspect ratio.
- An increase in the square root of either the density or the diameter yields a proportional decrease in aspect ratio.
- The shape factor of the extrudate plays a role. Quadrulobes hold up better than cylinders, although the difference is not very dramatic.

One could certainly argue that these three trends are intuitive, but it is because of the engineering analysis – based on basic laws – that the square-root relationship comes about. In Beeckman et al. [8, 22], I investigated extrudates dropped in ambient air onto a hard empty plate. The theoretical results were in good agreement with the experimental results, thus paving the way for applications to commercial plants.

I selected materials with a wide variety of properties in terms of shape, size, and strength. I also investigated different methods of preparation, such as different binders, as well as intermediate samples (samples of the same catalyst after drying or after calcination). I also conducted the drop test at different heights. In total, I evaluated about 25 catalysts, all of which are shown in Table 3.16.

My main goal was to develop and validate a general methodology and theoretical treatment that hopefully would have practical applications. For the dynamic modeling of the drop test, I applied the general model and determined the asymptotic parameters Φ_∞ and Φ_α via least-squares minimization of the predicted and observed aspect ratios. For materials where the starting aspect ratio was short,

Table 3.16 Materials used in impulsive breakage testing.

Test ID	Catalyst ID	Nature	Binder	Final *T* (°C)	Shape	Diameter (mm)	Drop height (m)
1	A	Zeolite	Alumina	120	Cylinder	1.56	0.20
2	A	Zeolite	Alumina	120	Cylinder	1.56	0.61
3	A	Zeolite	Alumina	120	Cylinder	1.56	1.83
4	A	Zeolite	Alumina	120	Cylinder	1.56	6.40
5	B1	Zeolite	Silica	120	Cylinder	1.50	1.83
6	B3	Zeolite	Silica	120	Cylinder	1.54	1.83
7	B5	Zeolite	Silica	120	Cylinder	1.53	1.83
8	C1	Zeolite	Silica	540	Cylinder	1.48	1.83
9	C2	Zeolite	Silica	540	Cylinder	1.51	1.83
10	C3	Zeolite	Silica	540	Cylinder	1.49	1.83
11	C4	Zeolite	Silica	540	Cylinder	1.54	1.83
12	C5	Zeolite	Silica	540	Cylinder	1.51	1.83
13	D1	Zeolite	Alumina	540	Quadrulobe	1.43	1.83
14	D2	Zeolite	Alumina	540	Quadrulobe	1.43	1.83
15	D4	Zeolite	Alumina	540	Quadrulobe	1.45	1.83
16	E1	Zeolite	Alumina	540	Quadrulobe	1.46	1.83
17	E2	Zeolite	Alumina	540	Quadrulobe	1.46	1.83
18	E3	Zeolite	Alumina	120	Quadrulobe	1.49	1.83
19	E4	Zeolite	Alumina	120	Quadrulobe	1.43	1.83
20	F	Amorphous	NA	540	Cylinder	2.70	1.83
21	G	Zeolite	NA	540	Quadrulobe	1.90	1.83
22	H	Amorphous	Alumina	540	Cylinder	0.83	1.83
23	L	Zeolite	Alumina	540	Trilobe	2.84	1.83
24	M	Zeolite	Alumina	540	Cylinder	0.95	1.83
25	N	Carbon	NA	200	Cylinder	1.40	1.83

Source: Courtesy of Wiley, doi: 10.1002/aic.15231.

I could not determine the parameter Φ_α; thus, I assumed they were ideal ($\Phi_\alpha = 2\Phi_\infty$) (see catalysts E1, E2, and E3 in Table 3.17).

I derived the catalyst strength properties with the method established in Section 3.3.1 on the bending strength of an extrudate and Section 3.4 on the experimental measurement of the mechanical strength of extrudates. Table 3.18 gives

Table 3.17 Catalyst aspect ratio Φ versus number of drops and parameter-fitting results.

Test ID	Catalyst ID	Number of repeated drops													Φ_α	Φ_∞
		Φ_0	1×	2×	3×	4×	5×	7×	8×	9×	11×	16×	21×	31×		
1	A	24	6.09		4.13		3.45	3.68		3.69	2.98	2.99	3.28	2.73	6.70	2.89
2	A	24	4.54		2.93		2.61	2.35		2.21	2.12				4.93	1.93
3	A	24	3.31		2.14		2.02	1.95		1.72	1.75				3.44	1.60
4	A	24	2.66		2.01		1.92	1.73		1.80	1.61				2.73	1.62
5	B1	7.92	3.73	2.92	2.70	2.53	2.40								4.46	1.94
6	B3	6.52	2.65	2.38	2.18	1.95	1.90								2.97	1.72
7	B5	3.96	2.06	1.90	1.65	1.63	1.46								2.45	1.34
8	C1	7.92	4.28	4.24	3.98	3.83	3.57								4.59	3.57
9	C2	6.94	5.30	4.33	4.19	3.83	3.57								9.46	2.57
10	C3	6.52	3.22	2.92	2.60	2.60	2.44								3.58	2.25
11	C4	7.03	3.22	2.92	2.77	2.45	2.63								3.46	2.36
12	C5	3.96	2.17	2.05	1.88	1.84	1.78								2.35	1.69
13	D1	4.37	3.28	3.08	2.86										3.86	2.53
14	D2	4.01	2.78	2.47	2.31										3.38	1.92
15	D4	4.46	3.54	3.16	3.00										4.93	2.39
16	E1	2.25	2.25	2.24		2.25			2.22			2.18			4.27	2.14
17	E2	2.21	2.22	2.22		2.21			2.21			2.16			4.26	2.13

(Continued)

Table 3.17 (Continued)

		Number of repeated drops														
Test ID	**Catalyst ID**	Φ_0	**1×**	**2×**	**3×**	**4×**	**5×**	**7×**	**8×**	**9×**	**11×**	**16×**	**21×**	**31×**	Φ_α	Φ_∞
18	E3	2.94	2.85	2.77		2.58			2.35			2.17			4.41	2.20
19	E4	4.19	3.41	3.18		2.86			2.61			2.54			4.99	2.37
20	F	2.66	1.89	1.67		1.45			1.28			1.28			2.60	1.16
21	G	3.71	2.92	2.73		2.50			2.36			2.18			3.92	2.12
22	H	7.48	4.10	3.92		3.69			3.46			3.17			4.48	3.28
23	L	2.28	1.92			1.80			1.80			1.65			2.13	1.69
24	M	5.92	4.63	3.73		2.99			2.88			2.61			7.72	2.22
25	N	57.41	41.15	36.35		32.09			26.46			19.80			87.26	17.96

Source: Courtesy of Wiley, doi: 10.1002/aic.15231.

Table 3.18 Catalyst properties.

Test ID	Catalyst ID	Shape	Drop height (m)	Φ_α (–)	Φ_∞ (–)	Density, ρ (kg/m^3)	Diameter, D (m)	Impact velocity, V (m/s)	Rupture force, Fr (N)	Shape factor, s (–)	Bridge width, w (m)	Modulus of rupture, σ (MPa)	Shape factor, ψ (–)	$\sigma/\psi\rho Dv$ (1/s)
1	A	Cylinder	0.20	6.70	2.89	1059	1.56E-03	1.88E+00	1.31E-01	2.55	3.18E-03	2.81E-01	2	4.54E+04
2	A	Cylinder	0.61	4.93	1.93	1059	1.56E-03	2.97E+00	1.31E-01	2.55	3.18E-03	2.81E-01	2	2.87E+04
3	A	Cylinder	1.83	3.44	1.60	1059	1.56E-03	4.00E+00	1.31E-01	2.55	3.18E-03	2.81E-01	2	2.14E+04
4	A	Cylinder	6.40	2.73	1.62	1059	1.56E-03	4.32E+00	1.31E-01	2.55	3.18E-03	2.81E-01	2	1.98E+04
5	B1	Cylinder	1.83	4.46	1.94	1242	1.50E-03	4.16E+00	3.87E-01	2.55	3.18E-03	9.20E-01	2	5.92E+04
6	B3	Cylinder	1.83	2.97	1.72	1178	1.54E-03	4.12E+00	3.20E-01	2.55	3.18E-03	7.16E-01	2	4.80E+04
7	B5	Cylinder	1.83	2.45	1.34	1126	1.53E-03	4.06E+00	1.11E-01	2.55	3.18E-03	2.49E-01	2	1.78E+04
8	C1	Cylinder	1.83	4.59	3.57	1150	1.48E-03	4.04E+00	6.63E-01	2.55	3.18E-03	1.66E+00	2	1.21E+05
9	C2	Cylinder	1.83	9.46	2.57	1180	1.51E-03	4.10E+00	6.49E-01	2.55	3.18E-03	1.52E+00	2	1.04E+05
10	C3	Cylinder	1.83	3.58	2.25	1049	1.49E-03	3.93E+00	2.93E-01	2.55	3.18E-03	7.14E-01	2	5.81E+04
11	C4	Cylinder	1.83	3.46	2.36	1087	1.54E-03	4.01E+00	4.21E-01	2.55	3.18E-03	9.40E-01	2	7.01E+04
12	C5	Cylinder	1.83	2.35	1.69	1037	1.51E-03	3.93E+00	1.93E-01	2.55	3.18E-03	4.55E-01	2	3.71E+04
13	D1	Quadrulobe	1.83	3.86	2.53	1010	1.43E-03	3.88E+00	2.94E-01	2.20	3.18E-03	7.03E-01	1.81	6.95E+04
14	D2	Quadrulobe	1.83	3.38	1.92	1067	1.43E-03	3.95E+00	9.79E-02	2.20	3.18E-03	2.34E-01	1.81	2.14E+04
15	D4	Quadrulobe	1.83	4.93	2.39	870	1.45E-03	3.69E+00	1.82E-01	2.20	3.18E-03	4.16E-01	1.81	4.93E+04
16	E1	Quadrulobe	1.83	4.27	2.14	1104	1.46E-03	4.03E+00	4.63E-01	2.20	3.18E-03	1.03E+00	1.81	8.77E+04
17	E2	Quadrulobe	1.83	4.26	2.13	1050	1.46E-03	3.96E+00	7.03E-01	2.20	3.18E-03	1.57E+00	1.81	1.42E+05

(Continued)

Table 3.18 (Continued)

Test ID	Catalyst ID	Shape	Drop height (m)	Φ_α (−)	Φ_∞ (−)	Density, ρ (kg/m^3)	Diameter, D (m)	Impact velocity, V (m/s)	Rupture force, Fr (N)	Shape factor, s (−)	Bridge width, w (m)	Modulus of rupture, σ (MPa)	Shape factor, ψ (−)	$\sigma/\psi\rho Dv$ (1/s)
18	E3	Quadrulobe	1.83	4.41	2.20	1138	1.49E-03	4.09E+00	2.71E-01	2.20	3.18E-03	5.75E+00	1.81	4.59E+04
19	E4	Quadrulobe	1.83	4.99	2.37	1249	1.43E-03	4.16E+00	3.74E-01	2.20	3.18E-03	8.89E-01	1.81	6.60E+04
20	F	Cylinder	1.83	2.60	1.16	2350	2.70E-03	5.32E+00	1.76E+00	2.55	6.35E-03	1.45E+00	2	2.16E+04
21	G	Quadrulobe	1.83	3.92	2.12	1094	1.90E-03	4.35E+00	5.43E-01	2.20	6.35E-03	1.10E+00	1.81	6.72E+04
22	H	Cylinder	1.83	4.48	3.28	1870	8.28E-04	3.91E+00	2.67E-01	2.55	2.38E-03	2.85E+00	2	2.35E+05
23	L	Trilobe	1.83	2.13	1.69	965	2.84E-03	4.52E+00	2.53E+00	2.28	6.35E-03	1.60E+00	1.65	7.82E+04
24	M	Cylinder	1.83	7.72	2.22	750	9.45E-04	2.82E+00	8.01E-02	2.55	2.38E-03	5.76E-01	2	1.44E+05
25	N	Cylinder	1.83	87.26	17.96	994	1.40E-03	3.77E+00	6.27E+00	2.55	6.40E-03	3.72E+01	2	3.55E+06

Source: Courtesy of Wiley, doi: 10.1002/aic.15231.

the flexural properties. Finally, I calculated the impact velocities for each drop test according to Eq. (3.54).

Figure 3.35 shows the fit of the data when employing the model: the asymptotic parameter values and the fit are quite reasonable. In addition, a graph of the asymptotic aspect ratio Φ_∞ as a function of the group $\sigma/\psi\rho Dv$ to the power of one-half (1/2) indicates a straight line through the origin, as shown in Figures 3.36 and 3.37.

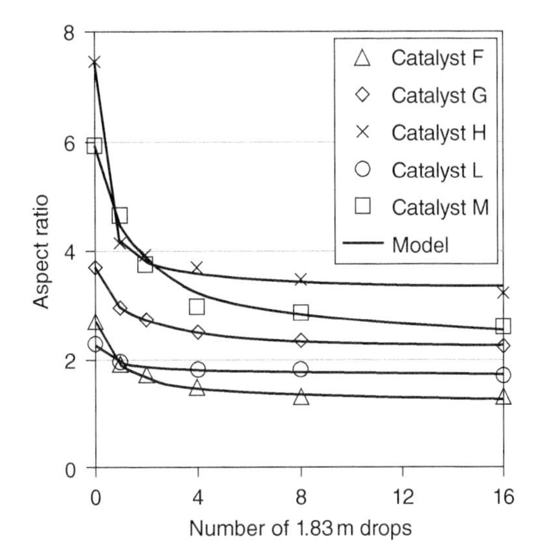

Figure 3.35 Aspect ratio model-fitting results. Source: Courtesy of Wiley, doi: 10.1002/aic.15231.

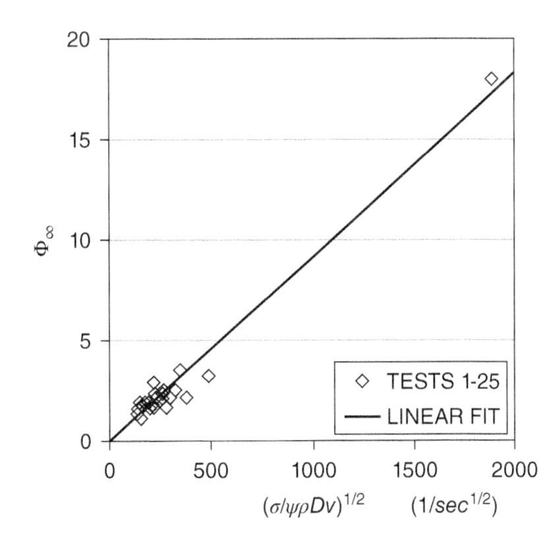

Figure 3.36 Asymptotic aspect ratio versus $\sqrt{\sigma/\psi\rho Dv}$ (full range). Source: Courtesy of Wiley, doi: 10.1002/aic.15231.

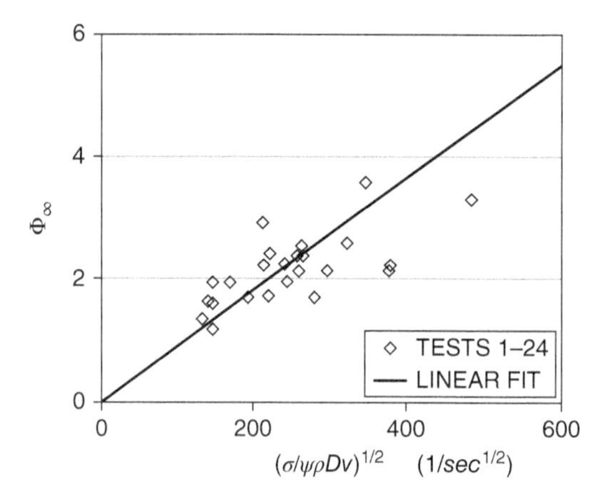

Figure 3.37 Asymptotic aspect ratio versus $\sqrt{\sigma/\psi\rho Dv}$ (reduced x-axis range). Source: Courtesy of Wiley, doi: 10.1002/aic.15231.

The slope of the straight line has a value of $\beta\Delta\tau/8$ and thus can help indicate the contact time of the collision. Assuming a collision factor of unity, one can calculate the value of $\Delta\tau$ at about 0.6 ms, which is indeed a short time interval. A value of 0.6 ms is longer than what high-speed photography indicates, yet considering all of the assumptions and treatments, it can be considered fair. Certainly, more in-depth study and a substantially faster frame speed could help improve this. Nonetheless, it is interesting to see that all catalysts seem to indicate that, in a first-order approximation, one can consider the product $\beta\Delta\tau$ a constant.

To estimate the square-root behavior, I plotted the asymptotic aspect ratio and the group $\sigma/\psi\rho Dv$ on a natural log scale. The slope of that line worked out at a value of 0.42, which is close to the one-half value from theory. Individual trends of the asymptotic aspect ratio with the specific variables in the group $\sigma/\psi\rho Dv$ are difficult to come by because often multiple variables change from experiment to experiment. Luckily, it is possible to isolate the effect of the impact velocity due to different drop heights and the effect of the modulus of rupture. Figures 3.38 and 3.39 show those trends.

As an example, I wondered what the dimensionless severity would be for a drop height of 1.83 m. The calculated value for the impact velocity in ambient air works out to be 4.0 m/s. With this 4.0-m/s value and the experimental value for the slope of the Φ_∞ versus $(\sigma/\psi\rho Dv\,)^{1/2}$ relationship, Eq. (3.69) finds:

$$S_\infty = \sqrt{8v/g\beta\Delta\tau} = \sqrt{4.0/9.81} \times (2000/18) = 71 \tag{3.69}$$

This dimensionless severity value of 71 can be used for drops from a height of 1.83 m on a hard empty plate for all catalysts or supports.

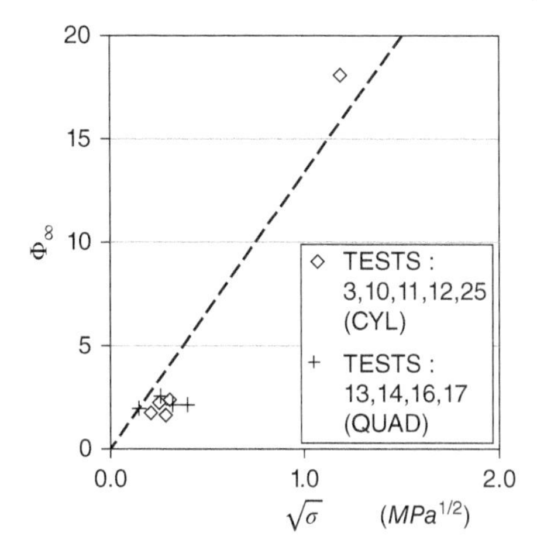

Figure 3.38 Trend of the asymptotic aspect ratio, Φ_∞, with the square root of the bending strength. Source: Courtesy of Wiley, doi: 10.1002/aic.15231.

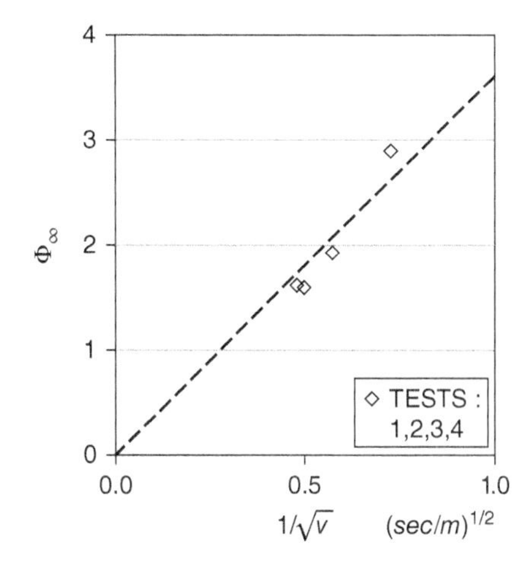

Figure 3.39 Trend of the asymptotic aspect ratio, Φ_∞, with the inverse square root of the impact velocity. Source: Courtesy of Wiley, doi: 10.1002/aic.15231.

To obtain the second asymptotic aspect ratio, Φ_α, I followed an earlier result from Beeckman et al. [8], where I found that Φ_α often is close to twice the asymptotic aspect ratio Φ_∞. Figure 3.40 demonstrates this. In addition, the effect of the impact velocity on Φ_α as shown in Figure 3.41 suggests that this reasoning is acceptable.

3.5.3.3.4 Worked Example: Catalysts in Space

Imagine three identical drop tests set up on Earth, Mars, and the Moon. How would the asymptotic aspect ratio Φ_∞ differ for these three tests in different

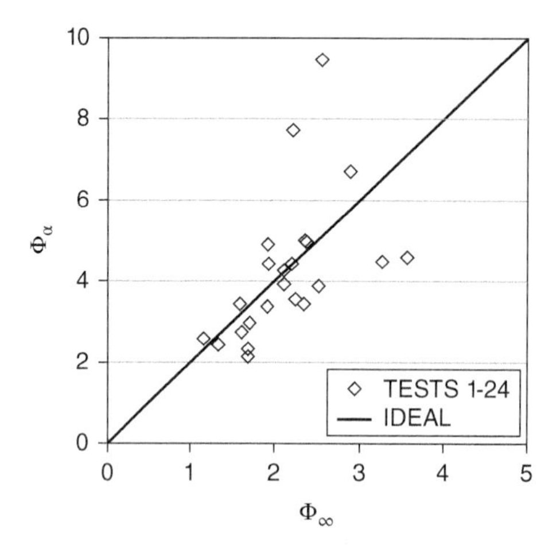

Figure 3.40 Correlation of the asymptotic aspect ratio, Φ_α, against Φ_∞. Source: Courtesy of Wiley, doi: 10.1002/aic.15231.

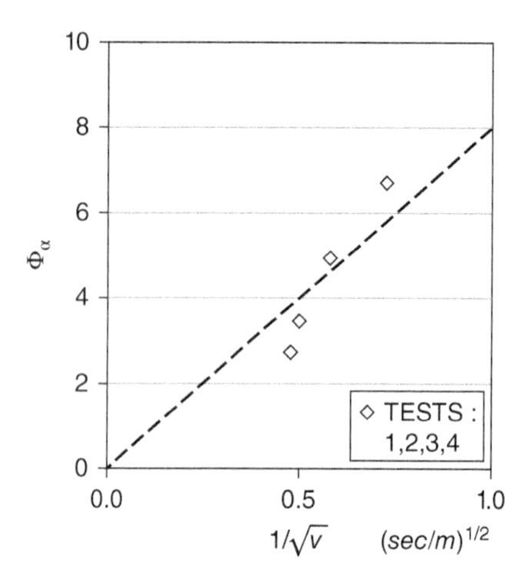

Figure 3.41 Trend of the asymptotic aspect ratio, Φ_α, with the square root of the impact velocity. Source: Courtesy of Wiley, doi: 10.1002/aic.15231.

gravitational fields? Assume short drop heights so that one can neglect the atmosphere and thus the drag in all three tests. Equation (3.70) expresses the impact velocity as:

$$v = \sqrt{2hg} \tag{3.70}$$

where h is the drop height and g is the local gravitational acceleration.

The contact time of the collision I assumed is unaffected by the gravitational force because the deceleration g of the particle from impact is much higher than the gravitational g. The dependence of Φ_∞ on the impact velocity then shows that the asymptotic aspect ratio is proportional to $g^{-0.25}$. A catalyst would yield $\Phi_\infty = 3.0$ on Earth, $\Phi_\infty = 3.8$ on Mars, and $\Phi_\infty = 4.7$ on the Moon.

The predictions for Φ_∞ could be simulated on Earth by adjusting the height of the drop test to simulate a lower g (this would of course be far less spectacular). On Earth, the height to simulate the lower gravitational field of the Moon would have to be 6.05 times lower, while to simulate the gravitational field of Mars, the height of the drop test on Earth would need to be 2.64 times lower.

3.5.3.4 Concluding Remarks on *Be*

To conclude and summarize this subject of the drop test, I have established a correlation that allows one to predict the asymptotic aspect ratio reached during collision with a surface of an extruded catalyst based on the modulus of rupture and the drop height. The impact surface is considered hard with essentially no yielding thereof based on the impact of the extrudate. The modulus of rupture in relation to catalyst extrudates is a relatively new strength property and has not been discussed in many external references.

My correlation leads to a new dimensionless group, Be_g, and a dimensionless severity of the test, S_∞. The group Be_g considers the aspects of catalyst strength, density, and shape, while the group S_∞ contains the impact velocity and the contact time (such as a soft landing versus a hard landing).

Later in this chapter, I will show how to apply the knowledge gained so far to commercial plants. This extends the concept of the severity for a particular interaction further to obtain the severity for an entire plant. With this information, one can compare plants in the catalyst manufacturing circuit and also predict plant severity as a function of the design architecture.

3.6 Breakage by Stress in a Fixed Bed

Let us further extend the breakage of extrudates caused by impact forces experienced during a collision to breakage experienced in a fixed bed caused by stress. A catalyst will experience stress in a fixed bed for varied and complex reasons. The weight of extrudates, especially at the bottom, could cause a great deal of stress. There are external references that provide a solid understanding of the factors at play. As the height of the extrudates in the bed increases, so does the stress due to the increased weight of layer upon layer. It has been shown, however, that the increase in stress decreases as the bed height increases because the gravitational forces spread out radially toward the walls of the bed. The frictional forces at the wall essentially hold the bed up and do not allow the gravitational forces to transpire all the way down.

The equilibrium of the gravitational forces and the frictional forces is labile, and a small disturbance can suddenly loosen the grip on the wall. For a short duration – typically fractions of a second to a few seconds – the bottom of the vessel experiences the entire weight of the column. The crushing forces induced by a disturbance, especially at the bottom of the bed, can be substantially higher.

In addition to the entire weight of the bed of catalyst, there is also the acceleration of the bed to consider during this short time, which creates extra stress at the bottom. Thereafter, the bed quickly finds a new labile equilibrium and goes on as if nothing had happened, except that the sudden jolt of stress at the bottom can create breakage accompanied by fines, causing operational difficulties later. The jolt can occur during upsets, startup, and shutdown, where the operation is not lined out nor in steady state. Keep this in mind when bringing reactor vessels online. The weight of the catalyst bed can also be substantially higher than the dry weight when the process fluids are liquids because of the absorption of liquids in the internal pore structure of the catalyst and the partial filling of the interparticle void space. In addition to stress caused by weight, stress can also be caused by process pressure drop over the catalyst bed during normal operation.

Therefore, because stress in a fixed bed can be quite variable, it is not easy to quantify it. To simplify the task, consider very shallow beds (say, a few centimeters) with substantial stress applied externally. One encounters such a scenario when performing a bulk crush strength measurement. This procedure can be particularly well controlled and has the added benefit of being an established ASTM procedure.

I will further restrict the study here to extrudates only, not spheres, granules, or shaped bodies made with a pelletizing press. Just as was the case in breakage by collision, a dimensionless group is at the core of the breakage phenomenon when basing the strength on the modulus of rupture. It will become clear that this attempt is only a first-order approximation and that much can be improved through further study. I expect that the field of breakage caused by stress in a fixed bed will become so much more than what it is now.

3.6.1 Experimental

The experimental procedure available in Beeckman et al. [26] is also available in video format as well, and a simple Google search for "Beeckman *JoVE*" or doi: 10.3791/57163 will get you there. I will summarize the research here. I tested a total of five different catalysts with substantial differences in properties such as shape, size, bending strength, and density. Some but not all of the catalysts contained zeolite crystal, while the binder used in the extrusion process was typically silica or alumina. I unfortunately cannot share the chemical composition of these catalysts, but this information has no impact on the methodology or analysis.

The aspect ratio is experimentally determined in much the same way as in Section 3.5.2.1; the video in Beeckman et al. [26] referenced in the previous paragraph clearly shows this procedure. In short, riffle a sample of about 200 extrudates from the entire catalyst specimen and perform an optical scan. The scan is then transferred to the Alias analysis software, which calculates every individual length-to-diameter ratio and determines the average or the aspect ratio of all of the extrudates. My coauthors in the *JoVE* publication then repeated this procedure for every starting catalyst, as well as for every catalyst sample exposed to individual crushing stresses.

My coauthors determined the bending strength of each catalyst with the Instron instrument, using the Bluehill software to analyze a total of 25 extrudates each time. They then determined the arithmetic average of the bending strength for the starting catalyst only. The bulk crush test itself is essentially the same as the ASTM D7084-04 test, except that the focus is on the aspect ratio after the crush test, while the ASTM test focuses on the fines produced during the crush test.

The test cell we used is a shallow hollow cylinder approximately 0.0635 m in diameter and 0.0254 m high. We packed the catalyst sample into the enclosure and evened out the surface with a ruler right to the rim. We placed the top plate and ball carefully on the bed in preparation for measurement.

The *JoVE* publication video shows the entire procedure. Beyond modeling the aspect ratio as a function of the strength and crushing stress, I make an attempt to model the formation of fines, dust, and chips – or what is commercially called "fines-make" – in this book. By the end of this section, I will show fairly good agreement of this simple model with the data.

3.6.2 Theoretical

Describing and quantifying how catalyst extrudates break when experiencing stress in a fixed bed is obviously not easy. First, even the simplest case – the breakage of a single extrudate – depends on the location of the supporting points of the extrudate in the bed and on the location and direction of the force(s) that make contact with the extrudate.

Perhaps a simplifying factor for the analysis is the fact that the amount that an extrudate bends when experiencing stress is actually very small; I think the extrudate rarely sees the establishment of new contact points because of the bending phenomenon. Second, do not forget that the catalyst has a distribution of extrudate lengths and hence, even if only applied in the middle, so does the force that breaks the extrudate because of the hyperbolic nature of the torque. This distribution of extrudate lengths will vary. Third, the distribution of the location and direction of the extrudates in the fixed bed is highly irregular.

Wooten [32] offers a good review and comparison of different types of loading techniques. Catalyst beds that are sock-loaded tend to be of a higher void fraction and show more of a house-of-cards structure. Catalyst beds loaded using the catalyst-oriented packing method are more densely packed; catalyst extrudates tend to come to rest on average in a more horizontal position than catalyst extrudates that were sock-loaded. In either case, it is difficult to describe the local arrangement of extrudates in a fixed bed.

I thought it most logical to start with a very simplified model and to compare a model prediction of the average aspect ratio after stress to an experimental data set of five catalysts; the agreement is good, even with this simple model. As extrudates break, they also produce fines, and this predicted production of fines also agrees well with the data.

While in this book my focus is on the prediction of the aspect ratio, it is important to predict fines caused by breakage in a fixed bed because the majority of ASTM D7084-04 test data in commercial facilities is based on fines.

My simple model first applies the force that breaks an extrudate of average length; the force is applied in the middle of the extrudate. This force is determined experimentally and leads to the modulus of rupture. The model assumes that, on average, catalyst extrudates are stacked in a preferred position of 45° during loading. The model also calculates the areal density of extrudates and the average number of extrudates per unit area that support the bed. Balancing the stress applied with the bending strength of the bed yields a dimensionless group that predicts the stress required to break the bed an infinitesimal amount, thus leading to an infinitesimal reduction in aspect ratio to achieve a new equilibrium.

3.6.2.1 Areal Densities of Catalyst Extrudates

In Beeckman et al. [25], I wrote more about breakage in fixed beds. First, there is a need to develop an expression for the areal density of catalyst extrudates in a horizontal layer holding up the weight of a catalyst bed, also called the load. The load may be the weight of the bed, it may be the process pressure drop, or it may, for instance, be the stress applied to the bed when running the bulk crush strength measurement used in the ASTM D7084-04 test. When looking at the load for a horizontal slice through the bed, this load is actually carried by a number of individual catalyst extrudates spread over that area of the slice. It would be an astonishing experiment to be able to visualize such a slice and the individual contact points. A slice in this context is actually the uneven surface formed by the stacked surface of the extrudates. Any horizontal top surface of a fixed bed can be considered the top layer or the start of the bed being built.

Each layer contains a number of extrudates per unit area; I assume that these extrudates hold up the bed of catalyst above it. A packing of spheres of diameter D or a packing of short cylinders with an aspect ratio of approximately unity would give a density proportional to $1/D^2$. Catalysts of this kind are obtained by bead granulation or bead dripping; short extrudates are made by pressing in a tablet press or by cutting at the die face. What we need is an expression for extrudates that are relatively long. Regardless of whether the catalyst sample has a length-to-diameter distribution, assume that extrudates are all of average length.

To obtain an expression for the number of extrudates in a layer:

- Consider a partial bed built by dropping catalysts at a slow rate such that they do not interfere with each other when coming to rest.
- Assume the bed is, on average, geometrically horizontal over the long range, but locally, over an area of only a few extrudates, the surface is quite uneven.
- Arbitrarily assign the top layer as layer 0. By definition, there are extrudates assigned to layer 0, but one does not know at this point how many there are or where they are.

To create the next layer, or layer 1:

- Drop an extrudate at a random location over the bed and let it come to rest. This is the first extrudate assigned to layer 1.
- Extrudates that are dropped later on, one at a time, which touch extrudates in layer 0 but not any previous single extrudate(s) assigned to layer 1, also belong to layer 1.
- Extrudates that touch previous extrudates assigned to layer 1 would belong to layer 2 and hence are removed from the bed.

After all possible locations are taken for layer 1, count all of the extrudates assigned to layer 1, which then determines the result. Thereafter, one can proceed further and continue with layer 2.

Not every layer will have the same number of extrudates given the random nature of this process, but one can still make a fairly good estimate. Each extrudate in a layer will claim some space in the top of the bed; assume that the average position is actually along the long diagonal of a cube and that this long diagonal harbors an extrudate of average length. A long diagonal in this context is defined as the diagonal that runs from corner to opposing corner of the cube.

Stacking the number of cubes orderly, one next to the other, but not necessarily at the same height in a square pattern with no open spaces between the cubes, counts as forming the number of extrudates in a layer that supports the bed. The packing of catalyst extrudates built in this way is likely more similar to sock-loaded than catalyst-oriented packing. Catalyst-oriented packing is more like a parallelepiped of short height for harboring extrudates of average length in a more horizontal position.

I am probably oversimplifying the stacking, but since I am not accounting for the length-to-diameter distribution, it makes no sense to envision anything more complicated at this point. The model predictions shown later in Section 3.6.2.5 will prove that the assumptions are fair.

Equation (3.71) expresses the areal density of a number of extrudates of average length L, stacked as long diagonals in a cube with the projection factor ζ:

$$\textit{Areal density}\ (m^{-2}) = \frac{1}{(\zeta L)^2} \tag{3.71}$$

with Eq. (3.72) expressing the projection factor ζ as:

$$\zeta = 1/\sqrt{3} \tag{3.72}$$

In Beeckman et al. [25], I showed that the void fraction of such an arrangement for an aspect ratio of 2.5 calculates at 35%, which is a very typical value. In fact, I think the void fraction measurement could be a good tool with which to test some of the more complex treatments when considering distributions in length and packing.

3.6.2.2 Balance of the Rupture Force with a Fixed-bed Static Load

After developing an approximate expression for the areal density of a bed of extrudates along a "horizontal" surface slice through the bed, it is now time to obtain an approximate value for the force that a single extrudate can take at rupture, also known as breakage. Clearly, the bed surface is a complex entity.

I assume here that a first-order approximation of the average position of an extrudate on the top of a bed is the long diagonal of a cube that sits horizontal and tiles the surface of the bed. From the well-known three-point bending test, Eq. (3.73) calculates the force to break or rupture the extrudate as:

$$F_r = \frac{\sigma D^3}{sL} \tag{3.73}$$

When the aspect ratio is small, then nonlinear behavior is introduced into Eq. (3.73), as explained by Wachtman et al. [33]. In this book, I will only use the correlation as expressed by Eq. (3.73) for slender beams.

The vertical component of this rupture force is the projection onto the vertical direction of the bed. Equation (3.74) calculates the fraction of the force, γ_a, obtained from the vertical projection of the long diagonal of the cube:

$$\gamma_a = \sqrt{2/3} \tag{3.74}$$

I also brought a bridging factor called γ_b into the analysis because I found that even a small pocket of extrudates or even a single extrudate can hold or will be forced to hold a fraction of the bed that is substantially larger than its own size.

This bridging factor could be the subject of very interesting numerical studies in the future. Because it is hard to establish, I set it equal to unity for simplicity.

Equation (3.75) now calculates the stress, P, that a bed of extrudates of a given aspect ratio can endure without rupture:

$$P = F_r \left[\frac{1}{\zeta L}\right]^2 \left[\frac{\gamma_a}{\gamma_b}\right] \tag{3.75}$$

Equation (3.76) rearranges Eq. (3.75) in a more manageable form to express the average aspect ratio, Φ, as a function of the stress and the bed properties:

$$\Phi = \Psi \, Be_r^{1/3} \tag{3.76}$$

where Ψ is a coefficient determined by the stacking of extrudates in the bed.

In a first-order approximation, Eq. (3.77) gives this coefficient as:

$$\Psi = \left[\frac{\gamma_a}{\zeta^2 \gamma_b}\right]^{1/3} \tag{3.77}$$

The value of Ψ calculated based on the parameter values in Eqs. (3.72) and (3.74) works out to be $6^{1/6}$, or about 1.35.

The stacking coefficient will probably not be constant during the breaking process; it seems intuitive that extrudates may prefer a more horizontal position as breakage occurs. However, at the same time, the broken extrudate pieces may also stack more vertically as they settle into crevices; this is difficult to predict. Only experimental data can shed light on this, and later I will show that a constant, Ψ, is probably a reasonable assumption.

The group Be_r is dimensionless and is given by Eq. (3.78):

$$Be_r = \sigma / sP \tag{3.78}$$

The subscript r is designed to draw attention to the rupture phenomenon. The physical meaning of the group is clear: it is the ratio of the rupture strength to the stress applied to the bed. This rupture strength, σ, is also the same as the tensile strength of the extrudate, as I have mentioned previously. The important point here is to assume that the breakage is properly modeled as the force needed to cause rupture by bending.

During bending, the stress both in the direction along the axis of the extrudate and on a surface element perpendicular to the axis will be highest in the underbelly of the extrudate. When the stress reaches the tensile strength, the extrudate breaks.

Let us recap the meaning and interpretation of the expression for Φ. When a fixed bed of extrudates is exposed to vertical stress, the extrudates will break to a smaller Φ value than the one loaded when the stress applied is too high, as per Eqs. (3.75) or (3.76). If the stress is too low, no breakage will occur. Thus, there is a critical pressure that a bed of a given Φ_0 can withstand without breakage, given by Eq. (3.79):

$$P_c = \sigma\,\Psi^3/\left[s\,\Phi_0{}^3\right] \tag{3.79}$$

The critical pressure that a bed of extrudates can withstand is thus proportional to the tensile strength of the extrudates and inversely proportional to the third power of the average aspect ratio of the bed.

Beds do not easily reach these critical pressures because of the wall effect: the weight of the bed and its stress is distributed toward the walls and is offset by wall friction. In the bulk crush strength test, the wall effect is not an issue because the bed is very shallow. Only in the case of a very shallow bed will the breakage be uniform throughout.

3.6.2.2.1 Catalyst Breakage Due to the Weight of a Shallow Catalyst Bed

Breakage in shallow beds due to stress by weight is typically negligible because the dry weight is just too low. Often, breakage during loading (collision) is more important. When considering pressure drop and process weight (due to wetting) as well, it is prudent to at least consider breakage. Equation (3.80) easily calculates the stress at the bottom from first principles:

$$P = H_B\,(1-\epsilon)\rho g = H_B\rho_B g \tag{3.80}$$

where ε is the bed void fraction, ρ is the particle density, ρ_B is the bed bulk density, g is the gravitational constant, and H_B is the bed depth.

Equation (3.80) can now be easily combined with the strength considerations established earlier, assuming that the catalyst at the bottom bears the entire weight of the bed without interacting with the wall. Equation (3.81) expresses the critical height of the bed that can withstand the weight without breakage:

$$H_c = \sigma\,\Psi^3/\left[s\rho_B g\,\Phi_0^3\right] \tag{3.81}$$

The critical height is proportional to the tensile strength of the extruded catalyst, inversely proportional to the bulk density of the bed, and inversely proportional to the third power of the aspect ratio of the charged catalyst.

The dependence of the critical height on bulk density and tensile strength is intuitive, but the third-power dependence on the aspect ratio is not; it is a result of this analysis.

3.6.2.2.2 Catalyst Breakage Due to the Weight of a Deep Catalyst Bed

When the bed height becomes appreciable compared to the diameter, the picture changes, as originally discovered by Janssen [34]. The picture is different in that the force caused by the weight of the catalyst radiates outward toward the wall and increases the friction forces against the wall. This friction can become so strong that it totally negates any extra weight loaded in the fixed bed. The extra weight simply no longer registers as extra stress at the bottom of the bed.

Figure 3.42 shows a stacking of just a few spheres in a narrow tube; one can see the frictional forces at the wall and between the particles. One can imagine a similar situation now with extrudates. To illustrate further, use a simple lab setup where a 1-m-long plastic tube 10 cm in diameter is mounted vertically; it sits about 2 cm above a weigh scale without touching. Initially, when pouring extruded catalyst into the tube from above, the weigh scale registers every gram of catalyst loaded into the tube. As the height of the catalyst bed rises, however, it becomes clear that the scale registers less and less of the loaded catalyst, and at some point registers nothing more of any further catalyst loaded at all. This is a labile equilibrium, however. A small tap on the column will break the frictional forces and cause the weigh scale to suddenly register the full weight of the column until it again enters a new labile equilibrium. This short time frame is measured in fractions of a second.

I mentioned at the beginning of this book that the labile equilibrium of weight and wall interaction can cause substantial breakage at the bottom of the bed and should be considered during unsteady-state operations such as startup and

Figure 3.42 Stacking of spheres in a column.

shutdown, when a fixed bed can experience a jolt when introducing feeds or changing feeds or feed rates.

Janssen's original model [34] accounts for wall friction, as expressed by Eq. (3.82):

$$\pi R^2 P(x + dx) = \pi R^2 P(x) + \pi R^2 g\rho_B dx - 2\pi R\mu_w P dx \tag{3.82}$$

where x is the distance from the top of the bed to a particular location in the bed and is counted positive downward. The friction coefficient, μ_w, quantifies the friction forces at the wall. I refer to Chapter 2 for more on friction and what it entails. In addition, consult Janssen [34] and Schulze [35] for more information about wall friction.

Rearranging the balance of Eq. (3.82) into Eq. (3.83) shows the stress gradient at every location:

$$\frac{dP}{dx} = g\rho_B - 2\mu_w P/R \tag{3.83}$$

Thus, for a small stress value, P, the increase in the load stress is proportional to the depth of the location. As the load stress increases, the second term on the right-hand side of Eq. (3.83) becomes larger and reduces the stress gradient to the point where it eventually becomes negligible.

Equation (3.84) is the solution to the differential equation in Eq. (3.83):

$$P = (g\rho_B R/2\mu_w)\left(1 - e^{-2\mu_w x/R}\right) \tag{3.84}$$

As mentioned, when x is sufficiently small, the stress caused by the load increases linearly with the depth of the location in the bed according to Eq. (3.85):

$$P_{x \ll R/2\mu_w} = g\rho_B x \tag{3.85}$$

In addition, due to wall friction, and when x is sufficiently large, the stress reaches a constant value given by Eq. (3.86):

$$P_{x \gg R/2\mu_w} = g\rho_B R/2\mu_w \tag{3.86}$$

With Eq. (3.86), one can quickly obtain experimental values for the wall friction coefficient for a column made of the same material as the wall in question based on the lined-out stress values.

3.6.2.3 Formation of Fines during Catalyst Breakage

As in the ASTM D7084-04 bulk crush test, a catalyst bed subjected to stress will generate chips and fines; the amount of fines, expressed as a percentage of the total catalyst weight, is also used as a measure of catalyst strength. The industry uses this measure of strength for scale-up reasons or for commercial applications.

Consider that every break of an extrudate generates an amount of fines. When pursuing a first-order type of approximation analysis, consider that all breaks are equivalent – independent of any distribution of extrudate length or the direction of the extrudate. Let a sample of catalyst have a starting aspect ratio equal to Φ_0.

After loading and performing a stress test, the sample will likely have a smaller aspect ratio; let this be equal to Φ_1. It is a simple calculation now to determine b_f, the number of breaks per unit weight required to change the aspect ratio from Φ_0 to Φ_1, according to Eq. (3.87):

$$b_f = \left(\frac{1}{\rho\Omega D}\right)(1/\Phi_1 - 1/\Phi_0) \tag{3.87}$$

where ρ is the particle density and Ω is the cross-sectional area.

All that is left to do now is to calculate the amount of fines generated for each break. One could obtain this number from a single experiment established in the ASTM D7084-04 test. In Beeckman et al. [25], I modeled the break phenomenon through an average break angle θ equal to 1 rad (or about 57°). Assuming that every break yields a loss of extrudates as fines equivalent to Φ_f and approximated as per Eq. (3.88):

$$\Phi_f = tan\theta \cong 0.65 \tag{3.88}$$

Using Eqs. (3.87) and (3.88) then yields Eq. (3.89) and gives the weight fraction of fines generated:

$$\omega_f = (1/tan\theta)(1/\Phi_1 - 1/\Phi_0) \tag{3.89}$$

Only use this weight fraction as per Eq. (3.89) when the pressure applied is above the critical pressure. Below or at the critical pressure, the weight fraction of fines is zero.

In Beeckman et al. [25], I calculated the slope of the fines generation curve and showed that this slope is proportional to the tensile strength and also to the second power of the loading aspect ratio. In addition, given the relationship between tensile strength, aspect ratio, and stress applied, the response of fines generation and the drop in aspect ratio with stress are now easily accessible.

3.6.2.4 A Consideration of Break Energy

Energy (work) is by definition the product of a force exerted on a body multiplied by the distance of travel of the force. During the bulk crush test, the applied force on the top surface exerted over the short distance traveled as it compresses the bed by breakage exerts work on the bed of catalyst. Below the critical pressure, there is no indentation of the bed, hence no work is delivered by the gradually increasing pressure.

Once the critical pressure is reached (and also beyond the critical pressure), the bed starts to indent due to catalyst breakage and subsequent rearrangement of the bed. For each pressure point reached, the bed breaks to an equilibrium value and reaches a specific indentation for that pressure. The force on top of the bed multiplied by the indentation is a work term that, when summed from the beginning, yields the total work done on the bed caused by breakage.

Each time the bed comes to equilibrium after breakage for a certain pressure, the aspect ratio drops and the critical pressure increases. Hence, pressure has to be increased in order to create further breakage.

The energy for breakage also includes the energy that goes into the accompanying formation of grit, fines, and dust. While frictional work is also likely during rearrangement, I will ignore its contribution here. This frictional work occurs in any location in the bed as well as at any location along the wall of the shallow bed. Frictional work in the bed is likely proportional to volume, while frictional work along the wall is likely proportional to the surface area of the vertical wall. It is possible that both frictional contributions could be separated from each other and separated from the work put into breakage by conducting experiments with different bed volumes and wall heights.

In this book, assume that all work done in testing is attributed to breakage. In a way, this depiction is similar to what occurs during compression of a gas, except that it is not reversible. Summing the work from the critical pressure (with indentation nil) to a certain pressure in small steps enables one to calculate the total work performed for catalyst sample breakage using Eq. (3.90):

$$\sum(F \times \Delta x_i) = \sum(P \times area \times \Delta x_i) = E_b \times (W/\rho\Omega D)(\,1/\Phi_1 - 1/\Phi_0\,) \tag{3.90}$$

where W is the weight of the catalyst sample in the test, E_b is the energy of a single break, and Δx_i is the finite-difference indentation going from one pressure point to the next.

As a last comment on work by breakage, it may also be interesting to determine whether it is possible to detect any heat release by breakage by carefully placing a thermocouple in the bed that is capable of registering only a few hundredths of a degree centigrade. In any case, more careful experimentation is probably a good next step.

3.6.2.5 Simulation of Fixed-bed Breakage via the Bulk Crush Strength Test

Table 3.19 lists catalysts A through E used in this study. I chose a broad spectrum of catalyst properties in order to test this theoretical approach. I selected three shapes – cylinders, quadrulobes, and trilobes – that represent a majority of the catalysts used in the industry.

Table 3.19 Catalysts and their properties employed in this study.

Catalyst	Shape	D (m)	Φ_0	s	ρ_P (kg/m^3)	σ (MPa)	P_c (kPa)
A	Quadrulobe	1.43E-03	3.18	2.20	1250	0.81	27.9
B	Cylinder	9.50E-04	5.92	2.55	750	1.38	6.4
C	Cylinder	8.30E-04	7.48	2.55	1870	2.83	6.5
D	Trilobe	2.89E-03	2.28	2.28	970	0.76	69.3
E	Cylinder	1.55E-03	3.54	2.55	NA	1.37	39.7

Source: Courtesy Wiley VCH, doi: 10.1002/ceat.201600550.

The initial aspect ratio ranged from 2.3 to 7.5, while catalyst particle densities ranged from 750 to 1870 kg/m^3. Catalyst extrudate diameters ranged over typical commercial values, while the tensile strength ranged from 0.76 to 2.8 MPa. My coauthors in Beeckman et al. [25, 26] performed all of the experimental work, loaded all of the catalysts, and subjected them to crushing pressures to various intermediate levels. They raised the pressure gradually and allowed the bed of catalyst to equilibrate at each pressure level. They also measured the level of indentation along with the fines and the aspect ratio.

Table 3.20 lists the data obtained during the testing. Using the properties in Table 3.19, I calculated the critical pressures; Figures 3.43–3.47 display the data. It is difficult to determine the critical pressure experimentally, although with enough trials, one can obtain a good average value. However, with limited data, one has to deal with inherent experimental scatter on the aspect ratio.

The solid lines in Figures 3.43–3.47 are predictions based on the model expressed by Eq. (3.76). I took all of the parameters in the model equation from Table 3.19 without further adjustments. In addition, I applied the theoretical value of 1.35 for the coefficient ψ. As can be seen, the agreement between the model and the experimental data is quite satisfactory. Upon reaching the critical pressure, the data shows a dramatic drop in aspect ratio, which the model captures well.

The critical pressure for catalyst A, for example, is 28 kPa. One could reach such a weight load if the bed were 8 m high, which is very substantial, but not uncommon. The calculated 8-m height does not take into account the wall friction for such a tall bed, so the stress will certainly be less. In the bulk crush test, the height is very limited and represents an almost differential slice of the bed. Thus, wall friction during the crush test plays no role.

The wide variety of critical pressures calculated in Table 3.19 shows that the starting aspect ratio plays a large role in the varying values of the critical pressure for each catalyst. After the pressure reaches higher values, the drop-off in pressure decreases with pressure, but is not negligible. Using the dimensionless group Be_r, it

Table 3.20 Catalyst bulk crush strength measurements.

Catalyst	Pressure (kPa)	Indent (m)	Weight (kg)	Fines (kg)	Φ (−)
A	6.9	0.00E+00	5.31E-02	0.00E+00	2.93
A	13.8	0.00E+00	5.13E-02	1.00E-04	3.50
A	20.7	0.00E+00	4.96E-02	3.00E-04	3.09
A	27.6	0.00E+00	5.04E-02	5.00E-04	3.20
A	41.4	1.59E-03	5.00E-02	8.90E-04	2.95
A	55.2	1.59E-03	5.15E-02	9.00E-04	2.89
A	82.7	2.38E-03	4.95E-02	4.06E-03	2.31
A	110.3	3.18E-03	5.13E-02	7.80E-03	2.11
A	165.5	4.76E-03	4.97E-02	1.24E-02	1.99
A	220.6	6.35E-03	4.88E-02	1.49E-02	1.64
B	13.8	0.00E+00	3.61E-02	1.00E-04	4.32
B	34.5	1.59E-03	3.43E-02	1.70E-04	3.70
B	69.0	2.38E-03	3.46E-02	7.00E-04	3.04
C	48.3	7.94E-04	7.41E-02	1.03E-02	2.95
C	69.0	1.59E-03	7.62E-02	8.00E-03	2.74
C	206.9	3.18E-03	7.67E-02	1.08E-02	2.53
D	344.8	8.06E-03	4.66E-02	8.30E-03	1.39
E	34.5	7.94E-04	4.48E-02	0.00E+00	3.12
E	69.0	1.59E-03	4.63E-02	2.16E-04	2.41
E	137.9	3.97E-03	4.31E-02	2.09E-03	2.27
E	275.8	4.76E-03	4.06E-02	6.31E-03	1.59
E	551.6	6.35E-03	3.43E-02	1.28E-02	1.77

Source: Courtesy Wiley VCH, doi: 10.1002/ceat.201600550.

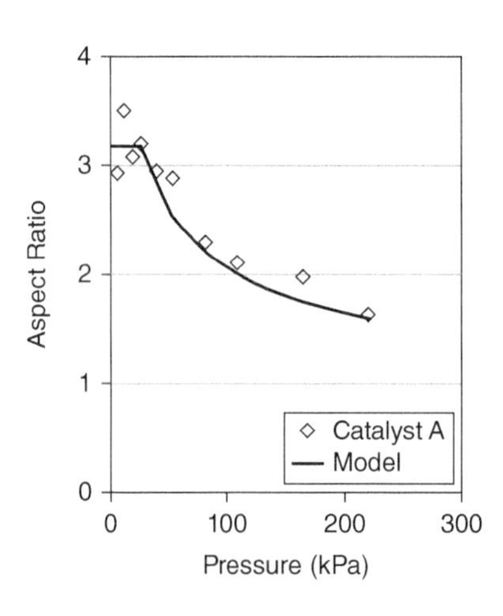

Figure 3.43 Predicted aspect ratio for catalyst A. Source: Courtesy of Wiley VCH, doi: 10.1002/ceat.201600550.

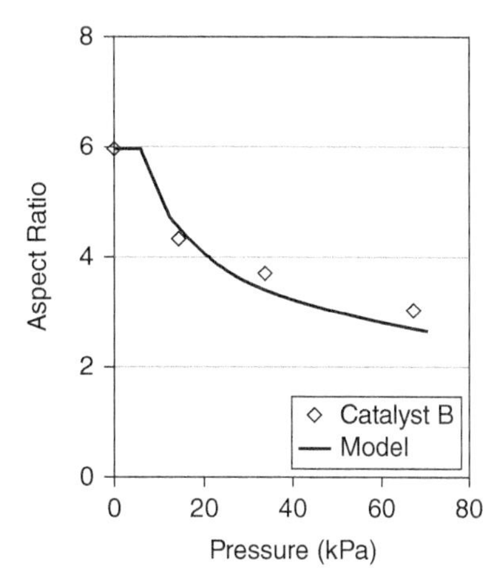

Figure 3.44 Predicted aspect ratio for catalyst B. Source: Courtesy of Wiley VCH, doi: 10.1002/ceat.201600550.

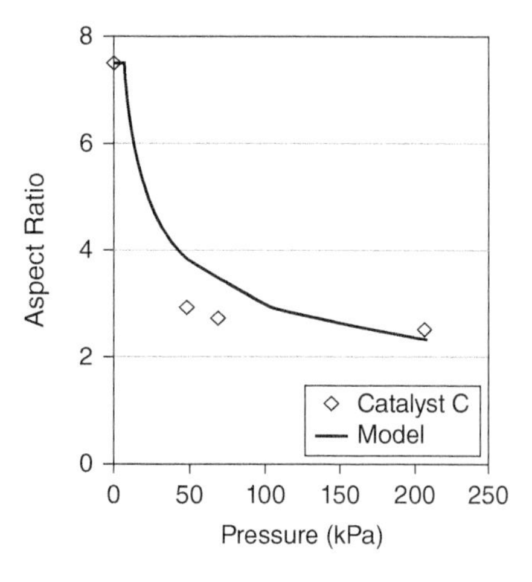

Figure 3.45 Predicted aspect ratio for catalyst C. Source: Courtesy of Wiley VCH, doi: 10.1002/ceat.201600550.

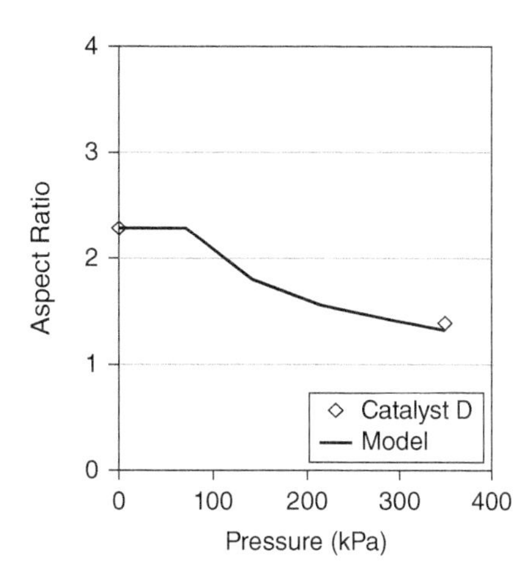

Figure 3.46 Predicted aspect ratio for catalyst D. Source: Courtesy of Wiley VCH, doi: 10.1002/ceat.201600550.

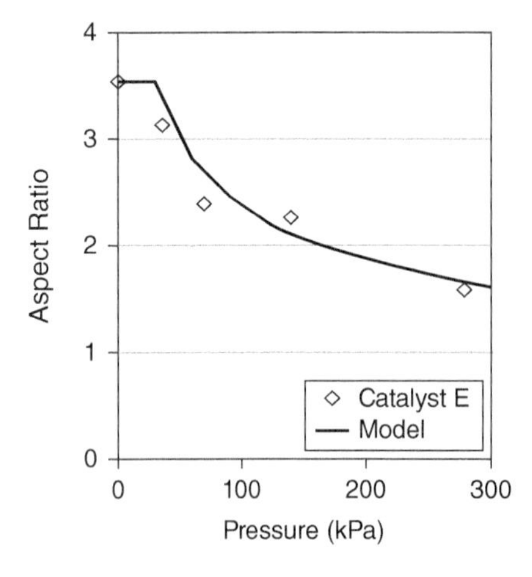

Figure 3.47 Predicted aspect ratio for catalyst E. Source: Courtesy of Wiley VCH, doi: 10.1002/ceat.201600550.

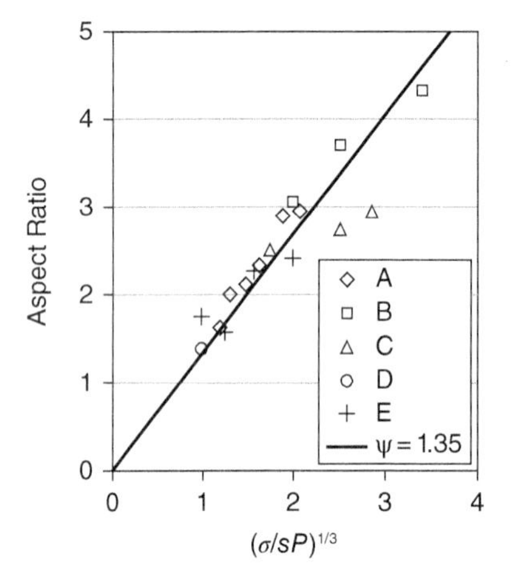

Figure 3.48 Dependence of the aspect ratio on $Be_r^{1/3}$. Source: Courtesy of Wiley VCH, doi: 10.1002/ceat.201600550.

is possible to combine the responses of the aspect ratios for all five catalysts in a single line that extends through the origin, as is shown in Figure 3.48.

I performed a statistical regression analysis on a linear model through the origin and obtained a slope of $1.25 \pm 10\%$ at a 95% confidence level. This slope is close to the 1.35 theoretical value, and from Figure 3.48 it is clear that catalyst C in particular shows two low data points that were likely causing this. However, there is no strong argument to be made on why to omit these data, since experimental error is part of the game.

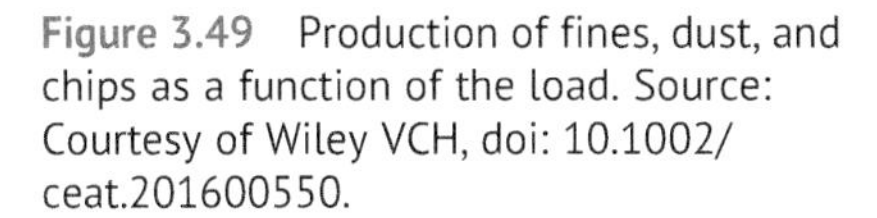

Figure 3.49 Production of fines, dust, and chips as a function of the load. Source: Courtesy of Wiley VCH, doi: 10.1002/ceat.201600550.

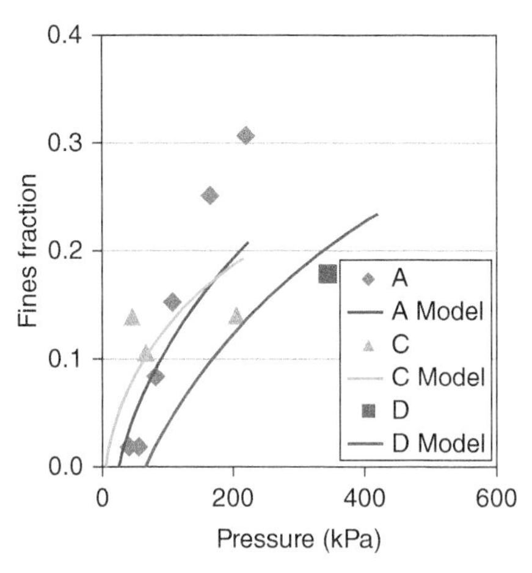

I also measured the fines for catalysts A, C, and D. The analysis that I presented in Beeckman et al. [25] and further illustrate in Figure 3.49 shows that it is possible to capture the breakage of extruded catalysts under a compressive load by using a dimensionless group Be_r, which is at the center of the description of this phenomenon.

A comparison of the bending strength of a fixed bed or the resilience of a bed against breakage with a compressive load on the bed leads to a dimensionless group that one can use to predict the reduction in aspect ratio. I determined that this reduction has a negative one-third power dependence with respect to the applied load. The agreement between the production of fines during breakage and the load further supports this methodology.

A wide range of catalyst properties illustrate all of the parameters that play a role in the process of catalyst breakage due to stress in a bed; only an engineering analysis is able to capture the influence of catalyst properties. The model equations describe the production of fines that accompany breakage with such specificity that one can observe a correlation with the well-established ASTM D7084-04 test already used to judge catalyst strength in the catalyst industry.

3.7 Breakage in Contiguous Equipment

3.7.1 Breakage from an Extrusion Line

Catalysts are manufactured according to a plant's architecture and its specific equipment setup. Although a plant's architecture is similar in broad lines from plant to plant, its equipment often differs substantially. Integration of equipment is heavily dictated by available space, footprint considerations, equipment accessibility, and personnel ingress and egress safety considerations.

For some catalyst properties such as surface area and pore volume, a plant's architecture is not a determining factor. However, the architecture of a plant does play an important role in the catalyst aspect ratio because breakage of catalyst in equipment and breakage during catalyst transfer from one piece of equipment to another depends on the severity of how the catalyst is handled.

For large catalyst manufacturing campaigns, catalyst samples are typically obtained with automatic samplers and collected in containers. It is important to obtain a representative sample of the catalyst container and to know the size of the batch or the lot it represents.

I will now describe the modeling effort to predict the aspect ratio of extruded catalysts manufactured in a contiguous and unchanging line of equipment. Assume that impulsive impact forces such as those encountered during a series of contiguous individual drops cause the catalyst breakage.

In cases where a catalyst originates directly from an extrusion line and the starting extrudate strands are fairly long, one can model the aspect ratio after passage through a single piece of equipment using the basic Eq. (3.65), which contains a dimensionless normalized strength group and a dimensionless normalized severity factor specific to that equipment. I thought it should be possible to characterize a line of contiguous equipment using an overall severity factor specific for the equipment line. I will show that one can obtain an overall severity factor from individual equipment severities through a set of Riccati-type nonlinear finite-difference equations.

The catalysts I examine in this book were obtained through the process of extrusion, and these catalysts typically have an aspect ratio well above two throughout the manufacturing line. For extruded catalysts, the diameter and the cross section of the catalyst are essentially constant, except for typically small changes attributed to the wear of the die plate.

The differences in aspect ratio among individual catalyst particles are essentially caused by differences in the actual length of the extrudates. For a large sample, the aspect ratio typically shows a Gaussian-type distribution. I did not consider granules or shaped catalysts made with a pelletizing press in this book. Pelletized catalysts are typically short and have an aspect ratio from 0.5 to 1.5.

Natural breakage occurs during catalyst handling at a manufacturing site during transport and handling at the destination site where the catalyst will be used. Forced breakage – also known as comminution of a catalyst – occurs during the sizing and sieving of catalysts at a manufacturing plant. Catalyst breakage is thus a balance between catalyst strength and the severity of catalyst handling. At a manufacturing plant, a structurally weak catalyst will break to a small aspect ratio, while a structurally strong catalyst will lead to a large aspect ratio.

My objective is to quantitatively relate the aspect ratio of a catalyst to its mechanical strength and the severity of handling on the manufacturing line. The evaluation of breakage of catalyst extrudates along a line of contiguous equipment from a commercial perspective is typically handled ad hoc. This book offers a methodology that combines information about the operational severity of individual pieces of equipment and catalyst strength in order to predict the aspect ratio of the catalyst as it proceeds through a contiguous line of equipment. Assume that all breakage is caused by an impact or collision of catalyst extrudates with a surface or a bed of catalyst.

3.7.1.1 Severity Functional for a Single Piece of Equipment

An avenue to describe the breakage of extrudates from impact in a given single operation is given by Eqs. (3.91) and (3.92) for the case of ideal materials:

$$\Phi_{j+1} = 2\Phi_{\infty} - \Phi_{\infty}^2/\Phi_j \quad \text{for } j = 0, 1, 2, \ldots \text{ and } \Phi_j \geq \Phi_{\infty} \tag{3.91}$$

$$\Phi_{j+1} = \Phi_j \quad \text{for } j = 0, 1, 2, \ldots \text{ and } \Phi_0 < \Phi_{\infty} \tag{3.92}$$

In Eqs. (3.91) and (3.92), Φ_0 is the starting aspect ratio, while Φ_j is the aspect ratio of the extrudate sample after it experienced j consecutive impacts. Φ_{∞} is the asymptotic aspect ratio obtained after a large number of identical impacts, and it is a function of both the strength of the material and the severity of the impact. Beyond this first asymptote, the aspect ratio also reaches a second asymptote after only a single impact for extrudates that are sufficiently long. For ideal materials, Eq. (3.93) shows that this second asymptote is equal to twice the asymptotic aspect ratio Φ_{∞}:

$$\lim_{\Phi_0 \to \infty} \Phi_1 = 2\Phi_{\infty} \tag{3.93}$$

In the model for breakage of ideal materials by impact, there is only one parameter to determine: the asymptotic aspect ratio Φ_{∞}. Experimentally, one can determine Φ_{∞} by repeatedly running the material through the particular unit operation until the asymptotic value is reached. Practically, one can run the extrudates a few times through the operation and then apply a model fit using Eq. (3.91) to get a best-fit estimate for the asymptote.

One can obtain a rough estimate for the asymptotic aspect ratio by passing the catalyst a single time through the equipment. Based on Φ_0 and Φ_1, which are the measured aspect ratios of the feed and the product, respectively, Eq. (3.94) is obtained from Eq. (3.91) applied for just one drop:

$$\Phi_1 = 2\Phi_{\infty} - \Phi_{\infty}^2/\Phi_0 \tag{3.94}$$

Solving Eq. (3.94) for the asymptotic aspect ratio yields Eq. (3.95):

$$\Phi_{\infty} = \Phi_0 \left(1 - \sqrt{1 - \Phi_1/\Phi_0}\right) \tag{3.95}$$

To relate the asymptotic aspect ratio to the impact, in Beeckman et al. [22], I applied a balance between the force due to collision and the force for rupture of the extrudate in bending mode. The force that an extrudate experiences due to collision is obtained from Newton's second law as the change in momentum divided by the contact time of the collision. The rupture force of the extrudate in bending mode is obtained from the modulus of rupture. Applying this force balance to the asymptotic aspect ratio allows one to write the severity of the impact and the strength of the extrudate as two separate quantities, as shown in Eqs. (3.96) and (3.97):

$$\Phi_{\infty} = (1/S_{\infty})\sqrt{Be_g} \tag{3.96}$$

$$Be_g = \sigma/\psi\rho Dg \tag{3.97}$$

where σ is the modulus of rupture of the extrudate, ψ is the extrudate shape factor, ρ is the extrudate density, D is the extrudate diameter, and g is the gravitational acceleration. Be_g is a dimensionless group normalized to the gravitational acceleration, thus the subindex g. S_{∞} is called the asymptotic severity of the impact and it is dimensionless. It too is normalized to the gravitational acceleration g and can be expressed as per Eq. (3.98):

$$S_{\infty} = \sqrt{8v/\beta g\Delta\tau}\, S_{\infty} = \sqrt{8v/\beta g\Delta\tau} \tag{3.98}$$

where v is the impact velocity and $\Delta\tau$ is the contact time of the collision.

The ratio of v to $\Delta\tau$ is a measure of the deceleration that an extrudate experiences upon collision with a surface. The severity is measured by the square root of the deceleration of the extrudate. The group Be_g contains information about the strength of the material, while the asymptotic severity S_{∞} contains information about the intensity of the collision. After determining the asymptotic aspect ratio Φ_{∞} from Eq. (3.95) and the catalyst properties required to determine Be_g, one can then easily obtain the asymptotic severity S_{∞}.

In Beeckman et al. [22], I showed experimentally that for a collection of 25 catalysts, the group $\sqrt{\beta\Delta\tau}$ is in the first approximation a constant that is independent of the nature of the extrudate. This collection of 25 catalysts consisted of silica-bound catalysts, alumina-bound catalysts, dried-only catalysts, catalysts calcined at about 800–1100 °F, zeolite-based catalysts, and carbon-based catalysts. The asymptotic severity is a useful tool for comparing the breakage of catalyst in a single piece of equipment or operation from impact, and it can be obtained from any catalyst or starting aspect ratio larger than Φ_{∞}.

3.7.1.2 Severity Derivation for Contiguous Equipment

Here, I show a methodology to combine equipment for impact breakage when the starting aspect ratio, Φ_0, is large. In an extrusion line, the catalyst strands are typically very long as they exit the extruder die. After extrusion, the catalyst strands are dried as they proceed through a moving belt drier. The starting point in the analysis is thus the catalyst strands as they lie on the moving belt at the exit of the drier. As the catalyst travels through the equipment line, it experiences a series of contiguous individual impacts as it drops from different heights; each drop has a particular severity determined by a plant's architecture and its equipment. The catalyst will experience an impact as it drops from the drier belt onto a conveyor, or from one conveyor onto another, or from a conveyor into a bucket elevator, for example. I suspect the methodology also applies to the sieving and sizing operation, because beyond the actual drop, the vibrational frequency and amplitude of the equipment are set and are independent of the grade of catalyst being manufactured (except for the mesh size of the screen, and here I neglect this difference).

During each drop or impact, the catalyst typically experiences a reduction in aspect ratio that can be large or small depending on the aspect ratio before the drop and the severity of the impact. It is also possible that the aspect ratio remains unaltered during impact, such as in cases when the severity of a particular step is too low to cause further breakage. For this analysis, assume that apart from the aspect ratio, the catalyst is unchanged as it travels through the line of equipment (i.e. that the group Be_g remains constant throughout the equipment line).

The drop from the drier bed trays onto a conveyor is designated with the letter A; it is the first breakage step by collision. When the starting aspect ratio Φ_0 is large, the aspect ratio Φ_A of the extrudates after the drop onto the conveyor is expressed by Eq. (3.99):

$$\Phi_A = 2\Phi_{\infty,A} - \Phi^2_{\infty,A}/\Phi_0 \cong 2\Phi_{\infty,A} \tag{3.99}$$

where $\Phi_{\infty,A}$ is the asymptotic aspect ratio of step A.

Suppose this operation were followed by a second impact designated with the letter B. Equations (3.100) and (3.101) then express the aspect ratio Φ_{AB} of the extrudates after steps A and B, respectively, as:

$$\Phi_{AB} = 2\Phi_{\infty,B} - \Phi^2_{\infty,B}/\Phi_A \text{ for } \Phi_A \geq \Phi_{\infty,B} \tag{3.100}$$

$$\Phi_{AB} = \Phi_A \text{ for } \Phi_A < \Phi_{\infty,B} \tag{3.101}$$

where $\Phi_{\infty,B}$ stands for the asymptotic aspect ratio of step B.

Equation (3.101) covers the case where the severity of step B is too low to cause a reduction in aspect ratio. Suppose step B were followed by yet another impact designated with the letter C. Equations (3.102) and (3.103) then express the aspect ratio Φ_{ABC} of the extrudates after operations A, B, and C:

$$\Phi_{ABC} = 2\Phi_{\infty,C} - \Phi^2_{\infty,C}/\Phi_{AB} \text{ for } \Phi_{AB} \geq \Phi_{\infty,C} \quad (3.102)$$

$$\Phi_{ABC} = \Phi_{AB} \text{ for } \Phi_{AB} < \Phi_{\infty,C} \quad (3.103)$$

where $\Phi_{\infty,C}$ is the asymptotic aspect ratio of step C.

Equation (3.103) covers the case where the severity of step C is too low to cause a further reduction in aspect ratio. Continuing the analysis of Eqs. (3.99)–(3.103) for each collision step enables one to account up to the final impact in the plant. Although the equations are simple, they are nonlinear and must be solved consecutively. Notice that the asymptotic aspect ratios $\Phi_{\infty,A}$, $\Phi_{\infty,B}$ and $\Phi_{\infty,C}$ are all proportional to the group $\sqrt{\sigma/\psi\rho Dg}$; therefore, all quantities Φ_A, Φ_{AB} and Φ_{ABC} will be proportional to $\sqrt{\sigma/\psi\rho Dg}$ as well. Equation (3.104) now calculates the final aspect ratio:

$$\Phi_{ABC\ldots} = (1/S_{ABC\ldots})\sqrt{\sigma/\psi\rho Dg} \quad (3.104)$$

where $\Phi_{ABC\ldots}$ is the aspect ratio of the final catalyst product and $S_{ABC\ldots}$ is the severity of that specific equipment lineup.

Remember that $S_{\infty,A}$ stands for the asymptotic severity of operation A because it is defined via the asymptotic aspect ratio $\Phi_{\infty,A}$, which is reached after many impacts. $S_{ABC\ldots}$ stands for the severity of a sequence of steps, $A \rightarrow B \rightarrow C \rightarrow \ldots$, each step executed only once and in that specific order.

For the case of just three impacts, $A \rightarrow B \rightarrow C$, it is possible to calculate severities S_A, S_{AB}, and S_{ABC} for catalyst extrudates to go through steps A, $A \rightarrow B$ and $A \rightarrow B \rightarrow C$ a single time, in that order, from the individual asymptotic severities $S_{\infty,A}$, $S_{\infty,B}$, and $S_{\infty,C}$ with Eqs. (3.105)–(3.109):

$$S_A = S_{\infty,A}/2 \quad (3.105)$$

$$S_{AB} = \left(2/S_{\infty,B} - S_A/S^2_{\infty,B}\right)^{-1} \quad \text{for } S_A \leq S_{\infty,B} \quad (3.106)$$

$$S_{AB} = S_A \quad \text{for } S_A > S_{\infty,B} \quad (3.107)$$

$$S_{ABC} = \left(2/S_{\infty,C} - S_{AB}/S^2_{\infty,C}\right)^{-1} \quad \text{for } S_{AB} \leq S_{\infty,C} \quad (3.108)$$

$$S_{ABC} = S_{AB} \quad \text{for } S_{AB} > S_{\infty,C} \quad (3.109)$$

The sequence of Eqs. (3.105)–(3.109) forms a familiar finite-difference Riccati equation. If more steps are in play, then one can follow the procedure I have outlined in an analogous way in order to obtain the overall severity. Regarding the topic of severity sequencing, I explained in Beeckman et al. [8] that the individual severities need to be combined in the proper order in which they appear in the lineup of the equipment because, as per Eq. (3.110):

$$S_{AB} \neq S_{BA} \tag{3.110}$$

If steps A, B, C, ..., were all identical and there were many of them, then the severity $S_{ABC...}$ would become equal to $S_{AAA....}$. Hence, for many identical consecutive impulsive breakage steps, it would become equal to the asymptotic severity $S_{\infty,\,A}$ by definition, as can be easily verified.

According to Eq. (3.104), a graph with the final aspect ratio $\Phi_{ABC...}$ of a given equipment lineup on the y-axis as a function of the group $\sqrt{\sigma/\psi\rho Dg}$ on the x-axis should now yield a straight line through the origin. The slope of this line is the inverse of the combined severity of the equipment lineup. With the help of this engineering analysis, it is not necessary to know all of the detailed steps in an existing plant, only that all of the steps lead to breakage by collision (also known as impulsive breakage of the catalyst). A graph depicting Φ as a function of $\sqrt{Be_g}$ (from one or perhaps a few grades of catalyst) could obtain the overall severity. One could then use this information to compare the severities of individual lineups in manufacturing plants or the severities among plants.

3.7.1.3 Application to a Commercial Plant

Table 3.21 lists the properties of eight extruded catalysts manufactured on the same line of contiguous equipment. In order to test the methodology over a wide range of catalyst properties, I varied the diameter and the density by approximately twofold while varying the modulus of rupture by approximately threefold.

All of the individual drops of catalysts lead to impulsive breakage and can be treated using the above analysis. Each data point represents the average of several

Table 3.21 Catalyst properties.

Catalyst	Shape	D (m)	ρ (kg/m^3)	ψ (–)	σ (MPa)	Be_g (–)	Φ (–)
A	Quadrulobe	1.50E-03	1023	1.81	6.73	2.47E+05	2.5
B	Quadrulobe	1.21E-03	925	1.81	5.46	2.76E+05	3.4
C	Cylinder	1.55E-03	900	2.00	8.83	3.23E+05	3.2
D	Quadrulobe	1.07E-03	573	1.81	7.56	6.96E+05	4.8
E	Quadrulobe	1.97E-03	880	1.81	4.21	1.36E+05	2.2
F	Cylinder	1.56E-03	950	2.00	12.0	4.13E+05	3.2
G	Cylinder	1.57E-03	810	2.00	5.58	2.23E+05	2.5
H	Cylinder	1.51E-03	690	2.00	6.25	3.07E+05	3.5

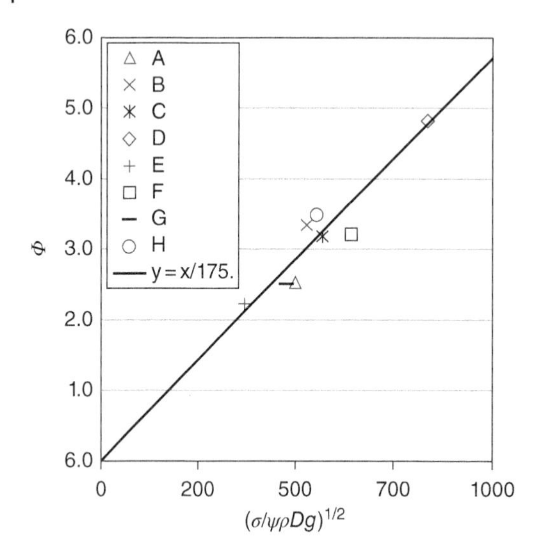

Figure 3.50 Aspect ratio correlation.

lots of catalyst and is of substantial tonnage, as is to be expected in a manufacturing plant. Due to confidentiality, detailed chemical properties and compositions of the catalysts cannot be disclosed, but they are not required for the analysis.

Figure 3.50 shows the aspect ratios of the different catalysts as a function of the group $\sqrt{Be_g}$. A linear line statistical analysis at the 95% confidence level shows that the intercept is not significantly different from zero; hence, a linear correlation with zero intercept is statistically justified. From the slope of the straight line, a value of 175 is obtained for the dimensionless severity of this equipment lineup, as is shown in the legend of Figure 3.50. This dimensionless slope is different for different plants or different equipment lines in a given plant, and it enables the quantification of one plant versus another as far as their severity is concerned. The model also enables the quantitative design and evaluation of a grassroots plant or alteration of the architecture of a given plant.

A lesson to learn from the group Be_g is that the aspect ratio of a catalyst increases with an increasing modulus of rupture σ, a lower density ρ, a smaller diameter D, and a smaller shape factor ψ of the extrudates. These trends are often difficult to capture from raw plant data, since multiple extrudate properties change from grade to grade. This engineering analysis helps describe impulsive breakage in both individual and contiguous equipment, as well as helping with the architectural design of commercial plants.

The modulus of rupture is able to tie the resilience to breakage of extruded catalysts to the severity of equipment handling in a commercial plant. The analysis shows that a dimensionless strength group is at the core of the breakage behavior of catalysts.

3.7.2 Breakage with a Variable-input Aspect Ratio

3.7.2.1 Theoretical

The mechanical strength of catalysts is an important property in manufacturing plants because it controls many facets of catalyst production. The aspect ratio of catalysts is a critical variable. The aspect ratio is governed by two groups: the dimensionless normalized group Be_g and the dimensionless severity factor S_g.

A catalyst support is prepared and stored in inventory at a plant or purchased from third parties; hence, the starting aspect ratio is a variable. One cannot assume that the aspect ratio is large, but one can assume that it is at least satisfying a minimum value set by specifications agreed upon per plant specifications or set in a purchasing contract. A catalyst support is typically exposed to treatments such as exchange, steaming, metal impregnation, drying, and calcination, which vary from grade to grade. Equipment may also entail impregnation vessels, rotary calciners, fixed-bed calciners, and sieving equipment. During these treatments, the bending strength of the catalyst may increase, but it could also weaken. Catalyst that is exposed to moisture will tend to weaken, while removal of moisture by calcination will tend to strengthen it. A three-point bending test using small quantities of catalyst prepared in a laboratory can prove these hypotheses on an ad hoc basis.

A catalyst can break upon impact from varying drop heights, but also during fixed-bed compression as a function of the variable bed height during various treatments, as well as during impact as a vessel is unloaded. Since the scenario is very complex, I combine the entire process into a single impact step, looking at the severity as a single parameter to be estimated. The variables are the starting aspect ratio, the bending strength of the final catalyst, and the severity of the operation. This may be an oversimplification of the process, and it may work satisfactorily for some plants but not for others. If a plant falls into the latter case, it may be interesting to organize the process into one or more steps in series as in Section 3.7.1.2. When several operations are contiguous with an arbitrary starting value of Φ_0, then one could consider using the asymptotic severities as parameters to be estimated and base the performance of the equipment line on those parameter values.

For a single impact step as I will assume here, however, Eq. (3.111) describes the breakage of extrudates in a given single operation (again for ideal materials):

$$\Phi_{j+1} = 2\Phi_\infty - \Phi_\infty^2/\Phi_j \text{ for } j = 0, 1, 2 \tag{3.111}$$

The asymptotic aspect ratio Φ_∞ of the single impact step is related to the strength of the material and the severity of the impact through Eq. (3.112):

$$\Phi_\infty = S_\infty^{-1} \times \sqrt{\sigma/\psi\rho Dg} = S_\infty^{-1} \times \sqrt{Be_g} \tag{3.112}$$

Overall, for a single-collision breakage step, one can now write Eqs. (3.113) and (3.114):

$$\Phi_C = 2\Phi_{\infty,C} - \Phi^2_{\infty,C}/\Phi_S \tag{3.113}$$

$$\Phi_{\infty,C} = S^{-1}_{\infty,C} \times \sqrt{Be_{g,C}} \tag{3.114}$$

where Φ_C is the final aspect ratio of the catalyst, Φ_S is the aspect ratio of the support, and $\Phi_{\infty,C}$ is the asymptotic aspect ratio of the catalyst.

The term $S_{\infty,C}$ stands for the normalized severity of that part of the plant that turns the support into the final catalyst. The group $Be_{g,C}$ stands for the dimensionless normalized group Be_g based on final catalyst properties.

Of course, if Eq. (3.115) holds:

$$\Phi_{\infty,C} \geq \Phi_S \tag{3.115}$$

then Eq. (3.116) follows:

$$\Phi_C = \Phi_S \tag{3.116}$$

When the starting aspect ratio, Φ_S, is arbitrary, then only in the case of a single operation A is it straightforward to make a graph that applies to arbitrary catalysts. Dividing both sides of Eq. (3.113) by $\sqrt{Be_g}$ yields Eq. (3.117):

$$\Phi_C/\sqrt{Be_g} = 2/S_{\infty,C} - S^{-2}_{\infty,C}/(\Phi_S/\sqrt{Be_g}) \tag{3.117}$$

Hence, graphing $\Phi_C/\sqrt{Be_g}$ as a function of $(\Phi_S/\sqrt{Be_g})^{-1}$ should yield a straight line with a slope and intercept that are both related to the asymptotic severity of the operation. Estimating this asymptotic severity from Eq. (3.117) obtains the best statistical fit.

The intercept equals $2/S_{\infty,C}$ while the slope equals a negative $S^{-2}_{\infty,C}$; hence, they are not independent. Thus, the typical best-fit two-parameter least squares is actually constrained and becomes a nonlinear single-parameter least-squares fit.

3.7.2.2 Experimental

Table 3.22 shows catalyst grades selected from two commercial plants A and B. The catalyst properties required in the analysis were mostly available from the existing plant database and included the shape, diameter, density, and starting and final aspect ratios. What remains is to determine the modulus of rupture of the samples. The final catalyst samples were obtained from the warehouse and were riffled before measuring the bending strength.

Table 3.22 Plant A and B aspect ratio properties for a variety of catalyst grades.

Grade	Plant	Shape	Ψ (shape)	Size	Diameter (in.)	Density (g/cc)	Support MOR (psi)	Support Φ	Catalyst MOR (psi)	Catalyst Φ	Final catalyst crush (lbf/in.)
A	A	Quadrulobe	1.81	1/16″	0.059	1.02	921	2.6	718	2.3	80
A	A	Quadrulobe	1.81	1/16″	0.059	1.01	1040	2.6	786	2.9	82
A	A	Quadrulobe	1.81	1/16″	0.059	1.02	883	2.3	817	2.3	66
A	A	Quadrulobe	1.81	1/16″	0.059	1.04	1060	2.6	NA	2.5	81
B	A	Quadrulobe	1.81	1/20″	0.048	0.92	666	3.4	775	3.2	57
B	A	Quadrulobe	1.81	1/20″	0.047	0.93	919	3.3	501	2.8	62
C	A	Cylinder	2	1/16″	0.061	0.9	1280	3.2	1750	3.2	161
D	A	Quadrulobe	1.81	1/20″	0.042	0.58	1170	5.5	1750	3.8	122
D	A	Quadrulobe	1.81	1/20″	0.042	0.57	1070	4.9	1240	4.3	142
D	A	Quadrulobe	1.81	1/20″	0.042	0.57	1050	4.1		5.0	108
E	A	Quadrulobe	1.81	1/10″	0.076	0.88	564	2.4	1060	2.1	NA
E	A	Quadrulobe	1.81	1/10″	0.078	0.88	591	2.1	757	2.1	NA
E	A	Quadrulobe	1.81	1/10″	0.079	0.88	509	2.3	664	2.2	NA
E	A	Quadrulobe	1.81	1/10″	0.078	0.88	778	2.1	NA	2.1	NA
F	A	Cylinder	2	1/16″	0.062	0.94	1670	2.8	1550	2.6	166
F	A	Cylinder	2	1/16″	0.061	0.96	1700	3.3	1780	2.8	165
F	A	Cylinder	2	1/16″	0.062	0.94	1600	3	1500	2.5	134
F	A	Cylinder	2	1/16″	0.061	0.96	2000	3.8	1760	2.6	183
G	A	Cylinder	2	1/16″	0.061	0.81	883	3.1	1260	2.2	110
G	A	Cylinder	2	1/16″	0.063	0.81	804	2.5	1410	2.3	73
G	A	Cylinder	2	1/16″	0.0625	0.81	772	2.38	1400	1.9	103

(Continued)

Table 3.22 (Continued)

Grade	Plant	Shape	Ψ (shape)	Size	Diameter (in.)	Density (g/cc)	Support MOR (psi)	Support Φ	Catalyst MOR (psi)	Catalyst Φ	Final catalyst crush (lbf/in.)
G	A	Cylinder	2	1/16″	0.061	0.81	776	2.1	1330	2.3	73
H	A	Cylinder	2	1/16″	0.06	0.69	836	4.3	842	3.1	109
H	A	Cylinder	2	1/16″	0.06	0.69	825	2.7	1380	2.9	149
H	A	Cylinder	2	1/16″	0.058	0.69	1060	3.5	1330	3.0	NA
A	B	Quadrulobe	1.81	1/16″	0.058	1.04	1310	2.6	1160	2.4	97
I	B	Quadrulobe	1.81	1/16″	0.062	0.91	1400	2.6	1570	2.7	153
I	B	Quadrulobe	1.81	1/16″	0.062	0.88	1450	3.0	1590	2.6	175
I	B	Quadrulobe	1.81	1/16″	0.062	0.9	1630	2.7	1690	2.7	147
I	B	Quadrulobe	1.81	1/16″	0.061	0.87	950	3.0	1290	2.8	155
I	B	Quadrulobe	1.81	1/16″	0.059	0.91	1510	2.3	1840	2.5	153
J	B	Cylinder	2	1/16″	0.061	0.93	1620	2.8	1430	2.5	116
J	B	Cylinder	2	1/16″	0.063	0.91	1250	3.3	1310	2.5	120
J	B	Cylinder	2	1/16″	0.063	0.94	1470	2.5	1620	2.6	118
J	B	Cylinder	2	1/16″	0.062	0.98	1400	3.0	1630	2.6	125
K	B	Cylinder	2	1/16″	0.061	1.12	2210	4.0	NA	2.0.3	102
K	B	Cylinder	2	1/16″	0.062	1.12	1930	3.5	1800	2.4	106
K	B	Cylinder	2	1/16″	0.062	1.15	2540	2.8	1940	2.5	112
K	B	Cylinder	2	1/16″	0.063	1.12	1410	3.5	NA	2.8	111
L	B	Quadrulobe	1.81	1/20″	0.047	0.69	1698	3.7	837		

1″ = 1 in. = 2.54 cm; 1 lb. = 0.454 kg; 145 psi = 1 MPa.

Figure 3.51 Plant A severity.

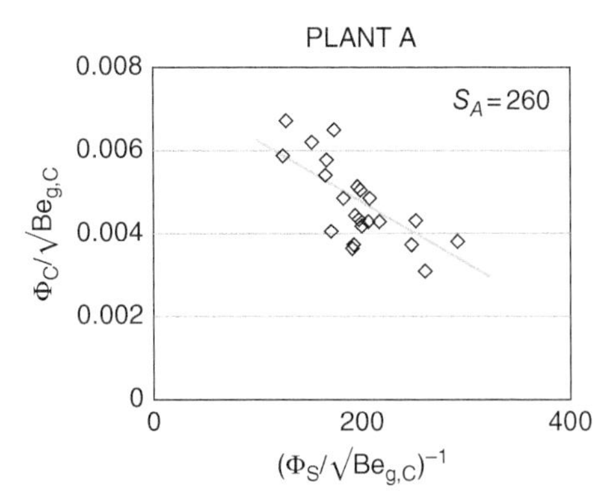

3.7.2.2.1 Aspect Ratio Comparison between the Support and the Final Catalyst

Both plants A and B practice complex and varied operations. Figure 3.51 and Table 3.22 show the aspect ratio for plant A as a function of the aspect ratio of the catalyst support. In an ideal world, all aspect ratios of the final catalyst should be below or at the parity line. Aspect ratios above the parity line likely mean that there is a sampling/measurement error for either the starting or final aspect ratio. Some grades come from a support with a good aspect ratio, which makes it particularly disappointing when their aspect ratio is much lower than the starting value after secondary operations. Figure 3.52 shows a similar aspect ratio comparison for plant B.

3.7.2.2.2 Bending Strength Comparison of Catalyst and Support

As mentioned, the only data left to obtain is the modulus of rupture of the final catalyst. Table 3.22 shows these catalyst properties. Depending on the grade, catalysts will receive different treatments; these treatments may change the flexural

Figure 3.52 Plant B severity.

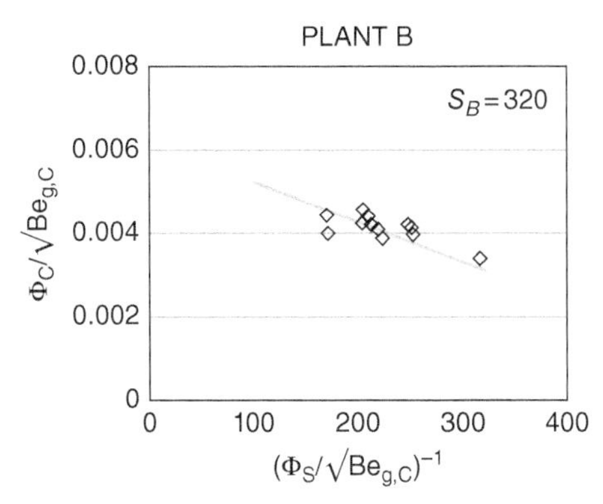

Table 3.23 Severities for plants A and B.

Severity for plant A	260
Severity for plant B	320

properties of the catalyst either for better or for worse. The bending strength can be checked in a laboratory, and one can measure the value after each intermediate preparation step. Intermediate steps can be ion exchange, impregnation, etc.

In the single-impact step approach and from Eq. (3.117), one can see that graphing the group $\Phi_C/\sqrt{Be_{g,C}}$ as a function of the group $\left(\Phi_S/\sqrt{Be_{g,C}}\right)^{-1}$ should yield a unique straight line irrespective of the various catalyst grades, since the slope or intercept only depends on plant severity. This line has a negative slope of absolute value $S_{\infty,C}^{-2}$ and a y-intercept value equal to $2/S_{\infty,C}$.

Figures 3.51 and 3.52 shows graphs for both plants. The linear line is fully determined by only one parameter, $S_{\infty,C}$, which is the dimensionless severity of that plant and should be obtained via nonlinear regression, as shown by the solid blue line. For both plants, the least-squares fit is fair, but it could be improved for this simplified approach of a single impact step.

It is good to see that the single impact step is able to predict the trend of the data for both plants well. As is shown in Table 3.23, the severity of plant B is higher than that of plant A. This is information that experienced technical staff and operating crews will know, but it has been difficult to quantify up to now. The general methodology developed in this book also affords the opportunity to compare the extrudate bending strength quality of each plant and to quantify severity differences.

3.8 Statistical Methods Applied to Manufacturing Materials

It is possible to handle the adequate mixing of materials that are difficult to discern visually by applying statistics to a chemical analysis of samples.

3.8.1 Dry Mixing of Powders

Samples that are taken from the vessel used to mix powders can be analyzed for one or more of the components, such as silica, alumina, etc. When taking representative samples across the batch, the analyses should show a constant and consistent level of the component in agreement with a standard material balance, expressed via the standard deviation. Equation (3.118) calculates the relative standard deviation σ_m:

$$\sigma_m = \frac{1}{\hat{C}}\sqrt{\frac{\sum (C_i - \hat{C})^2}{(n-1)}} \tag{3.118}$$

where C_i is the concentration in the analysis and $\hat{C}$ is the average concentration of the component found in the n samples obtained from Eq. (3.119):

$$\hat{C} = \frac{\sum C_i}{n} \tag{3.119}$$

If the relative standard deviation is below a 0.05 (5%) threshold, the material normally is considered well mixed, but the threshold value is a choice one must make. One will also have to consider the relative standard deviation σ_a of the analysis itself. The relative standard deviation of mixing cannot become less than the analytical relative standard deviation. The value of $\hat{C}$ should match the expected value from a material balance perspective; investigate and find the root cause if it does not.

3.8.2 Expected Values of a New Batch Compared to Historical Values

At times, a plant will begin remaking a catalyst grade that it has made in the past. The question naturally arises with the first batch made as to whether this batch is in line with what was manufactured previously. It is often advantageous to plot the historical values and their properties in order to see where the new batch falls in this cloud of data. If the new data falls well within the cloud, then there is no issue; production can proceed further. If the new data falls outside of the cloud, one will need to apply some solid statistical analysis in order to determine further action steps.

For example, based on historical levels, one can compute the averages and standard deviations of the properties of interest. With that knowledge, one can calculate a 90% confidence interval around the historical average in order for the new batch to be part of that distribution.

Nomenclature

a	Impact deceleration (m/s^2)
b_f	Number of breaks per unit weight
Be	Dimensionless group defined as $\sigma/\psi\rho Da$
Be_g	Dimensionless normalized group defined as $\sigma/\psi\rho Dg$
Be_r	Dimensionless group defined as σ/sP
C	Distance from the extreme fiber to the neutral axis (centroid) (m)
C_i	Concentration of colorant in sample i (au, arbitrary unit)

$\hat{C}$	Average concentration of colorant (au)
CSB	Crush strength of a spherical catalyst (N)
D	Extrudate diameter (m)
E_B	Breakage energy (J)
E	Young's modulus of elasticity (Pa)
F	Force (N)
F_i	Impulsive force (N)
F_r	Rupture force (N)
g	Gravitational acceleration (m/s^2)
G	Relative impact deceleration defined as $G = a/g$
h	Height (m)
$H_{k,j}$	Break function
H_B	Bed depth (m)
H_c	Critical bed depth (m)
i	Integer counter
I	Moment of inertia (m^4)
j	Integer counter
L	Average extrudate length (m)
n	Number of samples
m	Extrudate mass (kg)
M	Mass of the impact plate (kg)
MOR	Modulus of rupture (Pa)
N_j	Number of extrudates observed after j drops
P	Local stress in a bed (Pa)
P_c	Critical stress in a bed with aspect ratio Φ_0 (Pa)
R	Radius (m)
s	Shape factor ($s = cD^3/4I$)
S	Dimensionless severity of an operation
S_∞	Dimensionless asymptotic severity
SCS	Side crush strength of an extrudate (N/m)
u	Ratio of slope over intercept defined in Eq. (3.18)
v	Impact velocity (m/s)
v'	Post-collision velocity (m/s)
v_P	Velocity of the impact plate post-collision (m/s)
v_t	Terminal velocity (m/s)
w	Width between support points in the bending test (m)
W	Weight of a catalyst bed (kg)
W_a	Width of an anvil (m)
W_f	Catalyst fines weight (kg)
W_s	Distance between two support points (m)

x Coordinate
x_i Indent of a bed (m)
Z Defined in the text, Eq. (3.37)

Greek Symbols

α Parameter (Table 3.1)
β Collision interaction factor
γ Parameter defined in Eq. (3.9); $\gamma = (\Phi_\alpha - \Phi_\infty)/(\Phi_0 - \Phi_\infty)$
γ_a Average orientation factor of an extrudate in a bed
γ_b Bed bridging factor
Γ Forward fractional difference to reach the asymptotic aspect ratio Φ_∞
δ Deviation (m)
ε Bed void fraction (–)
θ Average break angle of an extrudate (rad)
κ Parameter defined in Eq. (3.56)
μ_w Wall friction coefficient (–)
ζ Projection factor (–)
ρ_g Density of ambient air (kg/m^3)
ρ Catalyst density (kg/m^3)
ρ_B Catalyst bed density (kg/m^3)
σ Tensile strength, also known as modulus of rupture (Pa)
σ_m Relative standard deviation of mixing (–)
σ_a Relative standard deviation for an analytical test (–)
$\Delta\tau$ Duration of an impact (s)
φ Defined as Ω/D^2 (in Table 3.1)
φ_{GR} Golden Ratio = $\left(1 + \sqrt{5}\right)/2$
Φ Aspect ratio of an extrudate
Φ_0 Extrudate aspect ratio at the start (before applying a load or before collision)
Φ_j Aspect ratio after j consecutive drops
Φ_∞ Asymptotic aspect ratio after many repeated drops
Φ_α Asymptotic aspect ratio for a single drop of sufficiently long extrudates
Φ_f Loss of catalyst to fines in a single break expressed as a loss in aspect ratio
χ Forward fractional change in aspect ratio in a drop
ψ Shape factor, ($\psi = s\Omega/D^2$) (in Table 3.1)
Ψ Proportionality factor for a fixed bed defined in Eq. (3.76)
ω_f Weight fraction of fines (–)
Ω Extrudate cross-sectional area (m^2)

Subscripts

A, B, C Designation of a particular impact for equipment A, B, or C
ABC Designation of single contiguous impacts $A \rightarrow B \rightarrow C$ in that order

References

1 Le Page, J.F. (1987). *Applied Heterogeneous Catalysis*. Paris, France: Institut Français du Pétrole Publications, Éditions Technip.
2 Woodcock, C.R. and Mason, J.S. (1987). *Bulk Solids Handling: An Introduction to the Practice and Technology*. New York: Chapman & Hall.
3 Bertolacini, R.J. (1989). *Mechanical and Physical Testing of Catalysts*, ACS Symposium Series. Washington D.C.: American Chemical Society.
4 Wu, D.F., Zhou, J.C., and Li, Y.D. (2006). Distribution of the mechanical strength of solid catalysts. *Chemical Engineering Research and Design* 84 (12): 1152–1157. https://doi.org/10.1205/cherd05015.
5 Li, Y., Wu, D., Chang, L. et al. (1999). A model for the bulk crushing strength of spherical catalysts. *Industrial and Engineering Chemical Research* 38 (5): 1911–1916. https://doi.org/10.1021/ie980360j.
6 Li, Y., Wu, D., Zhang, J. et al. (2000). Measurement and statistics of single pellet mechanical strength of differently shaped catalysts. *Powder Technology* 113: 176–184.
7 Staub, D., Meille, S., Corre, V. et al. (2015). Revisiting the side crushing test using the three-point bending test for the strength measurement of catalyst supports. *Oil & Gas Science and Technology* 70 (3): 475–486. https://doi.org/10.2516/ogst/2013214.
8 Beeckman, J.W., Fassbender, N.A., and Datz, T.E. (2016). Length to diameter ratio of extrudates in catalyst technology I. Modeling catalyst breakage by impulsive forces. *AICHE Journal* 62: 639–647. https://doi.org/10.1002/aic.15046.
9 Papadopoulos, D.G. (1998). Impact breakage of particulate solids. PhD thesis. University of Surrey.
10 Salman, A.D., Biggs, C.A., Fu, J. et al. (2002). An experimental investigation of particle fragmentation using single particle impact studies. *Powder Technology* 128: 36–46.
11 Subero-Couroyer, C., Ghadiri, M., Brunard, N., and Kolenda, F. (2005). Analysis of catalyst particle strength by impact testing: the effect of manufacturing process parameters on the particle strength. *Powder Technology* 160: 67–80.
12 Bridgwater, J. (2007). *Particle Breakage Due to Bulk Shear, Chapter 3, Handbook of Powder Technology*, 1e. Amsterdam, Netherlands: Elsevier B.V.

13 Bridgwater, J., Utsumi, R., Zhang, Z., and Tuladhar, T. (2003). Particle attrition due to shearing – the effects of stress, strain and particle shape. *Chemical Engineering Science* 58 (20): 4649–4665. https://doi.org/10.1016/j.ces.2003.07.007.

14 Li, Y., Li, X., Chang, L. et al. (1999). Understandings on the scattering property of the mechanical strength data of solid catalysts: a statistical analysis of iron-based high-temperature shift catalysts. *Catalysis Today* 51: 73–84. https://doi.org/10.1016/S0920-5861(99)00009-7.

15 Heinrich, S. (2016). Multiscale strategy to describe breakage and attrition behavior of agglomerates. 2016 Frontiers in Particle Science & Technology Conference, Houston, TX.

16 Wassgren, C. (2016). Discrete element method modeling of particle attrition. 2016 Frontiers in Particle Science & Technology Conference, Houston, TX.

17 Potapov, A. (2016). Approaches for accurate modeling of particle attrition in DEM simulations. 2016 Frontiers in Particle Science & Technology Conference, Houston, TX.

18 Hosseininia, E. and Mirgashemi, A. (2006). Numerical simulation of breakage of two-dimensional polygon-shaped particles using discrete element method. *Powder Technology* 166: 100–112. https://doi.org/10.1016/j.powtec.2006.05.006.

19 Potyondy, D. (2016). Bonded-particle modeling of fracture and flow. 2016 Frontiers in Particle Science & Technology Conference, Houston, TX.

20 Carson, J. (2016). Effective use of discrete element method (DEM) modeling of particle attrition applications and understanding unforeseen consequences with attrition reduction design techniques. 2016 Frontiers in Particle Science & Technology Conference, Houston, TX.

21 Beeckman, J. (2016). Modeling catalyst extrudate breakage by impulsive forces, 2016 Frontiers in Particle Science & Technology Conference, Houston, TX.

22 Beeckman, J.W., Fassbender, N.A., and Datz, T.E. (2016). Length to diameter ratio of extrudates in catalyst technology II. Bending strength versus impulsive force. *AICHE Journal* 62: 2658–2669. https://doi.org/10.1002/aic.15231.

23 Wu, D., Song, L., Zhang, B., and Li, Y. (2003). Effect of the mechanical failure of catalyst pellets on the pressure drop of a reactor. *Chemical Engineering Science* 58 (17): 3995–4004. https://doi.org/10.1016/S0009-2509(03)00286-0.

24 Timoshenko, S.P. and Goodier, J.N. (1951). *Theory of Elasticity*. New York: McGraw-Hill Book Co.

25 Beeckman, J.W., Fassbender, N.A., Cunningham, M., and Datz, T.E. (2017). Length to diameter ratio of extrudates in catalyst technology III. Catalyst breakage in a fixed bed. *Chemical Engineering and Technology* 40: 1844–1851. https://doi.org/10.1002/ceat.201600550.

26 Beeckman, J.W., Fassbender, N.A., Datz, T.E. et al. (2018). Predicting catalyst extrudate breakage based on the modulus of rupture. *Journal of Visualized Experiments* 135 https://doi.org/10.3791/57163.

27 Hertz, H. (1891). *Journal fur die Reine und Angewandte Mathematik* 92: 155.
28 Hertz, H. (1895). *About the Contact of Solid Elastic Bodies: Collected Works*, vol. 1. Leipzig.
29 Gugan, D. (2000). Inelastic collision and the Hertz theory of impact. *American Journal of Physics* 68: 920–924.
30 Leroy, B. (1985). Collision between two balls accompanied by deformation. *American Journal of Physics* 53 (4): 346–349.
31 Bokor, A. and Leventhall, H.G. (1971). The measurement of initial impact velocity and contact time. *Journal of Physics D: Applied Physics* 4: 160.
32 Wooten, J.T. (1998). Dense and sock catalyst loading compared. *Oil & Gas Journal* 96: 66–70.
33 Wachtman, J.B., Cannon, W.R., and Matthewson, M.J. (2009). *Mechanical Properties of Ceramics*, 2e. Hoboken, NJ: Wiley.
34 Janssen, H.A. (1895). Getreidedruck in Silozellen. *Zeitung des vereins deutscher ingenieure* 39: 1045–1049.
35 Schulze, D. (2008). *Powders and Bulk Solids – Behavior, Characterization, Storage and Flow*. Berlin, Germany: Springer.

4

Steady-state Diffusion and First-order Reaction in Catalyst Networks

4.1 Introduction

The classic continuum approach for describing reaction and diffusion in catalysts forms a unique field that is very mature and has developed over nearly a century. Starting with the groundbreaking work of Thiele [1], his analysis is still the workhorse in any academic catalyst engineering course or in commercial catalyst laboratories. The continuum approach has very elegant solutions, and the literature is vast and should be consulted prior to embarking on any kind of catalyst investigation. In the classic chemical engineering literature, pearls of analysis, comprehensive reviews, and textbooks can be found throughout that capture the many nuances of the treatment of catalysts according to their kinetics, diffusivity, shape, size, and texture. Let me mention Froment and Bischoff [2], Froment et al. [3], Satterfield [4], Aris [5], Hill [6], Hegedus and McCabe [7], Hegedus et al. [8], Chen et al. [9], Becker and Pereira [10], and Reyes and Iglesia [11].

For the mathematical analysis, assumptions are made upfront concerning the homogeneity and isotropic nature of the catalyst. Most often, one assumes total uniformity of properties in both locale and direction in the catalyst pellet. Often, the effective medium approximation (EMA), pioneered by Kirkpatrick [12] is applied as in Mo and Wei [13]. This technique replaces the irregular architecture of the catalyst with an effective medium homogeneous architecture to describe molecular transport. This technique is applied in many fields dealing with the transport through and effectiveness of disordered media. Finally, however, for verification or for validation of the assumptions, one still performs large-system stochastic calculations dealing with very large sparse matrices with state-of-the-art numerical solvers. The computational effort is substantial in order to get good accuracy, especially near percolation thresholds. In Beeckman [14], I show the enormous number of possibilities of percolation paths through a rectangular network with random blockages.

Catalyst Engineering Technology: Fundamentals and Applications,
First Edition. Jean W. L. Beeckman.

In the classic treatment, a material balance is formulated across two area slices that are perpendicular to the direction of interest and at an infinitesimal distance away from each other. By "direction of interest," I mean along the axis for a flat slab and radially for cylinders or spheres. Average constant properties in the areal slices are used to quantify molar fluxes and reaction rates. At times, a catalytic activity profile is introduced when the process or the application necessitates. Mathematically, the material balance then translates to an ensuing differential equation (DE) that governs the process. Solving the DE with the appropriate boundary conditions then allows one to obtain the reagent concentration profile inside the catalyst and the catalyst performance via the effectiveness factor.

Catalysts are essentially built up from a compaction of small particles that themselves are often obtained via precipitation and coprecipitation of even smaller particles. The size of these building blocks ranges from fractions of nanometers to microns. The void space between the particles in catalysts therefore consists of terrific networks [15] of nodes connected by pores (channels), and the question arises in my mind as to how the material balance closure from slice to slice ensures closure from node to node. Certainly, not all nodes are conveniently located in slices, and pores are longer than an infinitesimal distance. It is not clear how to close material balances in the very local node-to-node structure of the catalyst, and this is a bit of a drawback for the classic approach. The difficulty in closing local material balances stems from the fact that nothing is built into the classic approach that allows one to do that. Use is made of porosity and a tortuosity factor that are expected to translate the complex architecture of the catalyst into an effective diffusion coefficient. Porosimetry data also allows one to obtain the surface area of a catalyst. The linkage with the chemical reactivity of the surface then allows one to quantify the chemical reaction rate. Over the years, both the effective diffusion coefficient and the chemical reaction rate have been used very successfully in the analysis. Yet I have the lingering thought on what to do and how to ensure the local mass balance closure in each node, as well as what the advantages and impacts are of insisting on closure in the analysis.

To enhance clarity and transparency and not unduly burden the reader with complex formulae, I will look at a simplified engineering model. I consider the chemical reactivity limited to just a first-order reaction in the nodes while the channels that connect the nodes are envisioned to provide mass transfer only. For the catalyst geometric shape, I always use the flat slab geometry, but I believe cylinders and spheres and even arbitrary shapes can be treated along similar lines. The variety will come in the networks considered herein and how their treatment and solution may hopefully contribute to the field of reaction and diffusion in catalysts.

First, for those perhaps not familiar with the pertinent literature, I give the classic solution for the case of diffusion accompanied by a first-order reaction in the catalyst. The mathematical solution is well known, and I give it here (i) to make it easy for the novice reader and (ii) because the mathematical form of the solutions will be referenced later on when dealing with networks.

After the classic approach, I deal with regular networks that, as their name suggests, have a periodic structure. These layer-by-layer structures are repeated over and over again and are elegant tools for dealing with materials such as zeolites, and a generic kind will be dealt with later on. The approach allows one to write the solutions in matrix form, but functionally the expressions are very similar to the classic solutions. For a specific carving of the parameter space, I will show that the matrix solutions functionally are the same as the classic solutions. The solutions are analytic and hence the number of nodes is of no real concern. The solutions can be for two-dimensional or three-dimensional structures, as shown in Beeckman [16]. More complex structures may also be treated with this approach. For instance, for ZSM-5, the structure can be solved via the use of a double layer, although it is a bit more complex and is not treated herein. Properties of the structure are periodic and hence the solutions to the material balance closures can be accomplished by means of finite-difference matrix techniques.

Following the section on regular networks (Section 4.3.3), I deal with irregular or arbitrary networks, and the network approach gives the solution again in the form of matrix equations. The generality of these networks, however, does not allow one to say much as far as specific solutions go. Only specific solutions, still in matrix form, are obtained, but no specific solutions are found related to the properties of the matrix or the matrix form. As solution examples, I mention two: (i) the Schur complement that determines the molar fluxes of reagent that enter along the surface of the network and (ii) the weight factor matrix that determines the concentrations in the internal nodes as a function of the concentrations in the nodes located on the periphery of the network.

Finally, in Section 4.3.5, I deal with the treatment of what I think of as amorphous catalyst networks. I hope to convince the reader that the networks here considered are very realistic. Even though we take utmost care when preparing catalysts in the laboratory or a manufacturing plant such that they are very homogeneous, measurements of properties always leave us with distributions to deal with. The networks in catalyst particles are absolutely massive, and their local properties of surface area, mass transfer, and chemical reactivity can change abruptly from point to point. To model these abrupt changes and the variability in these properties, I introduce the concept of "white noise." I will show that while at first sight the size of the network and the variability look like significant problems for solving the network, in the end the massive size will prove to be an advantage, especially when having to deal with the large degree of point-to-point white noise variability in the network.

I think of the parameter space of the problem as threefold:

1) Diffusion: the distribution of the mass transfer coefficients from node to node.
2) Reaction: the distribution of nodal surface areas and rate constants.
3) Catalyst: the size and architecture of the network used to describe the catalyst pellet.

Typically, we only look at two groups of parameters, namely those linked to diffusion and those linked to reaction. In addition, in the literature, often the parameters linked to reaction and diffusion are considered to be constant along the direction of interest. The network size considered when solving DEs with constant coefficients is essentially infinite because we look at layers that are an infinitesimal distance away from each other.

When considering networks that have abrupt changes in the node-to-node properties over a very short but finite distance, I have found that it is important to have networks that are appropriately sized. Due to the large variability in the properties from node to node, it is necessary to consider several hundred to several thousand layers in the direction of interest from the periphery to the core of the network in order to represent a catalyst realistically. The current standard student and researcher PC equipped with the Mathcad, MATLAB, or other appropriate matrix numerical packages, however, can well handle the size of the networks required.

Later in this chapter, I will show that carving the threefold parameter space to only consider deep networks combined with reaction perturbation in Section 4.3.5 allows one to obtain very specific solutions to the problems of reaction and diffusion in catalysts, solutions that are "analytical but of a special kind". What exactly I mean by "deep networks" and "reaction perturbation" I will explain later on. For those who are familiar with diffusion and reaction in catalysts and effective medium theory, I hope that Section 4.3.5 yields interesting reading and offers a new perspective on the subject.

Further, theory and simulations show that for the cases studied, local arithmetic averages of catalyst properties can be applied, but they are not necessarily optimal. Indeed, many averages of properties can be defined at will, but how to find the one that is optimal (i.e. the average property that leads to the correct mathematical solution) is the question. In order not to disappoint the reader, I assure you that there is an answer to this question, but as in any good thriller – the answer comes at the end.

However, I have only made a start, and I leave many questions unanswered. Much work is left to be done.

4.2 Classic Continuum Approach

I refer to the literature as expounded upon in Section 4.1 for a comprehensive reading. Here, I treat as an example a catalyst with a flat slab geometry.

The solution to diffusion and first-order reaction is governed by a DE shown in Eq. (4.1):

$$D_e \frac{d^2C}{dx^2} = kSC \tag{4.1}$$

with boundary conditions:

$$C_{x=L} = C_L \tag{4.2}$$

$$\left(\frac{dC}{dx}\right)_{x=0} = 0 \tag{4.3}$$

where x is the coordinate of interest, C is the concentration of the reagent at location x, and L is the thickness of the slab of catalyst. D_e is the effective diffusion coefficient, k is the rate constant, and S is the specific surface area of the catalyst. The parameters D_e, k, and S are considered constant along the direction coordinate x. The effective diffusion coefficient D_e is further expressed in Eq. (4.4):

$$D_e = \left(D_m^{-1} + D_K^{-1}\right)^{-1}(\theta/\tau) \tag{4.4}$$

where D_m is the bulk molecular diffusion coefficient while D_K is the Knudsen diffusion coefficient. The porosity θ takes account of the limited areal open space within the catalyst available for diffusion. The tortuosity factor τ takes account of the need for a molecule to change direction when an obstacle (solid catalyst surface) is encountered. The porosity can be measured accurately experimentally with standard techniques. Tortuosity has a fairly wide acceptable range and, if at all possible, should be measured and confirmed via an independent means. Techniques based on a tracer pulse injected in front of a bed of catalyst and analyzed by gas chromatography at the exit of the bed can help in this endeavor. Modeling the shape of the pulse changed by the dispersion in the bed and the intraparticle diffusion then allows one to obtain the effective diffusion coefficient. A Wicke–Kallenbach diffusion cell, when appropriate, can help as well. Be careful when measuring and applying an effective diffusion coefficient through a catalyst when it is not determined in the form used during reaction. For instance, crushing the catalyst to powder and re-pelletizing to make it amenable for a diffusion cell can change the diffusive properties appreciably.

The solution of the DE in Eq. (4.1) is:

$$C(x) = C_L \frac{\cosh(\xi\phi)}{\cosh(\phi)} \tag{4.5}$$

where ξ is the dimensionless coordinate defined as x/L while ϕ is the Thiele modulus defined in Eq. (4.6):

$$\phi = L\sqrt{\frac{kS}{D_e}} \tag{4.6}$$

The molar flux F of reagent entering the slab is given in Eq. (4.7):

$$F = D_e\left(\frac{dC}{dx}\right)_{x=L} = C_L\left(\frac{D_e}{L}\right)\phi \tanh(\phi) \tag{4.7}$$

The effectiveness factor of the flat slab catalyst given in Eq. (4.8) yields a measure of how effectively the catalyst is used:

$$\eta = \frac{\tanh(\phi)}{\phi} \tag{4.8}$$

Later on, I will show how the solutions depicted in Eqs. (4.5)–(4.8) will come back for the appropriate network counterparts.

As an aside, consider the same flat slab arrangement with the exception that at the origin I now consider the slab to be open to the bulk fluid with a concentration $C_0 = 0$. The solution is obtained by the standard techniques and leads to Eq. (4.9):

$$C(x) = C_L\frac{\sinh(\xi\phi)}{\sinh(\phi)} \tag{4.9}$$

The flux of material entering the slab at $x = L$ is given in Eq. (4.10):

$$F_{x=L} = D_e\left(\frac{dC}{dx}\right)_{x=L} = D_e\left(\frac{C_L}{L}\right)\frac{\phi}{\tanh(\phi)} \tag{4.10}$$

The flux of material exiting the slab at $x = 0$ is given in Eq. (4.11):

$$F_{x=0} = D_e\left(\frac{dC}{dx}\right)_{x=0} = D_e\left(\frac{C_L}{L}\right)\frac{\phi}{\sinh(\phi)} \tag{4.11}$$

Equation (4.11) can lead to an interesting experimental arrangement, such as a Wicke–Kallenbach diffusion cell with one side flushed with reagent and the opposite side continuously flushed with an inert carrier.

To be complete, consider that the slab is open on both sides with the same concentration C_e imposed at each end. The concentration of reagent is then obtained from Eq. (4.12):

$$C(x) = C_e\left[\frac{\sinh(\xi\phi)}{\sinh(\phi)} + \frac{\sinh((1-\xi)\phi)}{\sinh(\phi)}\right] \tag{4.12}$$

4.3 The Network Approach

I look at a catalyst as a large network of nodes that all have catalytic activity. The nodes are connected by channels that account for all of the mass transfer between the nodes. Some nodes are in direct contact with the bulk fluid outside of the catalyst and are called external nodes. The concentration of a reagent in an external node is considered to be the same as the concentration in the bulk fluid, hence I do not consider external mass transfer limitations. The concentration in an external node is considered to be an independent variable. Most nodes are only accessible from the bulk fluid via other nodes and are called internal nodes. The concentration in an internal node is a dependent variable. It is an objective to determine the concentration in the internal nodes as a function of the concentrations in the external nodes and as a function of the network properties.

4.3.1 Mass Transfer between Nodes

In this work, mass transfer between nodes is characterized by a mass transfer coefficient ν.

The molar flux per unit time f from a node at a concentration C_1 to a neighboring node at a concentration C_2 is written out in Eq. (4.13):

$$f = \nu(C_1 - C_2) \tag{4.13}$$

For a regular network with cylindrical pore connections between nodes, the mass transfer coefficient ν is obtained from Eq. (4.14):

$$\nu = \frac{\pi r^2 D}{L} \tag{4.14}$$

where r is the pore radius, D is the diffusion coefficient, and L is the distance between the two nodes.

For an irregular network, I drop the cylindrical pore assumption totally and rather consider the mass transfer coefficient ν to belong to a distribution of interest that the reader is free to choose. The distribution can be chosen arbitrarily, but in worked examples I employ either a random uniform distribution between two limits ν_1 and $\nu_1 + \nu_2$ or a Gaussian distribution with a given mean μ and variance σ^2. The reason for this in the examples is simply that the Mathcad software conveniently generates a set of random numbers belonging to either one. For a uniform distribution, Mathcad uses the *rnd*(x) function, which generates a random number between 0 and x one at a time. For a Gaussian distribution, the *rnorm*(g, μ, σ) function generates a vector g of random numbers that have a Gaussian distribution. The value of ν is then either obtained from Eq. (4.15):

$$\nu = \nu_1 + rnd(\nu_2) \text{ for a uniform distribution} \tag{4.15}$$

or as element i in the vector g from Eq. (4.16):

$$\nu = g(i) \text{ for a Gaussian distribution} \tag{4.16}$$

4.3.2 First-order Reaction in a Node

In this work, I only consider a first-order reaction occurring in the nodes. In a regular network, I assume all nodes to be identical and write the rate of reaction R as per Eq. (4.17):

$$R = k^+ C \tag{4.17}$$

where k^+ is the rate constant while C is the concentration in the node.

For the nodes in an irregular network, I write the rate of reaction as per Eq. (4.18):

$$R = kSC \tag{4.18}$$

where k is a rate coefficient while S is the surface area allocated to the node. In the examples, I use values of S belonging to either a uniform distribution between two limits or a Gaussian distribution, but any distribution can be applied. The rate coefficient k is considered to be a constant, but here an arbitrary distribution can also be applied if one wishes.

The interplay of mass transfer and reaction in a nodal network is investigated here. For the rate of reaction, the choice for local nodal surface areas can always be offset by the appropriate choice of the rate coefficient in order to balance with the mass transfer coefficient. In this way, mass transfer limitations in the network can be evaluated theoretically, while the distributions for mass transfer coefficients and surface area can be chosen freely for the network under consideration.

4.3.3 Regular Networks

4.3.3.1 Introduction

This section elaborates on an approach for rigorously solving the ensemble of material balances in the nodes of a pore network with first-order kinetics and in the presence of mass transfer limitations. The networks considered have linear node connectivity in one dimension, square node connectivity in two dimensions, and cubic node connectivity in three dimensions. The approach yields explicit solutions for the mass transfer flux in the nodes on the perimeter of the pore network as a function of the concentrations in those nodes. Complex boundary conditions, such as those encountered with network perimeter blocking, are easily handled, and the computational efforts are significantly reduced. This book builds further on an earlier publication by Beeckman [16] and presents analytical solutions for first-order reactions with mass transfer limitations in a variety of pore networks. Since the discrete nature of the networks is kept intact (i.e. they are not replaced with their continuum counterparts), the solutions are pore detailed and also quantify unreacted reagent diffusing back out of the pore network. Matrix difference calculus appears to be the preferred vehicle for solving the ensemble of continuity equations in the nodes. As will be shown, the matrix solutions for these networks can be cast in a mathematical form that is similar to the solutions for their network continuum counterparts, while the classical Thiele modulus is replaced with a matrix of similar characteristics. This matrix, for instance, is singular (a zero Thiele modulus) when only diffusion is considered and becomes non-singular (a non-zero Thiele modulus) when reaction is included. This approach is probably well suited for solving the diffusion–reaction problem in zeolite structures due to the inherent periodicity encountered with zeolites.

4.3.3.2 A One-dimensional Network: A String of Nodes

Consider a string of nodes as shown in Figure 4.1 that have a chemical reactivity constant k^+ and are connected by channels with a mass transfer coefficient ν. The string has N internal nodes and two open nodes at each end of the string. At the far left of the string the concentration of reagent is set at zero, while at location $N+1$ on the far right the concentration is set at C_e. Making a material balance in each internal node leads to the matrix in Eq. (4.19):

$$\begin{vmatrix} k^+ + 2\nu & -\nu & 0 \\ -\nu & k^+ + 2\nu & -\nu \\ 0 & -\nu & k^+ + 2\nu \end{vmatrix} \begin{vmatrix} C_1 \\ C_i \\ C_N \end{vmatrix} = \begin{vmatrix} 0 \\ 0 \\ \nu C_e \end{vmatrix} \tag{4.19}$$

Let:

$$\kappa = k^+ + 2\nu \tag{4.20}$$

Equation (4.19) is then solved numerically per Eq. (4.21):

$$\begin{vmatrix} C_1 \\ C_i \\ C_N \end{vmatrix} = \begin{vmatrix} \kappa/\nu & -1 & 0 \\ -1 & \kappa/\nu & -1 \\ 0 & -1 & \kappa/\nu \end{vmatrix}^{-1} \begin{vmatrix} 0 \\ 0 \\ C_e \end{vmatrix} \tag{4.21}$$

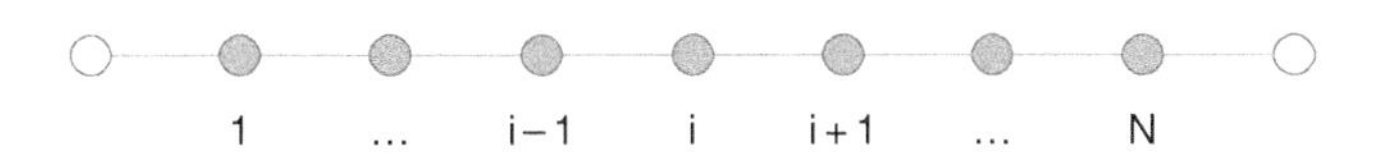

Figure 4.1 A string of nodes open on both sides.

There is an analytical solution that is obtained by standard finite-difference techniques as per Levy and Lessman [17]. Define φ per Eq. (4.22):

$$\varphi = (N + 1)a\cosh(\kappa/2\nu) \tag{4.22}$$

Then the analytical solution for the concentration in the internal node located at position i is obtained from Eq. (4.23):

$$C_i = C_e \frac{\sinh(\xi\varphi)}{\sinh(\varphi)} \quad i = 1, \dots, N \tag{4.23}$$

The dimensionless position ξ of the node at location i is given by Eq. (4.24):

$$\xi = \frac{i}{N + 1} \tag{4.24}$$

From Eqs. (4.9) and (4.23), it is clear that for a string of nodes, the factor φ formally plays the role of the Thiele modulus ϕ in the classic continuum case.

If the node on the far left would also have a concentration C_e imposed, then the solution is given in Eq. (4.25):

$$C_i = \left\{ \frac{\sinh(\xi\varphi)}{\sinh(\varphi)} + \frac{\sinh((1-\xi)\varphi)}{\sinh(\varphi)} \right\} C_e \tag{4.25}$$

again with φ playing the role of ϕ.

Consider now a string of nodes the same as in Figure 4.1, but with the channel to the left of node 1 closed off. A material balance in each node yields Eq. (4.26):

$$\begin{vmatrix} k^+ + \nu & -\nu & 0 \\ -\nu & k^+ + 2\nu & -\nu \\ 0 & -\nu & k^+ + 2\nu \end{vmatrix} \begin{vmatrix} C_1 \\ C_i \\ C_N \end{vmatrix} = \begin{vmatrix} 0 \\ 0 \\ \nu C_e \end{vmatrix} \tag{4.26}$$

The analytical solution for the concentrations in the nodes is then given by Eq. (4.27):

$$C_i = \frac{\cosh(\xi\varphi)}{\cosh(\varphi)} C_e \tag{4.27}$$

with:

$$\varphi = (N + 1/2)a\cosh(\kappa/2\nu) \tag{4.28}$$

while the dimensionless position ξ of a node is now given by Eq. (4.29):

$$\xi = \frac{i - 1/2}{N + 1/2} \tag{4.29}$$

Therefore, here again the factor φ plays the role of the classic Thiele modulus ϕ. The concentrations in the nodes of the string are formally identical to the classic continuum approach expressions and the only difference is in the formal expression for φ. To settle this, consider a carving of the parameter space and look at reaction as a perturbation of the diffusion field. For k^+ to act as a perturbation on v means that $k^+/v \ll 1$ and therefore one gets Eq. (4.30):

$$\lim_{k^+/v \to 0} (a\cosh(1 + k^+/2v)) = \sqrt{k^+/v} \tag{4.30}$$

Hence:

$$\varphi = (N + 1)\sqrt{\frac{k^+}{v}} \quad \text{for a string open on both ends} \tag{4.31}$$

and:

$$\varphi = (N + 1/2)\sqrt{\frac{k^+}{v}} \quad \text{for a string open on just one end} \tag{4.32}$$

The expression for φ in the case of perturbation in Eqs. (4.31) or (4.32) now also becomes the same as the classic continuum expression for the Thiele modulus from Eq. (4.6). Since k^+ is a perturbation of the diffusion field, it is necessary to select large enough values for N in order to get substantial mass transfer limitations; hence, one must consider a very long string.

4.3.3.3 Two-dimensional Networks with Square Connectivity

4.3.3.3.1 A Rectangular Network Open on All Sides

Consider a two-dimensional rectangular pore network as shown in Figure 4.2 and allow a first-order irreversible reaction $A \rightarrow B$ to proceed in the nodes. All pores that run parallel to the x-axis are assumed identical and have a mass transfer coefficient ν_x. Pores that run parallel to the y-axis have a mass transfer coefficient ν_y. Even for zeolite crystals, the pores connecting the nodes are most often not of a uniform cross section, but rather are channels with complex shape and electronic effects. One can therefore opt to obtain the mass transfer coefficient specific for the channel and the diffusing molecule based on *ab initio* calculations.

The position of each internal node in the pore network is defined by a set of integer coordinates (i, j) along the x-axis respective along the y-axis, with their ranges given by Eq. (4.33):

$$i = 1, 2, \ldots, N_x; \quad j = 1, 2, \ldots, N_y \tag{4.33}$$

for a total of $N_x \times N_y$ internal nodes. The concentration of component A in an internal node will be noted as $C_{i,j}$. The positions of the external nodes along each of the

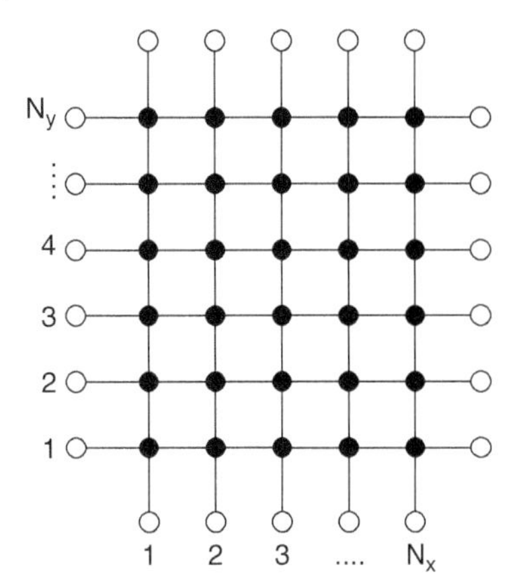

Figure 4.2 Node network open on all sides.

four sides of the pore network are defined similarly; for instance, the positions for the external nodes located on the x-axis are defined with their integer coordinates $(i, 0)$ and their range is given by Eq. (4.34):

$$i = 1, 2, \ldots, N_x \tag{4.34}$$

To fix our thoughts, I will first solve for the dependence of the concentrations in the internal nodes located per Eq. (4.35):

$$i = k;\ j = 1, 2, \ldots, N_y \tag{4.35}$$

as a function of the concentrations in the external nodes located per Eq. (4.36):

$$i = N_x + 1;\ j = 1, 2, \ldots, N_y \tag{4.36}$$

when all other external node concentrations are set equal to zero.

The concentrations in the internal nodes are written in column vector form as in Eq. (4.37):

$$\mathbf{C}(k) = \left|C_{k,\mathrm{N}_y} \quad C_{k,\mathrm{N}_y-1} \ \ldots \ C_{k,1}\right|^{\mathrm{T}} = \left(N_y \times 1\right) \text{ column vector} \tag{4.37}$$

with:

$$k = 1, 2, \ldots, N_x \tag{4.38}$$

The concentrations in the external nodes perpendicular to the x-axis for $i = N_x + 1$ are written similarly in column vector form as per Eq. (4.39):

$$\mathbf{C}(N_x + 1) = \left|C_{N_x+1,\mathrm{N}_y} \quad C_{N_x+1,\mathrm{N}_y-1} \ldots C_{N_x+1,1}\right|^{T} = \left(N_y \times 1\right) \text{ column vector} \tag{4.39}$$

A material balance in the nodes of layer $k > 1$ leads to Eq. (4.40):

$$-\mathbf{C}(k-1) + \mathbf{KC}(k) - \mathbf{C}(k+1) = \mathbf{0} \ \text{ for } \ k = 2, \ldots, N_x \tag{4.40}$$

The matrix $\mathbf{K}$ has dimensions $(N_y \times N_y)$ and is tri-diagonal, as shown in Eq. (4.41):

$$\mathbf{K} = \frac{1}{\nu_x}\begin{vmatrix} \kappa & -\nu_y & & & \\ -\nu_y & \kappa & -\nu_y & & \\ & . & . & . & \\ & & -\nu_y & \kappa & -\nu_y \\ & & & -\nu_y & \kappa \end{vmatrix} \tag{4.41}$$

with:

$$\kappa = k^{+} + 2(\nu_x + \nu_y) \tag{4.42}$$

Since $\mathbf{C}(0)$ was defined equal to zero, the material balance in the first layer of internal nodes reads as in Eq. (4.43):

$$\mathbf{KC}(1) - \mathbf{C}(2) = \mathbf{0} \tag{4.43}$$

Now define $\mathbf{\Psi}(k)$ as the $(N_y \times N_y)$ weight factor matrix that relates the concentrations in the nodes of layer $i = k$ to those in layer $i = k + 1$ for a network that extends from layer $k = 0$ to layer $k + 1$ for arbitrary $\mathbf{C}(k + 1)$. Per definition, this allows one to write Eq. (4.44):

$$\mathbf{C}(k) = \mathbf{\Psi}(k)\,\mathbf{C}(k + 1) \tag{4.44}$$

With the help of Eq. (4.43) for $k = 1$, I obtain Eq. (4.45):

$$\mathbf{\Psi}(1) = \mathbf{K}^{-1} \tag{4.45}$$

Applying Eq. (4.44) to the material balance Eq. (4.40) for $k > 1$ leads to Eq. (4.46):

$$[\mathbf{\Psi}(k-1)\mathbf{\Psi}(k) - \mathbf{K\Psi}(k) + \mathbf{I}]\mathbf{C}(k + 1) = \mathbf{0} \tag{4.46}$$

where $\mathbf{I}$ is a unit matrix of dimension $(N_y \times N_y)$. Since the vector $\mathbf{C}(k + 1)$ is arbitrary, Eq. (4.46) necessarily leads to the Riccati equation in Eq. (4.47):

$$\mathbf{\Psi}(k-1)\,\mathbf{\Psi}(k) - \mathbf{K\Psi}(k) + \mathbf{I} = \mathbf{0} \tag{4.47}$$

The infinite grid $(k \to \infty)$ is governed by Eq. (4.48):

$$\mathbf{\Psi}_{\infty}^{2} - \mathbf{K\Psi}_{\infty} + \mathbf{I} = \mathbf{0} \tag{4.48}$$

The classic canonical form of $\mathbf{K}$ is well known and is expressed in Eq. (4.49):

$$\mathbf{K} = \mathbf{U\Lambda U}^{T} \tag{4.49}$$

where $\mathbf{\Lambda}(N_y \times N_y)$ is the diagonal matrix of eigenvalues of $\mathbf{K}$ while the matrix $\mathbf{U}(N_y \times N_y)$ is the modal matrix of corresponding eigenvectors expressed in Eqs. (4.50) and (4.51):

$$(\mathbf{\Lambda})_{n,n} = \frac{1}{\nu_x}\left[\kappa - 2\nu_y \cos(\theta_n)\right] \text{ with n} = 1, 2, \ldots, N_y \tag{4.50}$$

$$(\mathbf{U})_{m,n} = \left(\sqrt{\frac{2}{N_y + 1}}\right) \sin\left[(N_y - m + 1)\theta_n\right] \text{ with } m, n = 1, 2, \ldots, N_y \tag{4.51}$$

and:

$$\theta_n = \frac{n\pi}{N_y + 1} \tag{4.52}$$

Note that the matrix of eigenvectors $\boldsymbol{U}$ is independent of the rate coefficient k^+, hence the first-order reaction does not impact the eigenvectors (i.e. the eigenvectors for diffusion only also apply for the case of diffusion and first-order reaction). The canonical representation for $\boldsymbol{\Psi}_\infty$ now becomes:

$$\boldsymbol{\Psi}_\infty = \mathbf{U}\boldsymbol{\Lambda}_\infty \mathbf{U}^T \tag{4.53}$$

with:

$$(\boldsymbol{\Lambda}_\infty)_{n,n} = \frac{1}{2}(\boldsymbol{\Lambda})_{n,n} - \sqrt{\frac{1}{4}(\boldsymbol{\Lambda})^2_{n,n} - 1} = e^{-\mathrm{acosh}\left[\frac{(\boldsymbol{\Lambda})_{n,n}}{2}\right]} \tag{4.54}$$

or also:

$$\boldsymbol{\Psi}_\infty = \frac{\mathbf{K}}{2} - \sqrt{\frac{\mathbf{K}^2}{4} - \mathbf{I}} = e^{-\mathrm{acosh}\left(\frac{\mathbf{K}}{2}\right)} \tag{4.55}$$

The Riccati equation in Eq. (4.47) together with the initial condition in Eq. (4.45) may now be solved to yield Eq. (4.56):

$$\boldsymbol{\Psi}(\mathrm{k}) = \left(\boldsymbol{\Psi}^{\mathrm{k}}_\infty + \boldsymbol{\Psi}^{-(\mathrm{k}+1)}_\infty\right)^{-1}\left(\boldsymbol{\Psi}^{\mathrm{k}-1}_\infty + \boldsymbol{\Psi}^{-\mathrm{k}}_\infty\right) \tag{4.56}$$

The weight factor matrix $\boldsymbol{\Omega}(\mathrm{k})$ that relates the concentrations in $\mathbf{C}(k)$ directly to those in $\mathbf{C}(N_x + 1)$ is then given by Eq. (4.57):

$$\boldsymbol{\Omega}(\mathrm{k}) = \boldsymbol{\Psi}(\mathrm{k})\boldsymbol{\Psi}(\mathrm{k}+1)\ldots\boldsymbol{\Psi}(\mathrm{N_x}) = \prod_{\mathrm{i}=\mathrm{k}}^{\mathrm{N_x}} \boldsymbol{\Psi}_{\mathrm{x}}(\mathrm{i}) \tag{4.57}$$

Inserting Eq. (4.56) into Eq. (4.57) and rearranging finally yields:

$$\boldsymbol{\Omega}(\mathrm{i}) = \frac{\sinh(\xi\boldsymbol{\Phi})}{\sinh(\boldsymbol{\Phi})} \text{ with } \xi = \frac{i}{N_x + 1} \tag{4.58}$$

ξ is the dimensionless position along the x-axis of the nodes in layer i.

While the $\boldsymbol{\Phi}$ matrix is given by Eq. (4.59):

$$\boldsymbol{\Phi} = (\mathrm{N_x} + 1)\ln\left(\boldsymbol{\Psi}^{-1}_\infty\right) = (\mathrm{N_x} + 1)\ln\left(\frac{\mathbf{K}}{2} + \sqrt{\frac{\mathbf{K}^2}{4} - 1}\right) \tag{4.59}$$

or by Eq. (4.60):

$$\mathbf{\Phi} = (\mathrm{N_x} + 1)\mathrm{acosh}\left(\frac{\mathbf{K}}{2}\right) \tag{4.60}$$

I now obtain the solution for the concentration in the nodes from Eq. (4.61):

$$\mathbf{C}(i) = \frac{\sinh\,(\xi\mathbf{\Phi})}{\sinh\,(\mathbf{\Phi})}\mathbf{C}(N_\mathrm{x} + 1) \tag{4.61}$$

Formally, the solution to the concentration in the nodes of the network is the same as is given in the classic continuum approach or also in the case of a string of nodes. The solution can be repeated easily if the nodes at $i = 0$ are now considered the independent variables (and the concentrations in the nodes at $N_x + 1$ are set equal to zero), which leads to Eq. (4.62):

$$\mathbf{C}(i) = \frac{\sinh\,((1-\xi))\mathbf{\Phi})}{\sinh\,(\mathbf{\Phi})}\mathbf{C}(0) \tag{4.62}$$

The same reasoning can be applied for the y-direction, and all solutions can then simply be added up to yield the final concentration in each node.

4.3.3.3.2 A Network Open on Opposing Sides Only

Envision a network as shown in Figure 4.3. Material balances in the nodes for constant i lead to the following tri-diagonal matrix $\mathbf{K}'$ shown in Eq. (4.63):

$$\mathbf{K}' = \frac{1}{\nu_\mathrm{x}}\begin{vmatrix} \kappa-\nu_\mathrm{y} & -\nu_\mathrm{y} & & & \\ -\nu_\mathrm{y} & \kappa & -\nu_\mathrm{y} & & \\ & \cdot & \cdot & \cdot & \\ & & -\nu_\mathrm{y} & \kappa & -\nu_\mathrm{y} \\ & & & -\nu_\mathrm{y} & \kappa-\nu_\mathrm{y} \end{vmatrix} \tag{4.63}$$

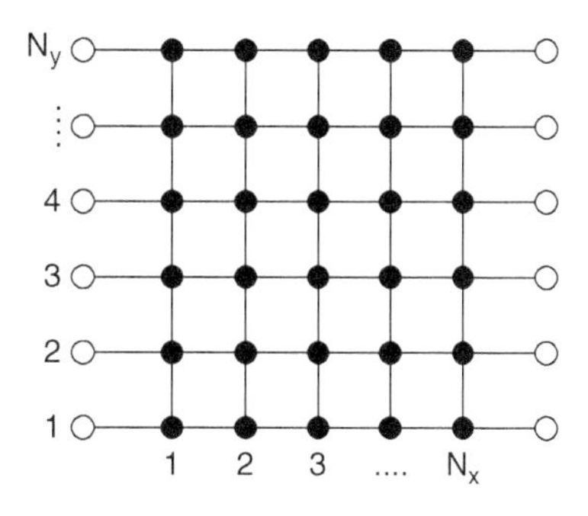

Figure 4.3 Node network open on opposing sides only.

The classic canonical form of $\mathbf{K}'$ is well known and is shown in Eq. (4.64):

$$\mathbf{K}' = \mathbf{U}'\mathbf{\Lambda}'\mathbf{U}'^{T} \tag{4.64}$$

where $\mathbf{\Lambda}'\,(\mathrm{N_y} \times \mathrm{N_y})$ is the diagonal matrix of eigenvalues of $\mathbf{K}'$ while $\mathbf{U}'\,(N_y \times N_y)$ is the modal matrix of corresponding eigenvectors given in Eqs. (4.65)–(4.67):

$$(\mathbf{\Lambda}')_{\mathrm{n,n}} = \frac{1}{\nu_{\mathrm{x}}}\left[\kappa - 2\nu_{\mathrm{y}}\cos\left(\theta_{\mathrm{n}}\right)\right] \text{ with } n = 1, 2, \ldots, N_y \tag{4.65}$$

$$(\mathbf{U}')_{\mathrm{m,n}} = \sqrt{\frac{1}{\mathrm{N_y}}} \text{ for } n = 1; m = 1, 2, \ldots, N_y \tag{4.66}$$

$$(\mathbf{U}')_{\mathrm{m,n}} = \left(\sqrt{\frac{2}{\mathrm{N_y}}}\right)\cos\left[\left(\mathrm{m} - \frac{1}{2}\right)\theta_{\mathrm{n}}\right] \text{ for } n > 1; m = 1, 2, \ldots, N_y \tag{4.67}$$

with:

$$\theta_{\mathrm{n}} = \frac{(\mathrm{n} - 1)\pi}{\mathrm{N_y}} \tag{4.68}$$

Again, the eigenvectors are independent of the rate constant. The solution to the concentration in the internal nodes is then as per Eq. (4.69):

$$\mathbf{C}(i) = \frac{\sinh\left((1 - \xi)\right)\mathbf{\Phi}')}{\sinh\left(\mathbf{\Phi}'\right)}\mathbf{C}(0) + \frac{\sinh\left(\xi\mathbf{\Phi}'\right)}{\sinh\left(\mathbf{\Phi}'\right)}\mathbf{C}(N_x + 1) \tag{4.69}$$

with:

$$\mathbf{\Phi}' = (\mathrm{N_x} + 1)\,\mathrm{acosh}\left(\frac{\mathbf{K}'}{2}\right) \tag{4.70}$$

and:

$$\xi = \frac{i}{N_x + 1} \tag{4.71}$$

4.3.3.3.3 A Network Open on One Side Only

Assume that only the external nodes located along $i = N_x + 1$ are open, as shown in Figure 4.4. With a matrix $\mathbf{K}'$ identical to the one in Section 4.3.3.3.2, the material balances can be written as per Eq. (4.72):

$$-\mathbf{C}(k - 1) + \mathbf{K}'\mathbf{C}(k) - \mathbf{C}(k + 1) = \mathbf{0} \text{ for } k = 2, \ldots, N_x \tag{4.72}$$

and since the nodes are considered blocked to the left of layer $k = 1$, I get:

$$(\mathbf{K}' - \mathbf{I})\mathbf{C}(1) - \mathbf{C}(2) = \mathbf{0} \tag{4.73}$$

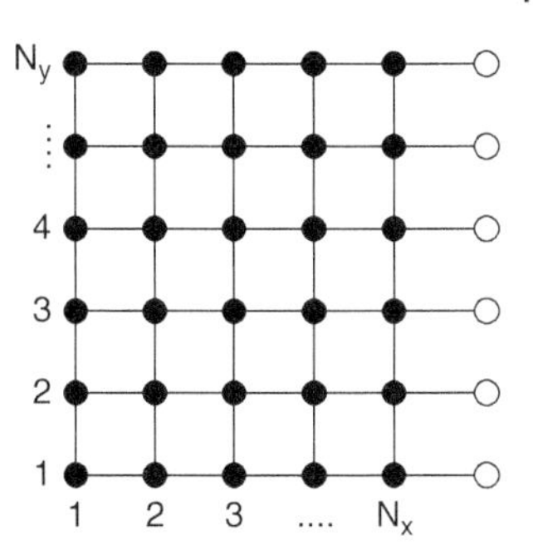

Figure 4.4 Node network open on one side only.

Following a very similar reasoning as earlier, it can be shown that the exact solution to the node concentrations is given by Eq. (4.74):

$$\mathbf{C}(i) = \frac{\cosh(\xi'\mathbf{\Phi}')}{\cosh(\mathbf{\Phi}')} C(\mathbf{N}_{\mathrm{x}} + 1) \tag{4.74}$$

with:

$$\mathbf{\Phi}' = \left(\mathrm{N}_{\mathrm{x}} + \frac{1}{2}\right) \mathrm{acosh}\left(\frac{\mathbf{K}'}{2}\right) \tag{4.75}$$

and:

$$\xi' = \frac{i - 1/2}{N_x + 1/2} \tag{4.76}$$

Clearly, ξ' is the dimensionless position along the x-axis of the nodes in layer i. The minimum eigenvalue of $\mathbf{\Phi}'$ is given by Eq. (4.77):

$$\lambda_{\min}(\mathbf{\Phi}') = (\mathrm{N}_x + 1/2)\,\mathrm{acosh}\left((\mathrm{k}^+ + 2\nu_x)/2\nu_x\right) \tag{4.77}$$

Notice that the minimum eigenvalue is independent of the mass transfer properties along the y-axis, so its value would be the same as for a single string of nodes parallel to the x-axis. One can indeed look at this as a system of stirred tank reactors in series. The exact mass transfer flux into the pore network may then be expressed in Eq. (4.78) as:

$$\mathbf{F}(N_x + 1) = \nu_x \left[\mathbf{I} - \frac{\cosh\left(\frac{N_x - 1/2}{N_x + 1/2}\mathbf{\Phi}'\right)}{\cosh(\mathbf{\Phi}')} \right] \mathbf{C}(N_x + 1) = \mathbf{HC}(N_x + 1) \tag{4.78}$$

with:

$$\mathbf{H} = \nu_{\mathrm{x}} \left[\mathbf{I} - \frac{\cosh\left(\frac{N_x - 1/2}{N_x + 1/2}\mathbf{\Phi}'\right)}{\cosh(\mathbf{\Phi}')} \right] \tag{4.79}$$

and:

$$\lambda_{\min}(\mathbf{H}) = \nu_{\mathrm{x}} \left[1 - \frac{\cosh\left(\frac{N_x - 1/2}{N_x + 1/2}\lambda_{\min}(\boldsymbol{\Phi}')\right)}{\cosh(\lambda_{\min}(\boldsymbol{\Phi}'))} \right] \tag{4.80}$$

When all of the external nodes are kept at the same concentration C_e, then the total rate of reaction R in the network is also the sum of all of the fluxes going into the network. This is obtained by summing all of the elements of the matrix $\mathbf{H}$ and can be written as per Eq. (4.81):

$$R = N_y \lambda_{\min}(\mathbf{H}) C_e \tag{4.81}$$

where $\lambda_{\min}$ is the smallest eigenvalue of the matrix $\mathbf{H}$. I will later show that the expression in Eq. (4.81) is very general, and will also come back for a specific carving of the parameter space in irregular networks (i.e. for real catalyst particles and the pore spaces therein).

The effectiveness factor for the network may now be obtained from Eq. (4.82):

$$\eta = \frac{R}{N_x N_y k^+ C_e} = \frac{\lambda_{\min}(\mathbf{H})}{N_x k^+} \tag{4.82}$$

Comparing Eqs. (4.5) and (4.74), it is observed that the vectors of concentrations can be functionally (mathematically) expressed in exactly the same way as for the classic continuum case. On the other hand, for the flux vector at the external nodes and the effectiveness factor, there are differences from the classic solution.

As in the case of a string of nodes, consider now a carving of the full three-dimensional parameter space, namely the reaction rate, the mass transfer rate, and the network size. Let the chemical reaction have a small rate constant compared to the mass transfer coefficient and consider networks with a large value of N_x (i.e. a deep network). The following approximations can be applied as per Eqs. (4.83)–(4.85):

$$\frac{N_x - 1/2}{N_x + 1/2} \cong 1 - \frac{1}{N_x} \tag{4.83}$$

and:

$$e^{\pm \frac{\boldsymbol{\Phi}'}{N_x}} \cong \mathbf{I} \pm \frac{\boldsymbol{\Phi}'}{N_x} \tag{4.84}$$

also:

$$\mathbf{I} - \frac{\cosh\left(\frac{N_x - 1/2}{N_x + 1/2}\boldsymbol{\Phi}'\right)}{\cosh(\boldsymbol{\Phi}')} \cong \boldsymbol{\Phi}' \tanh(\boldsymbol{\Phi}')/N_{\mathrm{x}} \tag{4.85}$$

Applying the approximations leads to Eqs. (4.86)–(4.91):

$$\mathbf{\Phi}' = \left(N_x + \frac{1}{2}\right)\sqrt{\frac{\mathbf{K}'}{2}} \tag{4.86}$$

$$\lambda_{\min}(\mathbf{\Phi}') \cong (N_x + 1/2)\sqrt{\frac{k^+}{\nu_x}} \tag{4.87}$$

Hence the minimum eigenvalue of the matrix $\mathbf{\Phi}'$ plays the role of the classic Thiele modulus. Also:

$$\mathbf{F}(N_x + 1) \cong \left(\frac{\nu_x}{N_x}\right)\mathbf{\Phi}' \tanh(\mathbf{\Phi}')\mathbf{C}(N_x + 1) = \mathbf{HC}(N_x + 1) \tag{4.88}$$

$$\mathbf{H} = \left(\frac{\nu_x}{N_x}\right)\mathbf{\Phi}' \tanh(\mathbf{\Phi}') \tag{4.89}$$

$$\lambda_{\min}(\mathbf{H}) = \left(\frac{\nu_x}{N_x}\right)\lambda_{\min}(\mathbf{\Phi}') \tanh(\lambda_{\min}(\mathbf{\Phi}')) \tag{4.90}$$

$$\eta \cong \frac{\tanh(\lambda_{\min}(\mathbf{\Phi}'))}{\lambda_{\min}(\mathbf{\Phi}')} \tag{4.91}$$

Now all of the classic expressions are quickly recognized. Hence, the classic solutions and the network approach merge together only for a specific carving of the parameter space (i.e. merging is valid for a deep network where a reaction acts as a mere perturbation of the diffusion field). Strong mass transfer limitations are obtained when considering very deep networks (VDNs) to offset the kinetic perturbation on the diffusion field.

4.3.3.3.4 Semi-infinite Networks

Consider a semi-infinite pore network located to the left of the y-axis with all of the external nodes located along the y-axis. Assume further that a concentration C_e is applied at the external node located in the origin while all other external node concentrations along the y-axis are assumed zero. By reworking the matrix solution for $\mathbf{\Psi}_\infty$ given in Eq. (4.48), it can be shown that the individual weight factors $\Psi_\infty(j)$ may be expressed as per Eq. (4.92):

$$\Psi_\infty(j) = \Psi_\infty(-j) = \frac{1}{\pi}\int_0^\pi \cos(jx)g(x)dx \text{ for } j \geq 0 \tag{4.92}$$

with:

$$g(x) = e^{-a\cosh[\alpha - \beta\cos(x)]} \tag{4.93}$$

$$\alpha = \frac{\kappa}{2\nu_x} \tag{4.94}$$

$$\beta = \frac{\nu_y}{\nu_x} \tag{4.95}$$

It is understood that the inverse of the hyperbolic cosine function, named $a\cosh$ in Eq. (4.93), yields the principal value of the function. It is recognized that the weight factors $\Psi_\infty(j)$ are the coefficients in the Fourier cosine series of $g(x)$. After some manipulations, it can be shown that Eq. (4.92) leads to Eq. (4.96):

$$\sum_{j=-\infty}^{j=+\infty} \psi_\infty(j) = e^{-a\cosh(\alpha-\beta)} = \begin{cases} 1 & \text{without reaction} \\ <1 & \text{with reaction} \end{cases} \tag{4.96}$$

The concentrations in the subsurface nodes ($i = -1$) may now be calculated from Eq. (4.97):

$$C_{-1,j} = C_e \Psi_\infty(j) \tag{4.97}$$

The concentrations in nodes located deeper in the network are found recursively from Eq. (4.98):

$$C_{i,j} = C_e \sum_{k=-\infty}^{k=+\infty} C_{i+1,k} \Psi_\infty(k-j) \tag{4.98}$$

The mass transfer fluxes at the network perimeter are obtained from Eqs. (4.99) and (4.100):

$$F_0 = \nu_x[1 - \Psi_\infty(0)]C_e \tag{4.99}$$

$$F_j = F_{-j} = -\nu_x \Psi_\infty(j) C_e \text{ for } j > 0 \tag{4.100}$$

The total amount $\mathfrak{R}_\infty$ of component A, which reacts in the semi-infinite network per unit time, is then obtained from Eq. (4.101):

$$\mathfrak{R}_\infty = F_0 + 2\sum_{j=1}^{j=\infty} F_j = \nu_x\left[1 - e^{-a\cosh(\alpha-\beta)}\right] C_e \tag{4.101}$$

$\mathfrak{R}_\infty$ may also be obtained by adding the contributions from all of the nodes and realizing that:

$$\sum_{j=-\infty}^{j=+\infty} C_{i,j} = e^{-a\cosh(\alpha-\beta)} \sum_{j=-\infty}^{j=+\infty} C_{i+1,j} \tag{4.102}$$

With reaction limited to the nodes only, I obtain Eq. (4.103):

$$\mathfrak{R}_\infty = \frac{k^+}{e^{a\cosh\left(1+\frac{k^+}{2\nu_x}\right)} - 1} C_e \tag{4.103}$$

A remarkable result then follows from Eq. (4.103) that $\mathfrak{R}_\infty$ is independent of the mass transport characteristics of the pores that run parallel to the y-axis. One can therefore choose $\nu_y = 0$, and a single infinitely long pore may now replace the network as far as the rate of disappearance of component A is concerned. For most practical applications, the rate constant k^+ in a node will be much smaller than the mass transfer coefficient ν_x (except possibly for cases with severe configurational constraints), and in that case one obtains the further simplification as per Eq. (4.104):

$$\mathfrak{R}_\infty \cong \sqrt{k^+ \nu_x} C_e \text{ for } \frac{k^+}{\nu_x} \ll 1 \tag{4.104}$$

For the case where the network is kinetically inert, the following may be shown in Eq. (4.105):

$$\psi_\infty(0) = \frac{2}{\pi}\left[(\beta + 1)\,\text{atan}\left(\frac{1}{\sqrt{\beta}}\right) - \sqrt{\beta}\right] \tag{4.105}$$

while the terms for $j \neq 0$ should be evaluated from Eq. (4.92).

4.3.3.4 Three-dimensional Regular Networks

4.3.3.4.1 Rectangular Parallelepiped

Consider a three-dimensional pore network in the form of a rectangular parallelepiped. The position of each internal node in the pore network is defined by the integer triplet (i, j, l) along the (x, y, z) axes, respectively, with their ranges given by Eq. (4.106):

$$i = 1, 2, \ldots, N_x \quad ; \ j = 1, 2, \ldots, N_y \quad ; \ l = 1, 2, \ldots, N_z \tag{4.106}$$

for a total of $N_x \times N_y \times N_z$ internal nodes. The positions of the external nodes along each of the six faces of the pore network are defined similarly; for instance, the positions for the external nodes in the $y = 0$ plane are defined with the integer triplet $(i, 0, l)$ with ranges:

$$i = 1, 2, \ldots, N_x \quad ; \ l = 1, 2, \ldots, N_z \tag{4.107}$$

The concentration of component A in an internal node will be noted as $C_{i,j,l}$. It is assumed that the diffusive properties of the pores that run parallel to the x-axis are identical. Similar statements hold for pores that run parallel to the y- and z-axes. It is further assumed that all nodes are identical in chemical reactivity and that each node acts as a continuous stirred tank reactor. To fix our thoughts, we will first search for how to express the concentrations in the internal nodes located in a plane perpendicular to the x-axis at position $i = k$ and characterized by:

$$i = k \quad ; \ j = 1, 2, \ldots, N_y \quad ; \ l = 1, 2, \ldots, N_z \tag{4.108}$$

as a function of the external node concentrations located at position in Eq. (4.109):

$$i = N_x + 1 \quad ; \quad j = 1, 2, ..., N_y \quad ; \quad l = 1, 2, ..., N_z \tag{4.109}$$

For this, it will be assumed that all of the other external node concentrations are set equal to zero.

The concentrations in the internal nodes in the plane at location $i = k$ are written in the form of a matrix whose elements are arranged according to the coordinate position of each node in that plane per Eq. (4.110):

$$\mathbf{C}(k) = \begin{vmatrix} C_{k,N_y,1} & C_{k,N_y,2} & . & C_{k,N_y,N_z} \\ . & . & . & . \\ C_{k,2,1} & C_{k,2,2} & . & C_{k,2,N_z} \\ C_{k,1,1} & C_{k,1,2} & . & C_{k,1,N_z} \end{vmatrix} \tag{4.110}$$

The matrix $\mathbf{C}$(k) will be vectorized and is noted as $vec\mathbf{C}(k)$ by moving column 2 of $\mathbf{C}(k)$ below column 1, then moving column 3 of $\mathbf{C}(k)$ below columns 1 and 2, etc. Define the matrix $\mathbf{\Psi}(k)$ with dimensions $(N_yN_z \times N_yN_z)$ as the weight factor matrix that relates the concentrations for the internal nodes in the plane at location i to those in the plane $i = k + 1$ for a network that starts at $i = 0$ and ends at $i = k + 1$ with the concentrations $C_{k+1,j,l}$ arbitrary. I then get Eq. (4.111) with $\mathbf{\Psi}(k)$ to be determined:

$$vec\mathbf{C}(k) = \mathbf{\Psi}(k) vec\mathbf{C}(k + 1) \tag{4.111}$$

I now introduce the tri-diagonal block matrix $\mathbf{K}$ with dimensions $(N_yN_z \times N_yN_z)$. Each block is of dimensions $(N_y \times N_y)$ and there are N_z blocks along each row and column of the matrix $\mathbf{K}$ as in Eq. (4.112).

$$\mathbf{K} = \frac{1}{\nu_x} \begin{vmatrix} \mathbf{L} & -\nu_z\mathbf{I}_y & & & \\ -\nu_z\mathbf{I}_y & \mathbf{L} & -\nu_z\mathbf{I}_y & & \\ & . & . & . & \\ & & -\nu_z\mathbf{I}_y & \mathbf{L} & -\nu_z\mathbf{I}_y \\ & & & -\nu_z\mathbf{I}_y & \mathbf{L} \end{vmatrix} \tag{4.112}$$

The matrix $\mathbf{I}_y(N_y \times N_y)$ is a unit matrix while the matrix $\mathbf{L}(N_y \times N_y)$ is tri-diagonal as per Eq. (4.113):

$$\mathbf{L} = \begin{vmatrix} \kappa & -\nu_y & & & \\ -\nu_y & \kappa & -\nu_y & & \\ & . & . & . & \\ & & -\nu_y & \kappa & -\nu_y \\ & & & -\nu_y & \kappa \end{vmatrix} \tag{4.113}$$

The value of κ is given by Eq. (4.114):

$$\kappa = k^{+} + 2(\nu_x + \nu_y + \nu_z) \tag{4.114}$$

Be aware that in the material balance equations any of the nodes located on the periphery are defined with a zero concentration (except for those with $k = N_x + 1$). The material balances in the internal nodes with $i = k > 1$ are expressed in Eq. (4.115):

$$-vec\mathbf{C}(k-1) + \mathbf{K}vec\mathbf{C}(k) - vec\mathbf{C}(k+1) = \mathbf{0} \tag{4.115}$$

where **I** is a unit matrix with dimensions $(N_yN_z \times N_yN_z)$. Remember that $vec\mathbf{C}(0)$ is chosen as zero, hence the material balance in the internal nodes for whom $k = 1$ leads to Eq. (4.116):

$$\mathbf{K}vec\mathbf{C}(1) - vec\mathbf{C}(2) = 0 \tag{4.116}$$

The material balance Eq. (4.115) leads to the Riccati-type Eq. (4.117):

$$\mathbf{\Psi}(k-1)\mathbf{\Psi}(k) - \mathbf{K}\mathbf{\Psi}(k) + \mathbf{I} = \mathbf{0} \tag{4.117}$$

In the first layer, I obtain Eq. (4.118):

$$\mathbf{\Psi}(1) = \mathbf{K}^{-1} \tag{4.118}$$

The infinite grid (in the x-direction) is governed by Eq. (4.119):

$$\mathbf{\Psi}_{\infty}^{2} - \mathbf{K}\mathbf{\Psi}_{\infty} + \mathbf{I} = \mathbf{0} \tag{4.119}$$

Define a tri-diagonal $(N_z \times N_z)$ matrix **S** such that all elements are zero except for those on the next upper and next lower location from the main diagonal, for which the elements are put equal to $-\nu_z$. The matrix **K** may then be written as per Eq. (4.120):

$$\mathbf{K} = \frac{1}{\nu_x}\left(\mathbf{S}\otimes\mathbf{I}_y + \mathbf{I}_z\otimes\mathbf{L}\right) = \frac{1}{\nu_x}(\mathbf{S}\oplus\mathbf{L}) \tag{4.120}$$

The canonical representations of **S** and **L** are well known and are expressed in Eqs. (4.121)–(4.126):

$$\mathbf{S} = \mathbf{U}_S\mathbf{\Lambda}_S\mathbf{U}_S^T \tag{4.121}$$

$$\mathbf{L} = \mathbf{U}_L\mathbf{\Lambda}_L\mathbf{U}_L^T \tag{4.122}$$

with:

$$(\mathbf{U}_S)_{m,n} = \left(\sqrt{\frac{2}{N_z+1}}\right)\sin\left[\frac{m(N_z-n+1)\pi}{(N_z+1)}\right] \text{ for } m, n = 1, 2, \ldots, N_z \tag{4.123}$$

$$(\mathbf{\Lambda}_S)_{\mathrm{m,m}} = -2v_z \cos\left(\frac{m\pi}{N_z + 1}\right) \text{ for } m = 1, 2, ..., N_z \tag{4.124}$$

while:

$$(\mathbf{U}_L)_{\mathrm{m,n}} = \left(\sqrt{\frac{2}{N_y + 1}}\right) \sin\left[\frac{m(N_y - n + 1)\pi}{(N_y + 1)}\right] \text{ for } m, n = 1, 2, ..., N_y \tag{4.125}$$

$$(\mathbf{\Lambda}_L)_{\mathrm{m,m}} = \kappa - 2\sigma_y \cos\left(\frac{m\pi}{N_y + 1}\right) \text{ for } m = 1, 2, ..., N_z \tag{4.126}$$

The canonical representation of K is then obtained as per Eqs. (4.127)–(4.129):

$$\mathbf{K} = \mathbf{U}_K \mathbf{\Lambda}_K \mathbf{U}_K^T \tag{4.127}$$

with:

$$\mathbf{\Lambda}_K = \frac{1}{v_x}(\mathbf{\Lambda}_S \oplus \mathbf{\Lambda}_L) \tag{4.128}$$

and:

$$\mathbf{U}_K = \mathbf{U}_S \otimes \mathbf{U}_L \tag{4.129}$$

The reader may also opt to determine the eigenvalues and eigenvectors of **K** directly with the appropriate software package and bypass the somewhat tedious nature of the analytical solutions for these properties.

The canonical representation for $\mathbf{\Psi}_\infty$ is then found as per Eqs. (4.130)–(4.132):

$$\mathbf{\Psi}_\infty = \mathbf{U}_K \mathbf{\Lambda}_\infty \mathbf{U}_K^T \tag{4.130}$$

with:

$$(\mathbf{\Lambda}_\infty)_{\mathrm{m,m}} = \frac{1}{2}(\mathbf{\Lambda}_K)_{\mathrm{m,m}} - \sqrt{\frac{1}{4}(\mathbf{\Lambda}_K)_{\mathrm{m,m}}^2 - 1} \tag{4.131}$$

for:

$$m = 1, 2, ..., N_y N_z \tag{4.132}$$

The Riccati-type Eq. (4.117) together with the initial condition in Eq. (4.118) may now be solved to yield Eq. (4.133):

$$\mathbf{\Psi}(k) = \left(\mathbf{\Psi}_\infty^k + \mathbf{\Psi}_\infty^{-(k+1)}\right)^{-1} \left(\mathbf{\Psi}_\infty^{k-1} + \mathbf{\Psi}_\infty^{-k}\right) \tag{4.133}$$

The weight factor matrix $\mathbf{\Omega}(i)$ that relates the concentrations in $vec\mathbf{C}(i)$ to $vec\mathbf{C}$ $(N_x + 1)$ is then given by Eq. (4.134):

$$\boldsymbol{\Omega}(i) = \prod_{m=i}^{N_x} \boldsymbol{\Psi}(m) \tag{4.134}$$

Inserting Eq. (4.133) into Eq. (4.134) and rearranging finally yields:

$$\boldsymbol{\Omega}(i) = \frac{\sinh(\xi\boldsymbol{\Phi})}{\sinh(\boldsymbol{\Phi})} \text{ with } \xi = \frac{i}{N_x + 1} \tag{4.135}$$

with:

$$\boldsymbol{\Phi} = (N_x + 1)\ln\left(\boldsymbol{\Psi}_{\infty}^{-1}\right) = (N_x + 1)a\cosh(\mathbf{K}/2) \tag{4.136}$$

and the final solution for the concentrations is then given by Eq. (4.137):

$$vec\mathbf{C}(i) = \frac{\sinh(\xi\boldsymbol{\Phi})}{\sinh(\boldsymbol{\Phi})} vec\mathbf{C}(N_x + 1) \text{ with } \xi = \frac{i}{N_x + 1} \tag{4.137}$$

The solutions for the internal concentrations when the concentrations in the other planes are allowed in turn to be the independent variables are obtained in a similar way and can all be added up to yield the total concentration in a particular internal node.

4.3.3.4.2 The Catalytic Box

As a special case, consider a pore network open only along the plane perpendicular to the x-axis at position $N_x + 1$ (i.e. a catalytic box). It is assumed that all pores ending in the closed planes are closed off. In analogy with the two-dimensional case, I now introduce the matrix $\mathbf{K}'$ in Eq. (4.138):

$$\mathbf{K}' = \frac{1}{\nu_x} \begin{vmatrix} \mathbf{L}'' & -\nu_z\mathbf{I}_y & & & \\ -\nu_z\mathbf{I}_y & \mathbf{L}' & -\nu_z\mathbf{I}_y & & \\ & \cdot & \cdot & \cdot & \\ & & -\nu_z\mathbf{I}_y & \mathbf{L}' & -\nu_z\mathbf{I}_y \\ & & & -\nu_z\mathbf{I}_y & \mathbf{L}'' \end{vmatrix} \tag{4.138}$$

The matrix $\mathbf{I}_y(N_y \times N_y)$ is a unit matrix while the matrices $\mathbf{L}'(N_y \times N_y)$ and $\mathbf{L}''(N_y \times N_y)$ are tri-diagonal as per Eqs. (4.139) and (4.140):

$$\mathbf{L}' = \begin{vmatrix} (\kappa - \nu_y) & -\nu_y & & & \\ -\nu_y & \kappa & -\nu_y & & \\ & \cdot & \cdot & \cdot & \\ & & -\nu_y & \kappa & -\nu_y \\ & & & -\nu_y & (\kappa - \nu_y) \end{vmatrix} \tag{4.139}$$

$$\mathbf{L}'' = \mathbf{L}' - \nu_z\mathbf{I}_y \tag{4.140}$$

The value of κ is still given by Eq. (4.141):

$$\kappa = \mathrm{k}^{+} + 2(\nu_{\mathrm{x}} + \nu_{\mathrm{y}} + \nu_{\mathrm{z}}) \tag{4.141}$$

The material balances are written as per Eq. (4.142):

$$-\mathbf{C}(k-1) + \mathbf{K}'\mathbf{C}(k) - \mathbf{C}(k+1) = \mathbf{0} \text{ for } k = 2, \ldots, N_x \tag{4.142}$$

and since the nodes are considered blocked to the left of layer $k = 1$, I get:

$$(\mathbf{K}' - \mathbf{I})\mathbf{C}(1) - \mathbf{C}(2) = \mathbf{0} \text{ for } k = 1 \tag{4.143}$$

Following very similar reasoning as demonstrated earlier in Section 4.3.3.3.3, it can be shown that the exact solution for the node concentrations is given by:

$$vec\mathbf{C}(i) = \frac{\cosh(\xi\boldsymbol{\Phi}')}{\cosh(\boldsymbol{\Phi}')} vec\mathbf{C}(N_{\mathrm{x}} + 1) \tag{4.144}$$

with:

$$\xi = \frac{i - 1/2}{N_x + 1/2} \tag{4.145}$$

and

$$\boldsymbol{\Phi}' = \left(\mathrm{N}_{\mathrm{x}} + \frac{1}{2}\right) \operatorname{acosh}\left(\frac{\mathbf{K}'}{2}\right) \tag{4.146}$$

It remains to evaluate the mass transfer flux *into* the pore network from each external node. For this, define the matrix $\mathbf{F}(N_x + 1)$ structurally in the same way as $\mathbf{C}(N_x + 1)$, but with every element now standing for the mass transfer flux into the pore network from that external node. $\mathbf{F}(N_x + 1)$ may then be expressed as per Eq. (4.147):

$$vec\mathbf{F}(N_x + 1) = \mathbf{H}vec\mathbf{C}(N_x + 1) \tag{4.147}$$

with:

$$\mathbf{H} = \nu_{\mathrm{x}} \left[\mathbf{I} - \frac{\cosh\left(\dfrac{N_x - 1/2}{N_x + 1/2}\boldsymbol{\Phi}'\right)}{\cosh(\boldsymbol{\Phi}')} \right] vec\mathbf{C}(N_x + 1) \tag{4.148}$$

The effectiveness factor for the network may now be obtained from Eq. (4.149):

$$\eta = \frac{\lambda_{\min}(\mathbf{H})}{\mathrm{k}^{+} N_x} \tag{4.149}$$

where $\lambda_{\min}$ is the smallest eigenvalue of the matrix H.

4.3.3.4.3 Semi-infinite Three-dimensional Space

Consider a three-dimensional pore network located along the negative part of the x-axis with all of the external nodes located in the plane $i = 0$. Apply a concentration C_e at the origin while all other external nodes are kept at a concentration $C = 0$. The rate of disappearance of component A by reaction in the network may then be shown to be as per Eq. (4.150):

$$\mathfrak{R}_{\infty} = \frac{k^{+}}{e^{\operatorname{acosh}\left(1+\frac{k^{+}}{2v_x}\right)} - 1} C_e \tag{4.150}$$

The same conclusion as in the two-dimensional case follows that the rate is independent of the mass transfer characteristics in the y- or z-direction, hence the network can be replaced by a single string of nodes along the x-axis.

4.3.3.5 Worked Example: p-Xylene Selectivity Boost

4.3.3.5.1 Generalities

The precursor for polyester resin production is p-xylene. In the chemical industry, p-xylene can be produced either by the alkylation of toluene with methanol or by the disproportionation of toluene. Under kinetic control, a mixture of xylenes is produced in near equilibrium. Every mole of toluene reacted yields roughly 0.25 mol of o-xylene, 0.50 mol of m-xylene, and 0.25 mol of p-xylene. Because of the shape differences between the o-, m-, and p-xylene molecules, the diffusivities of these compounds, inside the catalyst can be drastically changed, and also by altering the ingress or egress of the catalyst. This scenario is elegantly covered by Wei [18] by treating the catalyst as a single pore where the diffusivities of the xylene isomers are drastically different, with a very high diffusivity for p-xylene compared to the two other xylenes. It is my intent here to cover a simplified version of the kinetics applied to a two-dimensional network. This network scenario was not covered by Wei, and perhaps the ease of application of the approach shows in the coming analysis. For simplicity, a network open on one side only as given in Section 4.3.3.3.3 will be treated.

The simplified reaction scheme adopted here is:

$$\text{Toluene} \rightarrow \text{Xylene (25\% p-xylene, 75\% l-xylene)} \tag{4.151}$$

where l-xylene stands for o-xylene and m-xylene lumped together in a single component. The disappearance of toluene and methanol is kinetically controlled because the rate of diffusion of toluene and methanol is very fast, leading to essentially no internal mass transfer limitations. Each node of the network is assumed to yield a constant total rate r of the production of xylenes. The rate of formation toward p-xylene is written as rS_k while for the lumped component representing o- and m-xylene the rate is written as $r(1 - S_k)$. The quantity S_k represents the fractional kinetic selectivity toward p-xylene.

Besides the formation of the xylenes at the active sites, there is also isomerization of the xylenes. The rate constant for isomerization of the lumped xylene to p-xylene is shown as:

$$l\text{-}xylene \xrightarrow{k} p\text{-}xylene \tag{4.152}$$

with an equilibrium constant K_{eq} given by:

$$K_{eq} = \frac{C_p^{eq}}{C_l^{eq}} \tag{4.153}$$

where C_l^{eq} and C_p^{eq} represent the equilibrium concentrations for l-xylene and p-xylene, respectively. The reverse isomerization rate constant is given by k/K_{eq}. According to Wei, the equilibrium concentration for l-xylene is approximately threefold higher than the equilibrium concentration for p-xylene.

When the catalyst is prepared such that p-xylene diffuses much faster than o- and m-xylene, Wei has already reported selectivation of the catalyst for p-xylene. Here, I will look at a similar case, but for a catalyst network instead of a single pore. I will also further investigate the effect of blocking off many of the pore openings on p-xylene selectivity for a simple network.

4.3.3.5.2 Numerical Solution

Now let me show a way to describe the formation and isomerization of xylenes in a network of nodes. First, consider a network as shown in Figure 4.5. The blue nodes are considered nodes where toluene and methanol react to form xylene according to the reaction in Eq. (4.151) and where isomerization is occurring according to the reaction in Eq. (4.152). The white nodes are those where a certain concentration C_p^e for p-xylene and C_l^e for l-xylene can be imposed arbitrarily. The superscript e stands for external because the white nodes are assumed to be in intimate contact with the external "bulk" fluid. No reaction or isomerization is considered in the white nodes. The blue nodes are considered not to be in contact with the bulk fluid. In order to calculate the steady-state flux vector in the white nodes that represents the molar flux of p-xylene and l-xylene going into (or out off) the network, consider the following material balances in the blue nodes:

For p-xylene:

$$\text{Production in node } j : rS_k \text{ a constant rate of production} \tag{4.154}$$

$$\text{Isomerization in node } j : kC_{l,j} - \left(k/K_{eq}\right)C_{p,j} \tag{4.155}$$

Diffusion to neighboring nodes:

$$j = 1 \quad -D_p\left(C_{p,1} - C_{p,2}\right) - D_p\left(C_{p,1} - C_{p,1}^e\right) \tag{4.156}$$

Figure 4.5 First-generation pore network representation.

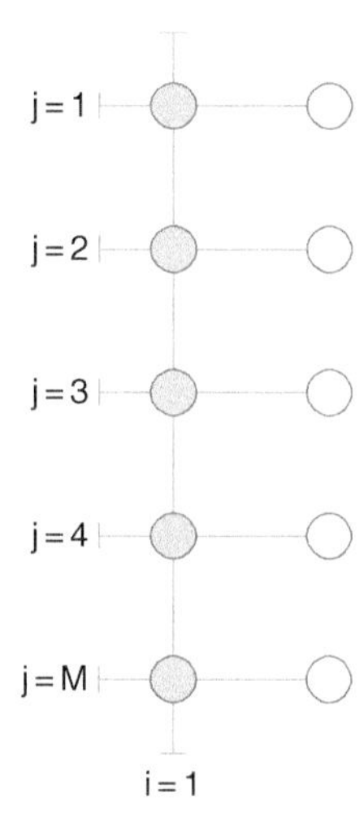

$$1 < j < M \quad -D_p\left(C_{p,j} - C_{p,j-1}\right) - D_p\left(C_{p,j} - C_{p,j+1}\right) - D_p\left(C_{p,j} - C^e_{p,j}\right) \tag{4.157}$$

$$j = M \quad -D_p\left(C_{p,M} - C_{p,M-1}\right) - D_p\left(C_{p,M} - C^e_{p,M}\right) \tag{4.158}$$

For l-xylene:

$$\text{Production in node } j: r(1 - S_k) \text{ a constant rate of production} \tag{4.159}$$

$$\text{Isomerization in node } j: \left(k/K_{eq}\right)C_{p,j} - kC_{l,j} \tag{4.160}$$

Diffusion to neighboring nodes:

$$j = 1 \quad -D_l(C_{l,1} - C_{l,2}) - D_l\left(C_{l,1} - C^e_{l,1}\right) \tag{4.161}$$

$$1 < j < M \quad -D_l\left(C_{l,j} - C_{l,j-1}\right) - D_l\left(C_{l,j} - C_{l,j+1}\right) - D_l\left(C_{l,j} - C^e_{l,j}\right) \tag{4.162}$$

$$j = M \quad -D_l(C_{l,M} - C_{l,M-1}) - D_l\left(C_{l,M} - C^e_{l,M}\right) \tag{4.163}$$

This can then conveniently be written in matrix form:

$$\text{For p-xylene}: \left[D_p\mathbf{B} + \left(k/K_{eq}\right)\mathbf{I}\right]\mathbf{C}_p - k\mathbf{C}_l = D_p\mathbf{C}^e_p + rS_k\mathbf{e} \tag{4.164}$$

$$\text{For l-xylene}: -\left(k/K_{eq}\right)\mathbf{C}_p + [D_l\mathbf{B} + k\mathbf{I}]\mathbf{C}_l = D_l\mathbf{C}^e_l + r(1 - S_k)\mathbf{e} \tag{4.165}$$

where $\mathbf{e}$ is a unit vector while $\mathbf{B}$ is a tri-diagonal matrix with diagonal elements equal to three, except for the very corners, where the elements are equal to two. The next upper and lower diagonals of $\mathbf{B}$ have elements that are all equal to negative unity (−1). The set of matrices in Eqs. (4.164) and (4.165) can then be written as in Eq. (4.166):

$$\begin{vmatrix} D_p\mathbf{B} + (k/K_{eq})\mathbf{I} & -k\mathbf{I} \\ -(k/K_{eq})\mathbf{I} & D_l\mathbf{B} + k\mathbf{I} \end{vmatrix} \begin{vmatrix} \mathbf{C}_p \\ \mathbf{C}_l \end{vmatrix} = \mathbf{D}_X \begin{vmatrix} \mathbf{C}_p \\ \mathbf{C}_l \end{vmatrix}^e + r\mathbf{S} \tag{4.166}$$

The vector **S** is the kinetic selectivity vector with its elements equal to S_k in the top half and equal to $(1 - S_k)$ in the bottom half. The matrix $\mathbf{D}_X$ is diagonal with elements equal to D_p in the top half and D_l in the bottom half of the diagonal.

Let:

$$\mathbf{A} = \begin{vmatrix} D_p\mathbf{B} + (k/K_{eq})\mathbf{I} & -k\mathbf{I} \\ -(k/K_{eq})\mathbf{I} & D_l\mathbf{B} + k\mathbf{I} \end{vmatrix} \tag{4.167}$$

The solution is then expressed in Eq. (4.168):

$$\begin{vmatrix} \mathbf{C}_p \\ \mathbf{C}_l \end{vmatrix} = \mathbf{A}^{-1}\mathbf{D}_X \begin{vmatrix} \mathbf{C}_p \\ \mathbf{C}_l \end{vmatrix}^e + r\mathbf{A}^{-1}\mathbf{S} \tag{4.168}$$

or:

$$\begin{vmatrix} \mathbf{C}_p \\ \mathbf{C}_l \end{vmatrix} = \mathbf{\Omega}_1^e \begin{vmatrix} \mathbf{C}_p \\ \mathbf{C}_l \end{vmatrix}^e + \mathbf{\Omega}_1^n \tag{4.169}$$

with the matrix:

$$\mathbf{\Omega}_1^e = \mathbf{A}^{-1}\mathbf{D}_X \tag{4.170}$$

and the vector:

$$\mathbf{\Omega}_1^n = r\mathbf{A}^{-1}\mathbf{S} \tag{4.171}$$

Equation (4.169) allows one to express the concentrations for both p-xylene and l-xylene as a linear function of the arbitrary external node concentrations $\mathbf{C}_p^e$ and $\mathbf{C}_l^e$. The flux vectors $\mathbf{F}_p^e$ and $\mathbf{F}_l^e$ of molar fluxes counted positive when entering each node from the bulk fluid are then written in vector form as:

$$\mathbf{F}_p^e = \mathbf{D}_p\left(\mathbf{C}_p^e - \mathbf{C}_p\right) \tag{4.172}$$

$$\mathbf{F}_l^e = \mathbf{D}_l\left(\mathbf{C}_l^e - \mathbf{C}_l\right) \tag{4.173}$$

or in matrix form as:

$$\begin{vmatrix} \mathbf{F}_p \\ \mathbf{F}_l \end{vmatrix}^e = \mathbf{D}_X\left(\mathbf{I} - \mathbf{\Omega}_1^e\right) \begin{vmatrix} \mathbf{C}_p \\ \mathbf{C}_l \end{vmatrix}^e - \mathbf{D}_X\mathbf{\Omega}_1^n \tag{4.174}$$

As in Wei [18], the production rates in every node are here considered constants. They do not necessarily have to be, and they could be selected arbitrarily as well by making the appropriate selection for k in each node of the material balance equations. For instance, they could be considered zero in the case of poisoning of the node or they could be selected according to some distribution. For the case of very low conversion of toluene to xylene (at the reactor entrance) where no p-xylene and l-xylene is present in the bulk fluid, the flux vectors simply become linear combinations of the production rates only. However, once we start to consider nodes that are located deeper in the structure, it is necessary to consider the full dependence as shown in Eq. (4.174), since there is no essential difference between going deeper in the reactor with p-xylene and l-xylene present in the bulk or solving a crystal for nodes that are located deeper in the crystal at the start of the reactor. Consider now a second column of nodes that is added to the previous one as shown in Figure 4.6. The material balances in each of the nodes in this second column ($i = 2$) is then written similarly as for the first column above, but the balances are now corrected for the fluxes going into the first column of nodes based on the p-xylene and l-xylene in those nodes and the respective production rates:

For p-xylene:

$$\text{Production in node} j : rS_k \tag{4.175}$$

$$\text{Isomerization in node} j : kC_{l,j} - \left(k/K_{eq}\right)C_{p,j} \tag{4.176}$$

Diffusion to neighboring nodes:

$$j = 1 \quad -D_p\left(C_{p,1} - C_{p,2}\right) - D_p\left(C_{p,1} - C^e_{p,1}\right) - F_{p,1} \tag{4.177}$$

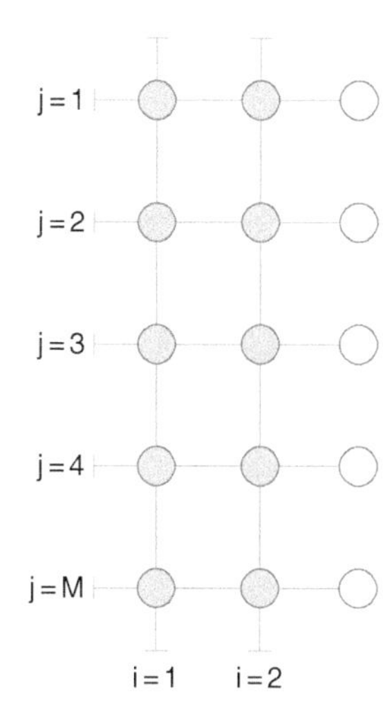

Figure 4.6 Second-generation pore network representation.

$$1 < j < M \quad -D_p(C_{p,j} - C_{p,j-1}) - D_p(C_{p,j} - C_{p,j+1}) - D_p\left(C_{p,j} - C^e_{p,j}\right) - F_{p,j} \tag{4.178}$$

$$j = M \quad -D_p(C_{p,M} - C_{p,M-1}) - D_p\left(C_{p,M} - C^e_{p,M}\right) - F_{p,M} \tag{4.179}$$

For l-xylene:

$$\text{Production in node } j : r(1 - S_k) \tag{4.180}$$

$$\text{Isomerization in node } j : \; -kC_{1,j} + (k/K_{eq})C_{p,j} \tag{4.181}$$

Diffusion to neighboring nodes:

$$j = 1 \quad -D_l(C_{1,1} - C_{l,2}) - D_l(C_{l,1} - C^e_{l,1}) - F_{l,1} \tag{4.182}$$

$$1 < j < M \quad -D_l(C_{l,j} - C_{l,j-1}) - D_l(C_{l,j} - C_{l,j+1}) - D_l\left(C_{l,j} - C^e_{l,j}\right) - F_{l,j} \tag{4.183}$$

$$j = M \quad -D_l(C_{l,M} - C_{l,M-1}) - D_p(C_{l,M} - C^e_{l,M}) - F_{l,M} \tag{4.184}$$

This can then conveniently be written in matrix form as:

$$\mathbf{A}\begin{vmatrix} \mathbf{C}_p \\ \mathbf{C}_l \end{vmatrix} = \mathbf{D}_X\begin{vmatrix} \mathbf{C}_p \\ \mathbf{C}_l \end{vmatrix}^e + r\mathbf{S} - \begin{vmatrix} \mathbf{F}_p \\ \mathbf{F}_l \end{vmatrix} \tag{4.185}$$

The flux vector here at the far right of Eq. (4.185) represents the flux from the nodes where the material balance is written for toward the nodes deeper into the network located to the left of them, as can be seen in Figure 4.6. This flux is formally represented by Eq. (4.174), but it is now to be written as per Eq. (4.186):

$$\begin{vmatrix} \mathbf{F}_p \\ \mathbf{F}_l \end{vmatrix} = \mathbf{D}_X(\mathbf{I} - \mathbf{\Omega}^e_1)\begin{vmatrix} \mathbf{C}_p \\ \mathbf{C}_l \end{vmatrix} - \mathbf{D}_X\mathbf{\Omega}^n_1 \tag{4.186}$$

Combining Eqs. (4.185) and (4.186) yields:

$$\mathbf{A}\begin{vmatrix} \mathbf{C}_p \\ \mathbf{C}_l \end{vmatrix} = \mathbf{D}_X\begin{vmatrix} \mathbf{C}_p \\ \mathbf{C}_l \end{vmatrix}^e + r\mathbf{S} - \mathbf{D}_X(\mathbf{I} - \mathbf{\Omega}^e_1)\begin{vmatrix} \mathbf{C}_p \\ \mathbf{C}_l \end{vmatrix} + \mathbf{D}_X\mathbf{\Omega}^n_1 \tag{4.187}$$

and is solved as per Eq. (4.188):

$$\begin{vmatrix} \mathbf{C}_p \\ \mathbf{C}_l \end{vmatrix} = (\mathbf{A} + \mathbf{D}_X(\mathbf{I} - \mathbf{\Omega}^e_1))^{-1}\left(\mathbf{D}_X\begin{vmatrix} \mathbf{C}_p \\ \mathbf{C}_l \end{vmatrix}^e + r\mathbf{S} + \mathbf{D}_X\mathbf{\Omega}^n_1\right) \tag{4.188}$$

or:

$$\begin{vmatrix} \mathbf{C}_p \\ \mathbf{C}_l \end{vmatrix} = \mathbf{\Omega}^e_2\begin{vmatrix} \mathbf{C}_p \\ \mathbf{C}_l \end{vmatrix}^e + \mathbf{\Omega}^n_2 \tag{4.189}$$

with:

$$\boldsymbol{\Omega}_2^e = \left(\mathbf{A} + \mathbf{D}_X\left(\mathbf{I} - \boldsymbol{\Omega}_1^e\right)\right)^{-1}\mathbf{D}_X \tag{4.190}$$

$$\boldsymbol{\Omega}_2^n = \left(\mathbf{A} + \mathbf{D}_X\left(\mathbf{A} - \boldsymbol{\Omega}_1^e\right)\right)^{-1}\left(r\mathbf{S} + \mathbf{D}_X\boldsymbol{\Omega}_1^n\right) \tag{4.191}$$

Equations (4.189)–(4.191) allow one to calculate the concentrations of p-xylene and l-xylene in the nodes located just below the periphery of the network (i.e. just to the left of the white nodes).

This set of two matrix equations can now be calculated over and over again for networks that extend deeper and deeper, and it also allows one to express the flux vectors for both p-xylene and l-xylene as a function of the external node concentrations, the production rates of the xylenes, and the isomerization rates in the nodes. The general iterative equations for the subsurface concentrations and the flux vectors from the external nodes into the network are then:

$$\begin{vmatrix} \mathbf{C}_p \\ \mathbf{C}_l \end{vmatrix}_N = \boldsymbol{\Omega}_N^e \begin{vmatrix} \mathbf{C}_p \\ \mathbf{C}_l \end{vmatrix}^e + \boldsymbol{\Omega}_N^n \tag{4.192}$$

$$\begin{vmatrix} \mathbf{F}_p \\ \mathbf{F}_l \end{vmatrix}^e = \mathbf{D}_X\left(\mathbf{I} - \boldsymbol{\Omega}_N^e\right)\begin{vmatrix} \mathbf{C}_p \\ \mathbf{C}_l \end{vmatrix}^e - \mathbf{D}_X\boldsymbol{\Omega}_N^n \tag{4.193}$$

$$\boldsymbol{\Omega}_N^e = \left(\mathbf{A} + \mathbf{D}_X\left(\mathbf{I} - \boldsymbol{\Omega}_{N-1}^e\right)\right)^{-1}\mathbf{D}_X \quad N = 2, 3, \ldots \tag{4.194}$$

$$\boldsymbol{\Omega}_N^n = \left(\mathbf{A} + \mathbf{D}_X\left(\mathbf{I} - \boldsymbol{\Omega}_{N-1}^e\right)\right)^{-1}\left(r\mathbf{S} + \mathbf{D}_X\boldsymbol{\Omega}_{N-1}^n\right) \quad N = 2, 3, \ldots \tag{4.195}$$

together with the starting values:

$$\boldsymbol{\Omega}_1^e = \mathbf{A}^{-1}\mathbf{D}_X \tag{4.196}$$

$$\boldsymbol{\Omega}_1^n = r\mathbf{A}^{-1}\mathbf{S} \tag{4.197}$$

The above relationships are convenient ways to perform the numerical calculations for any depth of the network.

4.3.3.5.3 An Analytical Solution

There is an analytical solution as well, and for this, start by writing the recurrence relation in Eq. (4.194) in the following way:

$$\boldsymbol{\Omega}_{N-1}^e\boldsymbol{\Omega}_N^e - \mathbf{K}\boldsymbol{\Omega}_N^e + \mathbf{I} = \mathbf{0} \quad N = 2, 3, \ldots \tag{4.198}$$

$$\boldsymbol{\Omega}_1^e = (\mathbf{K} - \mathbf{I})^{-1} \tag{4.199}$$

with:

$$\mathbf{K} = \mathbf{I} + \mathbf{D}_X{}^{-1}\mathbf{A} \tag{4.200}$$

There is an asymptotic solution available that would be reached when the network (crystal) extends infinitely deep as expressed in Eq. (4.201):

$$(\mathbf{\Omega}^e_\infty)^2 - \mathbf{K}\mathbf{\Omega}^e_\infty + \mathbf{I} = \mathbf{0} \tag{4.201}$$

After some manipulations along the same lines as shown in Section 4.3.3.3.3, one finds the general solution as:

$$\mathbf{\Omega}^e_N = \frac{\cosh(\xi_N \mathbf{\Phi}_N)}{\cosh(\mathbf{\Phi}_N)} \tag{4.202}$$

$$N = 1, 2, \ldots, \infty \tag{4.203}$$

with the dimensionless position ξ_N of the subsurface nodes given by:

$$\xi_N = \frac{N - 1/2}{N + 1/2} \tag{4.204}$$

and:

$$\mathbf{\Phi}_N = (N + 1/2)a\cosh(\mathbf{K}/2) \tag{4.205}$$

In addition, I remind the reader that what is sought here is the solution for the concentrations in the first set of nodes that are just below the surface nodes (i.e. the subsurface nodes for any depth of the network). Be aware that the matrix $\mathbf{K}$ is not symmetrical, hence its canonical decomposition leads to:

$$\mathbf{K} = \mathbf{U}\mathbf{\Lambda}\mathbf{U}^{-1} \tag{4.206}$$

The eigenvectors contained as columns in matrix $\mathbf{U}$ are not orthogonal. The eigenvalues are all real and are contained in the diagonal matrix $\mathbf{\Lambda}$. As is well known in the literature, the evaluation of the function of the matrix as needed in Eq. (4.205) is obtained from:

$$f(\mathbf{K}) = \mathbf{U}f(\mathbf{\Lambda})\mathbf{U}^{-1} \tag{4.207}$$

where the function f operates on each diagonal entry of $\mathbf{\Lambda}$.

The matrix $\mathbf{\Omega}^n_N$ remains to be evaluated. For this we consult Levy and Lessman [17] on general first-order finite-difference equations and try to make the extension to matrices, keeping in mind that they are nonsymmetrical. With the help of Eq. (4.202), for $\mathbf{\Omega}^e_N$ one gets:

$$\mathbf{\Omega}^n_N = \left(\mathbf{A} + \mathbf{D}_X\left(\mathbf{I} - \frac{\cosh(\xi_{N-1}\mathbf{\Phi}_{N-1})}{\cosh(\mathbf{\Phi}_{N-1})}\right)\right)^{-1}(r\mathbf{S} + \mathbf{D}_X\mathbf{\Omega}^n_{N-1}) \tag{4.208}$$

$$N = 2, 3, \ldots \infty \tag{4.209}$$

which leads to:

$$\left(\mathbf{K} - \frac{\cosh(\xi_{N-1}\mathbf{\Phi}_{N-1})}{\cosh(\mathbf{\Phi}_{N-1})}\right)\mathbf{\Omega}^n_N = r\mathbf{D}_X{}^{-1}\mathbf{S} + \mathbf{\Omega}^n_{N-1} \tag{4.210}$$

Define B_N as per Eq. (4.211):

$$\mathbf{B}_N = \mathbf{K} - \frac{\cosh(\xi_N \mathbf{\Phi}_N)}{\cosh(\mathbf{\Phi}_N)} \tag{4.211}$$

This then leads to:

$$\mathbf{B}_{N-1}\mathbf{\Omega}_N^n = r\mathbf{D}_X^{-1}\mathbf{S} + \mathbf{\Omega}_{N-1}^n \tag{4.212}$$

The solution to the homogeneous part of Eq. (4.212) is obtained by omitting the term $r\mathbf{D}_X^{-1}\mathbf{S}$ (note the order of multiplication) and leads to Eq. (4.213):

$$\mathbf{\Omega}_{N+1}^n = \left(\prod_{i=1}^{N} \mathbf{B}_i^{-1}\right)\mathbf{\Omega}_1^n = \mathbf{B}_N^{-1}\mathbf{B}_{N-1}^{-1}\ldots\mathbf{B}_1^{-1}\mathbf{\Omega}_1^n = (\mathbf{B}_1\mathbf{B}_2\ldots\mathbf{B}_N)^{-1}\mathbf{\Omega}_1^n = \mathbf{Z}_N^{-1}\mathbf{\Omega}_1^n \tag{4.213}$$

with:

$$\mathbf{Z}_N = \mathbf{B}_1\mathbf{B}_2\ldots\mathbf{B}_N \tag{4.214}$$

Left-multiplication of Eq. (4.212) with $\mathbf{Z}_{N-1}$ yields:

$$\mathbf{Z}_N\mathbf{\Omega}_{N+1}^n = \mathbf{X}_{N-1}r\mathbf{D}_X^{-1} + \mathbf{Z}_{N-1}\mathbf{\Omega}_N^n \tag{4.215}$$

Equation (4.215) leads to the following difference equation:

$$\mathbf{\Delta}\left(\mathbf{Z}_N\mathbf{\Omega}_{N+1}^n\right) = \mathbf{Z}_N\mathbf{\Omega}_{N+1}^n - \mathbf{Z}_{N-1}\mathbf{\Omega}_N^n = \mathbf{Z}_{N-1}r\mathbf{D}_X^{-1} \tag{4.216}$$

Apply Eq. (4.215) for N ranging from as shown down to the value of two:

$$\mathbf{B}_1\mathbf{\Omega}_2^n = \mathbf{rD_X}^{-1}\mathbf{S} + \mathbf{\Omega}_1^n \tag{4.217}$$

$$\mathbf{B}_1\mathbf{B}_2\mathbf{\Omega}_3^n = \mathbf{B}_1 r\mathbf{D}_X^{-1}\mathbf{S} + \mathbf{B}_1\mathbf{\Omega}_2^n \tag{4.218}$$

$$\mathbf{B}_1\mathbf{B}_2\mathbf{B}_3\mathbf{\Omega}_4^n = \mathbf{B}_1\mathbf{B}_2 r\mathbf{D}_X^{-1}\mathbf{S} + \mathbf{B}_1\mathbf{B}_2\mathbf{\Omega}_3^n \tag{4.219}$$

...

$$\mathbf{B}_1\mathbf{B}_2\mathbf{B}_3\ldots\mathbf{B}_{N-1}\mathbf{\Omega}_N^n = \mathbf{B}_1\mathbf{B}_2\ldots\mathbf{B}_{N-2}r\mathbf{D}_X^{-1}\mathbf{S} + \mathbf{B}_1\mathbf{B}_2\ldots\mathbf{B}_{N-2}\mathbf{\Omega}_{N-1}^n \tag{4.220}$$

Adding up all of the equations yields the following solution:

$$\mathbf{\Omega}_N^n = \mathbf{Z}_{N-1}^{-1}\left\{\mathbf{\Omega}_1^n + \left[\mathbf{I} + \left(\sum_{j=1}^{N-2}\mathbf{Z}_j\right)\right]r\mathbf{D}_X^{-1}\mathbf{S}\right\} \quad N = 3, \ldots \infty \tag{4.221}$$

together with the following starting values:

$$\mathbf{\Omega}_1^n = r\mathbf{A}^{-1}\mathbf{S} \tag{4.222}$$

$$\mathbf{\Omega}_2^n = \mathbf{B}_1^{-1}\left(r\mathbf{D}_X^{-1}\mathbf{S} + \mathbf{\Omega}_1^n\right) \tag{4.223}$$

This second starting value is only required due to the definition of the sum in the general solution of Eq. (4.221). Further algebraic manipulations lead to some simplifying properties:

$$\mathbf{B}_N = \frac{\cosh\left(\dfrac{N+3/2}{N+1/2}\mathbf{\Phi}_N\right)}{\cosh\left(\mathbf{\Phi}_N\right)} \tag{4.224}$$

and hence:

$$\mathbf{Z}_N = \frac{\cosh\left(\dfrac{(N+3/2)\mathbf{\Phi}_N}{(N+1/2)}\right)}{\cosh\left(\dfrac{3/2}{(N+1/2)}\mathbf{\Phi}_N\right)} \tag{4.225}$$

I was not able to find a more compact expression for $\sum_{j=1}^{N-2} \mathbf{Z}_j$ than the actual summation. However, since the summation only works on each of the eigenvalues, this is not a strong detriment.

4.3.3.5.4 *Selectivity for a Network Open on One Side*

The molar fluxes going into the network at the perimeter formed by the open nodes are given by Eq. (4.193). I will look here at the reactor entrance where no p-xylene and l-xylene is yet present in the bulk fluid. The iterative scheme shown in Eqs. (4.192)–(4.197) is applied and is calculated with the Mathcad software. Based on the results shown in Table 4.1, the picture that emerges for the selectivity toward p-xylene as a function of the isomerization rate constant shows interesting features. Without isomerization, the selectivity is the same as the kinetic selectivity, and the diffusive coefficients play no role. As the isomerization

Table 4.1 p-Xylene selectivity as a function of the isomerization rate constant *k*.

($r = 2$, $S_k = 0.25$, $K_{eq} = 1/3$, $D_p = 10$, $D_l = 1$, $N = 51$, $M = 51$; consistent defined units)	
k	S_{px}
0	0.250
10^{-6}	0.251
10^{-4}	0.303
10^{-2}	0.685
1	0.764
∞	0.769

rate increases, the selectivity increases for p-xylene since it is being formed, hence p-xylene is being favored. If the rate of isomerization is very fast or if there are enough sites available in the network (i.e. the network is deep enough to make isomerization substantial), then the network operates at or near equilibrium conditions. With the zero bulk concentrations for the p-xylene and l-xylene, I then get for the flux of p-xylene and l-xylene exiting the network:

$$F_p = MC_p^{eq} D_p \tag{4.226}$$

and:

$$F_l = MC_l^{eq} D_l \tag{4.227}$$

and the selectivity S_{pX} for p-xylene is given by Eq. (4.228):

$$S_{pX} = \frac{F_p}{F_p + F_l} \tag{4.228}$$

and reaches the limit value shown in Eq. (4.229):

$$\left(S_{pX}\right)_{\lim} = \frac{1}{1 + \dfrac{D_l}{K_{eq} D_p}} \tag{4.229}$$

Table 4.2 shows the limit value of p-xylene selectivity that can be reached as a function of the ratio of diffusivities for the xylenes.

4.3.3.5.5 *Selectivity for a Network Accessible via a Single Opening*

Left to consider now in Figure 4.7 is the solution to the special case where there is only one node left open to communicate to the bulk fluid while all of the other nodes on the periphery are blocked. The total number of nodes on the periphery is considered uneven, and it is the center node on the boundary that is considered open. Many other scenarios can be evaluated, but for this particular case the effort here is focused on the center node. In order to calculate the selectivity, I need the

Table 4.2 Limiting p-xylene selectivities for different ratios of p-xylene to l-xylene.

D_p/D_l	$(S_{pX})_{\lim}$
1	0.250
5	0.625
10	0.769
100	0.971
1000	0.997

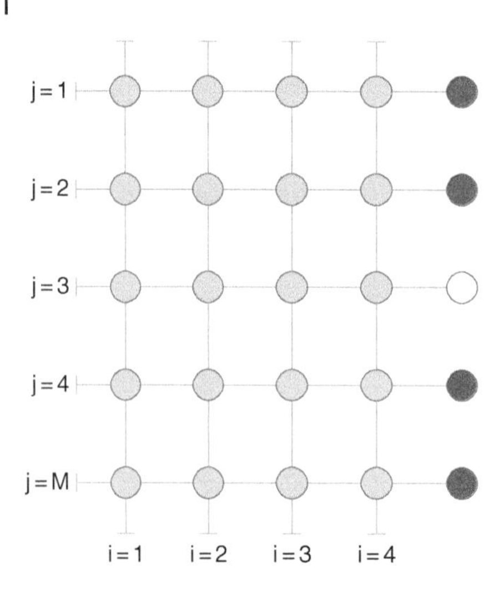

Figure 4.7 General picture for a network that is nearly fully blocked.

full expression of the flux vector, since it will be necessary to calculate the concentrations in all of the blocked nodes at the internal side of the blocked node (i.e. the network side). This allows one to impose zero flux into those blocked nodes. It will be assumed that all blocked nodes no longer have toluene to xylene activity nor isomerization activity. To solve the general equation in Eq. (4.193), I impose concentrations for p-xylene and l-xylene on the network side of the blocked periphery nodes that lead to zero flux from those nodes into the network. The conditional form of the matrix in Eq. (4.193) can then be written for the blocked nodes and the one open node as per Eq. (4.230):

$$\begin{vmatrix} \mathbf{F}_p \\ \mathbf{F}_l \end{vmatrix}^e = \begin{vmatrix} \mathbf{0} \\ f_p \\ \mathbf{0} \\ \mathbf{0} \\ f_l \\ \mathbf{0} \end{vmatrix}^e = \boldsymbol{D}_X\left(\boldsymbol{I} - \boldsymbol{\Omega}_N^e\right) \begin{vmatrix} \mathbf{c}_p \\ \mathbf{0} \\ \mathbf{c}'_p \\ \mathbf{c}_l \\ \mathbf{0} \\ \mathbf{c}'_l \end{vmatrix}^e - \mathbf{D}_X \boldsymbol{\Omega}_N^n \tag{4.230}$$

Equation (4.230) yields six matrix equations, four to solve for the unknown concentration vectors $\mathbf{c}_p^e$, $\mathbf{c}_p'^e$, $\mathbf{c}_l^e$, and $\mathbf{c}_l'^e$ in the blocked nodes and two to solve for the fluxes f_p and f_l exiting the network through the center node on the periphery. The four concentration vectors contain the concentrations of the respective components on the inside of the network. Because of symmetry, the vector of unknown concentrations $\mathbf{c}'_p$ can be expressed as:

$$\mathbf{c}'_p = \mathbf{I}'\mathbf{c}_p \tag{4.231}$$

where the matrix $\mathbf{I}'$ is a matrix with unity only on the diagonal running bottom-left to top-right and all other elements are equal to zero. A similar statement holds for $\mathbf{c}'_l$. Hence, the four vectors of unknown concentrations are now reduced to only two. In order not to make things cumbersome, one can apply a permutation matrix $\mathbf{P}$ to switch just the two equations for the fluxes from the open node to the bottom, and that leads to Eq. (4.232) (remember and apply that $\mathbf{P}^{-1} = \mathbf{P}$):

$$\mathbf{P}\begin{vmatrix} \mathbf{F}_p \\ \mathbf{F}_l \end{vmatrix}^e = \mathbf{P}\begin{vmatrix} \mathbf{0} \\ f_p \\ \mathbf{0} \\ \mathbf{0} \\ f_l \\ \mathbf{0} \end{vmatrix}^e = \begin{vmatrix} \mathbf{0} \\ \mathbf{0} \\ \mathbf{0} \\ \mathbf{0} \\ f_p \\ f_l \end{vmatrix}^e = \mathbf{PD}_X\left(\mathbf{I} - \mathbf{\Omega}_N^e\right)\mathbf{P}\begin{vmatrix} \mathbf{c}_p \\ \mathbf{c}'_p \\ \mathbf{c}_l \\ \mathbf{c}'_l \\ \mathbf{0} \\ \mathbf{0} \end{vmatrix}^e - \mathbf{PD}_X\mathbf{\Omega}_N^n \tag{4.232}$$

Equation (4.232) can now be conveniently partitioned and solved for the unknown concentrations and fluxes via the Schur complement.

$$\mathbf{P}\begin{vmatrix} \mathbf{0} \\ f_p \\ \mathbf{0} \\ \mathbf{0} \\ f_l \\ \mathbf{0} \end{vmatrix}^e = \begin{vmatrix} \mathbf{0} \\ \mathbf{0} \\ \mathbf{0} \\ \mathbf{0} \\ f_p \\ f_l \end{vmatrix}^e = \mathbf{PD}_X\left(\mathbf{I} - \mathbf{\Omega}_N^e\right)\mathbf{P}\begin{vmatrix} \mathbf{c}_p \\ \mathbf{c}'_p \\ \mathbf{c}_l \\ \mathbf{c}'_l \\ \mathbf{0} \\ \mathbf{0} \end{vmatrix}^e - \mathbf{PD}_X\mathbf{\Omega}_N^n = \begin{vmatrix} \mathbf{G}_1 & \mathbf{G}_2 \\ \mathbf{G}_3 & \mathbf{G}_4 \end{vmatrix}\begin{vmatrix} \mathbf{c}_p \\ \mathbf{c}'_p \\ \mathbf{c}_l \\ \mathbf{c}'_l \\ \mathbf{0} \\ \mathbf{0} \end{vmatrix}^e - \begin{vmatrix} \mathbf{Q}_1 \\ \mathbf{Q}_2 \end{vmatrix} \tag{4.233}$$

$$\begin{vmatrix} \mathbf{c}_p \\ \mathbf{c}'_p \\ \mathbf{c}_l \\ \mathbf{c}'_l \end{vmatrix} = \mathbf{G}_1^{-1}\mathbf{Q}_1 \tag{4.234}$$

$$\begin{vmatrix} f_p \\ f_l \end{vmatrix} = \mathbf{G}_3\mathbf{G}_1^{-1}\mathbf{Q}_1 - \mathbf{Q}_2 \tag{4.235}$$

Following Wei, the total molar rate of production of xylenes is equal to the sum NMr of all of the rates in the nodes, where M is the number of nodes in a column of the network and N is the number of columns of the network. The selectivity S_{pX} toward p-xylene is then obtained from Eq. (4.236):

$$S_{pX} = f_p/NMr = f_p/\left(f_p + f_l\right) \tag{4.236}$$

Table 4.3 Comparison of p-xylene selectivity for a single open node versus a fully open boundary as a function of the network size.

($r = 2$, $S_k = 0.25$, $K_{eq} = 1/3$, $D_p = 10$, $D_l = 1$, $k = 10^{-4}$, $(S_{pX})_{\text{lim}} = 0.769$; consistent defined units)		
$M \times N$	S_{pX} (center opening only)	S_{pX} (all nodes open)
5×5	0.253	0.251
11×11	0.264	0.253
21×21	0.303	0.260
51×51	0.475	0.303
101×101	0.647	0.402
201×201	0.733	0.548

Table 4.3 shows the results as the network perimeter is blocked to just one opening and compared with a fully open boundary as a function of the size of the network. The selectivity increases and shows a substantial boost with networks that hardly have any openings to the bulk fluid compared to networks with all nodes on the periphery open. Of course, these selectivity values are at the reactor entrance, and they will decrease as the bulk fluid progresses through the reactor, as shown by Wei with a single-pore model. The discrete network approach clearly shows some practical advantages when perimeter blocking is considered, since the cumbersome boundary conditions otherwise encountered with the continuum approach are relatively easy to handle with matrix calculus.

Consider a network with a fully open boundary. Figure 4.8 gives the selectivities that are reached with zero bulk concentrations of xylenes for different kinetic selectivities for p-xylene and l-xylene. The depth of the network is a variable and is shown on the horizontal axis. If the kinetic selectivity S_k equals unity such that only p-xylene is produced at the active site with none toward l-xylene, then the selectivity decreases steadily with the network size toward the asymptotic value. Obviously, here there is no need to fix something that is not broken (i.e. the selectivity is already at 100% to begin with and all mass transfer limitations should be avoided). Second, if the kinetic selectivity S_k is such that only l-xylene is produced with none toward p-xylene, then the selectivity increases steadily with the network size toward the asymptotic value. If the kinetic selectivity S_k equals ¼, as used in the earlier cases, then the selectivity increases monotonically to the asymptotic value, but with a higher starting value. The calculations shown in Figure 4.8 have been obtained using the Mathcad software. The dashed upward-trending curve is for l-xylene formation only at the site while the dashed downward-trending curve is for p-xylene formation only at the site.

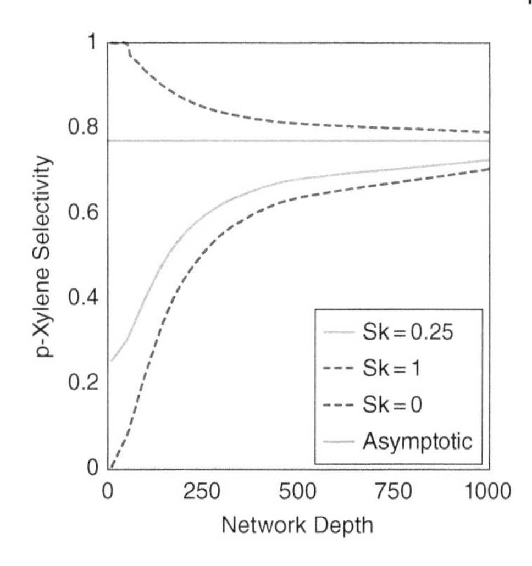

Figure 4.8 p-Xylene selectivity as a function of kinetic selectivity and network size.

4.3.4 Irregular Networks

4.3.4.1 Generalities

To represent the pore space in a catalyst, I here consider networks of a general nature, but perhaps with a very simplified arrangement. The catalyst is thought of as an ensemble of nodes, not necessarily all alike in their properties. Nodes are considered to be small volumes with a given concentration of the reactant that locally contain and represent the pore space and also contain all of the necessary aspects (e.g. surface area, chemistry of the surface, metal loading, etc.) for a chemical reaction to proceed. Nodes are sprinkled throughout the space of the catalyst particle and are as plentiful and abundantly located as one finds necessary. Some nodes are by definition connected by a "channel" that allows mass transfer to neighboring nodes to occur. Each channel is characterized not by its shape, but simply by a mass transfer coefficient. The properties of the nodes and the properties of the channels can be according to any distribution envisioned, but a Gaussian distribution is a likely candidate for amorphous catalysts. A substantial difference between this pore space arrangement and what is assumed in periodic networks or in the continuum approach is that here the pore structure properties are allowed to be variable everywhere. In periodic networks, which are typically not isotropic, the network properties are constant and simple only along a limited number of principal directions. In the classic continuum case, it is typically assumed that the catalyst is isotropic (i.e. the properties are not direction sensitive and are constant throughout the catalyst particle). As I did with regular networks in Section 4.3.3, all material balances in each node of the arbitrary network will be closed explicitly.

4.3.4.2 A Corollary

Before I go on, I have found the following corollary to come in handy in order to show certain properties of relationships. Consider the following identity applied to an arbitrary column vector $\mathbf{u}$ of dimensions $(m \times 1)$:

$$diag(\mathbf{u})\mathbf{e}_m = \mathbf{u} \tag{4.237}$$

where $\mathbf{e}_m$ is a unit column vector of dimensions $(m\times1)$ (i.e. each element is equal to unity), while $diag$ is an operator that turns the vector u into a diagonal $m \times m$ matrix such that:

$$diag(\mathbf{u})_{i,i} = \mathbf{u}_i \quad i = 1, 2, ..., m \tag{4.238}$$

To Eq. (4.237) one can evidently add a product of a matrix $\mathbf{A}$ and a column vector $\mathbf{e}$ that satisfies:

$$\mathbf{A}\mathbf{e}_m = \mathbf{0} \tag{4.239}$$

but is otherwise arbitrary. Hence, one gets:

$$(\mathbf{A} + diag(\mathbf{u}))\mathbf{e}_m = \mathbf{u} \tag{4.240}$$

or when the inverse exists:

$$\mathbf{e}_m = (\mathbf{A} + diag(\mathbf{u}))^{-1}\mathbf{u} \tag{4.241}$$

Applying Eq. (4.241) for:

$$\mathbf{u} = \mathbf{G}_{m,n}\mathbf{e}_n \tag{4.242}$$

where $\mathbf{G}_{m,n}$ is an arbitrary rectangular matrix of proper dimension yields:

$$\mathbf{e}_m = (\mathbf{A} + diag(\mathbf{G}\mathbf{e}_n))^{-1}\mathbf{G}\mathbf{e}_n \tag{4.243}$$

Equation (4.243) expresses that the sum of the elements in each row of the matrix $(\mathbf{A} + diag(\mathbf{G}\mathbf{e}_n))^{-1}\mathbf{G}$ equals unity. Applying the expression for a more general $\mathbf{u}_m$ such as:

$$\mathbf{u}_m = \mathbf{G}_{m,n}\mathbf{e}_n + \mathbf{w}_m \tag{4.244}$$

where $\mathbf{w}_m$ is an arbitrary column vector leads to:

$$\mathbf{e}_m = (\mathbf{A} + diag(\mathbf{G}\mathbf{e}_n) + diag(\mathbf{w}_m))^{-1}(\mathbf{G}\mathbf{e}_n + \mathbf{w}_m) \tag{4.245}$$

Therefore, summing the elements of each row of $(\mathbf{A} + diag(\mathbf{G}\mathbf{e}_n) + diag(\mathbf{w}_m))^{-1}\mathbf{G}\mathbf{e}_n$ and adding to it the respective sum of the elements of each row of $(\mathbf{A} + diag(\mathbf{G}\mathbf{e}_n) + diag(\mathbf{w}_m))^{-1}\mathbf{w}_m$, equals unity. Equation (4.245) will be useful for showing that the diffusion fluxes into a network from the periphery of the network equal the total rate of reaction in the nodes.

In addition, let:

$$[\mathbf{A} + diag(\mathbf{G}\mathbf{e}_n)]\mathbf{e}_m = \mathbf{A}'\mathbf{e}_m \neq \mathbf{0} \tag{4.246}$$

It then follows that:

$$\mathbf{e}_m = (\mathbf{A}' + diag(\mathbf{w}_m))^{-1}(\mathbf{A}'\mathbf{e}_m + \mathbf{w}_m) \tag{4.247}$$

Another peculiarity one easily obtains follows. Let:

$$\mathbf{u} = \alpha\mathbf{I}\mathbf{e}_m \text{ with } \alpha \neq 0 \tag{4.248}$$

One gets:

$$\mathbf{e}_m = \alpha(\mathbf{A} + \alpha\mathbf{I})^{-1}\mathbf{e}_m \tag{4.249}$$

and:

$$\lim_{\alpha \to 0} \mathbf{e}_m = \mathbf{e}_m = \left\{ \lim_{\alpha \to 0} \left[\alpha(\mathbf{A} + \alpha\mathbf{I})^{-1}\right] \right\}\mathbf{e}_m \tag{4.250}$$

When the matrix **A** is symmetrical, then the following applies:

$$\lim_{\alpha \to 0} \{\alpha(\mathbf{A} + \alpha\mathbf{I})^{-1}]\} = \mathbf{U}/m \tag{4.251}$$

where the matrix **U** is a matrix with all elements equal to unity.

4.3.4.3 Diffusion in an Arbitrary Network

Consider an arbitrary set of nodes. Each of the nodes is here considered to act as a continuous stirred tank reactor and each has a certain concentration. Each of the nodes has nearest-neighbor nodes, and exchange of mass among them occurs via diffusion. The rate of mass transfer between neighboring nodes is considered proportional to the concentration difference between the two nodes. This rate of mass transfer is also proportional to a mass transport coefficient ν specific for the transport between these two nodes. That is, the molar flux from the node at concentration C_i toward a nearest-neighbor node at concentration C_j can be written as:

$$f = \nu(C_i - C_j) \tag{4.252}$$

A number m of these nodes are called "internal" because they only exchange mass with other nodes of the network. However, there are nodes of the network – n in total – that are called "external" because by definition they are also able to exchange mass with the bulk fluid outside that surrounds the network.

4.3.4.3.1 The Material Balance Matrix for Diffusion

Networks are solved by making material balances in each node, imposing boundary conditions, and solving the ensuing set of equations simultaneously. The material balance for each internal node i can be represented by equating the sum of all fluxes leaving that node toward all of the other network nodes to zero:

$$\sum_{j=1, j \neq i}^{\substack{internal,\\ external}} \nu_{i,j}(C_i - C_j) = 0 \tag{4.253}$$

The summation can be divided up into two sums: one for the connections to other internal nodes and one for the connections to the external nodes. The sum concerning the connections to other internal nodes can be conveniently organized by working with an upper diagonal matrix **A**. The matrix **A** contains in row 1 the mass transfer coefficients of the node numbered $j = 1$ to the other internal nodes. Then on row 2, the matrix **A** contains the mass transfer coefficients of the node numbered $j = 2$ to the other internal nodes (except node 1). Then on row 3, the matrix **A** contains the mass transfer coefficients of the node numbered $j = 3$ to the other internal nodes (except nodes 1 and 2), and so forth. The full summation for making a material balance can then be obtained by making the matrix **P**:

$$\mathbf{P} = diag\left(\left(\mathbf{A} + \mathbf{A}^T\right)\mathbf{e}_m\right) - \left(\mathbf{A} + \mathbf{A}^T\right) \tag{4.254}$$

The operator *diag* converts a column vector into a diagonal matrix with preservation of the value and sequence of the elements. For example:

$$diag\left(\begin{vmatrix} a \\ b \\ c \end{vmatrix}\right) = \begin{vmatrix} a & 0 & 0 \\ 0 & b & 0 \\ 0 & 0 & c \end{vmatrix} \tag{4.255}$$

It is also clear from Eq. (4.254) that:

$$\mathbf{P}\mathbf{e}_m = \mathbf{0} \tag{4.256}$$

The sum of the mass transfer coefficients of each internal node to any of the external nodes is organized with the help of a matrix **G**. Each row is specific for an internal node and contains all of the mass transfer coefficients linking the internal node to all of the external nodes.

The material balance for each external node e can be represented by equating the sum of all fluxes leaving that node toward all of the other network nodes to the flux that must enter the node from the bulk fluid outside of the network:

$$\sum_{j=1, j \neq e}^{\substack{internal,\\ external}} \nu_{e,j}(C_e - C_j) = F_e \tag{4.257}$$

The same methodology can be used to organize and sum the connections of the external nodes to any of the other external nodes, and this leads to a matrix **Q**. The connections of the external nodes to any internal nodes is conveniently organized by the matrix $\mathbf{G}^T$, the transpose of **G**. The flux F_e is the flux entering from the bulk fluid into the network through this external node. If F_e is positive, this means that material actually enters the network from the bulk fluid, while if F_e is negative, this means that there is material exiting the network. This totality of equations can now conveniently be written in matrix form by considering the overall material balance matrix **M**:

$$\mathbf{M} = \begin{vmatrix} \mathbf{P} + diag(\mathbf{G}\mathbf{e}_n) & -\mathbf{G} \\ -\mathbf{G}^T & \mathbf{Q} + diag(\mathbf{G}^T\mathbf{e}_m) \end{vmatrix} \tag{4.258}$$

P and **Q** are typically sparse matrices that are formed from all of the diffusional fluxes among the internal nodes respective of the external nodes. The off-diagonal elements of **P** and **Q** are mostly zero or negative, and they contain the mass transfer coefficients between the nodes. Their diagonal elements contain the sum of all of the mass transfer coefficients from that node to all of the other network nodes if they exist (i.e. to either the internal nodes or the external nodes). The matrix **G** is typically a rectangular matrix and contains the mass transfer coefficients from each internal node to each of the external nodes.

4.3.4.3.2 General Solution

The overall matrix equation for the network is then given as:

$$\begin{vmatrix} \mathbf{P} + diag(\mathbf{G}\mathbf{e}_n) & -\mathbf{G} \\ -\mathbf{G}^T & \mathbf{Q} + diag(\mathbf{G}^T\mathbf{e}_m) \end{vmatrix} \times \begin{vmatrix} \mathbf{C}_i \\ \mathbf{C}_e \end{vmatrix} = \begin{vmatrix} \mathbf{0} \\ \mathbf{F}_e \end{vmatrix} \tag{4.259}$$

Partitioning of the matrix allows one to derive the following two matrix equations:

$$[\mathbf{P} + diag(\mathbf{G}\mathbf{e}_n)]\mathbf{C}_i - \mathbf{G}\mathbf{C}_e = \mathbf{0} \tag{4.260}$$

$$-\mathbf{G}^T\mathbf{C}_i + [\mathbf{Q} + diag(\mathbf{G}^T\mathbf{e}_m)]\mathbf{C}_e = \mathbf{F}_e \tag{4.261}$$

These matrix equations are easily solved and allow one to express the concentrations in the internal nodes and the fluxes from the bulk fluid into the external nodes, to be solved as shown in Eqs. (4.262) and (4.263):

$$\mathbf{C}_i = (\mathbf{P} + diag(\mathbf{G}\mathbf{e}_n))^{-1}\mathbf{G}\mathbf{C}_e = \mathbf{\Omega}_i\mathbf{C}_e \tag{4.262}$$

$$\mathbf{F}_e = [\mathbf{Q} + diag(\mathbf{G}^T\mathrm{e}_m) - \mathbf{G}^T(\mathbf{P} + diag(\mathbf{G}\mathbf{e}_n))^{-1}\mathbf{G}]\mathbf{C}_e = \mathbf{H}_e\mathbf{C}_e \tag{4.263}$$

$\mathbf{\Omega}_i$ is a rectangular matrix of weight factors that show how the concentrations in the internal nodes depend on the concentrations arbitrarily set in the external nodes. Each row of $\boldsymbol{\Omega}_i$ corresponds to a specific internal node. Each element of that row is a weight factor that, when multiplied by the concentration in the corresponding external node, represents the contribution of that external node to the internal node concentration. Summing these products over all of the external nodes then yields the total concentration in that internal node. From Eq. (4.243), in the corollary, it is immediately shown that each row of coefficients of $\boldsymbol{\Omega}_i$ sums to unity, in agreement with the fact that no material is lost due to the absence of reaction. The square matrix $\mathbf{H}_e$ shows how the fluxes going into the network depend on the concentrations arbitrarily set in the external nodes. Each row of $\mathbf{H}_e$ yields a specific solution to diffusion. The diagonal element of that row (positive) represents the flux entering the network when the concentration in that node is assumed to be unity, with all other external nodes set at zero, while the off-diagonal elements of that row are the (negative) fluxes exiting the network. The sum of all of the fluxes in a row of course being zero in order to close the material balance also means that the matrix $\mathbf{H}_e$ is singular. From Eq. (4.263), it is clear that the matrix $\mathbf{H}_e$ is what is called the Schur complement of the matrix $[\mathbf{P} + diag(\mathbf{G}\mathbf{e}_n)]$ in the full material balance matrix $\mathbf{M}$.

4.3.4.3.3 An Alternative Solution Method for Diffusion

There is an alternative solution method for the diffusion equations above that allows one to circumvent the singularity of the matrix $\mathbf{M}$ that prohibits a direct solution. Rearrange Eqs. (4.260) and (4.261) according to:

$$\begin{vmatrix} \mathbf{P} + diag(\mathbf{G}\mathbf{e}_n) & \mathbf{0} \\ \mathbf{G}^T & \mathbf{I}_e \end{vmatrix} \times \begin{vmatrix} \mathbf{C}_i \\ \mathbf{F}_e \end{vmatrix} = \begin{vmatrix} \mathbf{G} \\ \mathbf{Q} + diag(\mathbf{G}^T\mathbf{e}_m) \end{vmatrix} \mathbf{C}_e \tag{4.264}$$

Equation (4.264) can then be solved directly via:

$$\begin{vmatrix} \mathbf{C}_i \\ \mathbf{F}_e \end{vmatrix} = \begin{vmatrix} \mathbf{P} + diag(\mathbf{G}\mathbf{e}_n) & \mathbf{0} \\ \mathbf{G}^T & \mathbf{I}_e \end{vmatrix}^{-1} \begin{vmatrix} \mathbf{G} \\ \mathbf{Q} + diag(\mathbf{G}^T\mathbf{e}_m) \end{vmatrix} \mathbf{C}_e \tag{4.265}$$

The lower triangular block matrix on the right-hand side of Eq. (4.265) can be inverted along the standard route, and this leads to:

$$\begin{vmatrix} \mathbf{C}_i \\ \mathbf{F}_e \end{vmatrix} = \begin{vmatrix} (\mathbf{P} + diag(\mathbf{G}\mathbf{e}_n))^{-1} & \mathbf{0} \\ -\boldsymbol{\Omega}_i^T & \mathbf{I}_e \end{vmatrix} \begin{vmatrix} \mathbf{G} \\ \mathbf{Q} + diag(\mathbf{G}^T\mathbf{e}_m) \end{vmatrix} \mathbf{C}_e \tag{4.266}$$

or finally also:

$$\begin{vmatrix} \mathbf{C}_i \\ \mathbf{F}_e \end{vmatrix} = \begin{vmatrix} \mathbf{I}_i & \mathbf{0} \\ -\mathbf{G}^T & \mathbf{I}_e \end{vmatrix} \begin{vmatrix} (\mathbf{P} + diag(\mathbf{G}\mathbf{e}_n))^{-1} & \mathbf{0} \\ \mathbf{0} & \mathbf{I}_e \end{vmatrix} \begin{vmatrix} \mathbf{G} \\ \mathbf{Q} + diag(\mathbf{G}^T\mathbf{e}_m) \end{vmatrix} \mathbf{C}_e \tag{4.267}$$

4.3.4.3.4 On the Structure of the Matrix G

The matrix **G** that represents the mass transfer coefficients of all internal nodes to the external nodes can in practical cases be partitioned. Only subsurface nodes are connected to the external nodes, hence:

$$\mathrm{G} = \begin{vmatrix} \mathbf{0} \\ \mathbf{g} \end{vmatrix} \tag{4.268}$$

The matrix **g** is much smaller than the matrix **G**. The matrix **g** may be sparse, but every row has at least one non-zero element. The reasoning can be repeated to the next set of subsurface nodes, and so on.

4.3.4.3.5 The Complementary Situation

The general problem can further be worked by considering the complementary situation: reversing the roles of internal and external nodes. This then leads to:

$$\begin{vmatrix} \mathbf{P} + diag(\mathbf{G}\mathbf{e}_n) & -\mathbf{G} \\ -\mathbf{G}^T & \mathbf{Q} + diag(\mathbf{G}^T\mathbf{e}_m) \end{vmatrix} \times \begin{vmatrix} \mathbf{C}_i \\ \mathbf{C}_e \end{vmatrix} = \begin{vmatrix} \mathbf{F}_i \\ \mathbf{0} \end{vmatrix} \tag{4.269}$$

The matrix equation is again easily solved and yields:

$$\mathbf{C}_e = \boldsymbol{\Omega}_e\mathbf{C}_i \tag{4.270}$$

$$\mathbf{F}_i = \mathbf{H}_i\mathbf{C}_i \tag{4.271}$$

with:

$$\boldsymbol{\Omega}_e = \left(\mathbf{Q} + diag\left(\mathbf{G}^T\mathbf{e}_m\right)\right)^{-1}\mathbf{G}^T \tag{4.272}$$

$$\mathbf{H}_i = \mathbf{P} + diag(\mathbf{G}\mathbf{e}_n) - \mathbf{G}\left(\mathbf{Q} + diag\left(\mathbf{G}^T\mathbf{e}_m\right)\right)^{-1}\mathbf{G}^T \tag{4.273}$$

In practical cases, it is clear that only the arbitrary choice of the subsurface node concentrations has an impact on the surface node concentration. Hence, the matrix $\boldsymbol{\Omega}_e$ can be partitioned as:

$$\boldsymbol{\Omega}_e = \begin{vmatrix} \mathbf{0} & \boldsymbol{\Omega}_{e,ss} \end{vmatrix} \tag{4.274}$$

where, again, the matrix $\boldsymbol{\Omega}_{e,ss}$ is much smaller than the matrix $\boldsymbol{\Omega}_e$.

The following generalization can now easily be shown:

$$\begin{vmatrix} \mathbf{H}_i & \mathbf{0} \\ \mathbf{0} & \mathbf{H}_e \end{vmatrix} = \mathbf{M}\begin{vmatrix} \mathbf{I}_\mathrm{i} & \boldsymbol{\Omega}_\mathrm{i} \\ \boldsymbol{\Omega}_\mathrm{e} & \mathbf{I}_\mathrm{e} \end{vmatrix} \tag{4.275}$$

where $\mathbf{I}_i$ and $\mathbf{I}_e$ are identity matrices of the proper dimensions. This general expression can be considered a particular factorization of the matrix **M** and will come back herein for other cases. This factorization is, to myself at least, different from the typical factorizations that are available in the literature and in mathematical numerical packages.

4.3.4.3.6 The Meaning of $(\mathbf{P} + diag(\mathbf{Ge}_n))^{-1}$

The matrices $(\mathbf{P} + \boldsymbol{diag}(\mathbf{Ge}_n))^{-1}$ or $(\mathbf{Q} + diag(\mathbf{G}^T\mathbf{e}_m))^{-1}$ appear to play an important role. I will now show that the physical meaning of either of these two matrices represents the solution to a number of diffusional source point problems. For this, for the matrix $(\mathbf{P} + \boldsymbol{diag}(\mathbf{Ge}_n))^{-1}$, for instance, proceed as follows: beyond the external nodes, select an arbitrary internal node, call it node i, and declare it open. This means that this node is allowed to have a flux from the bulk fluid either enter or exit the node. Define the concentration in that internal node to be equal to C^* and set all external node concentrations equal to zero. I will show that the flux going into the internal node i is given by:

$$flux_i = C^*/(\mathbf{P} + diag(\mathbf{Ge}_n))_{i,i}^{-1} \tag{4.276}$$

while at the same time the concentrations in the other internal nodes j will be shown to be:

$$Conc_j = flux_i \times (\mathbf{P} + diag(\mathbf{Ge}_n))_{i,j}^{-1} \tag{4.277}$$

Hence, the matrix $(\mathbf{P} + diag(\mathbf{Ge}_n))^{-1}$ fully solves as many internal diffusional problems as there are internal nodes. The proof for this is rather straightforward. For ease of writing, let the node i in the material balance matrix $\mathbf{M}$ be located right next to the external nodes. For any other internal node, the matrix $\mathbf{M}$ can be so rearranged as to get that result. To start, the matrix $\mathbf{M}$ can then be specified as:

$$\mathbf{M} = \begin{vmatrix} \mathbf{A}' & -\mathbf{g} & -\mathbf{G}' \\ -\mathbf{g}^T & a & -\mathbf{p}^T \\ -\mathbf{G}'^T & -\mathbf{p} & diag(\mathbf{G}^T\mathbf{e}_m) \end{vmatrix} \tag{4.278}$$

with:

$$\mathbf{P} + diag(\mathbf{Ge}_n) = \begin{vmatrix} \mathbf{A}' & -\mathbf{g} \\ -\mathbf{g}^T & a \end{vmatrix} \tag{4.279}$$

and:

$$\mathbf{G} = \begin{vmatrix} \mathbf{G}' \\ \mathbf{p}^T \end{vmatrix} \tag{4.280}$$

The element a in matrix M is the element in the position of the i-th node and it has been assumed to be located right next to the set of external nodes. The matrix equation is now:

$$\begin{vmatrix} \mathbf{A}' & -\mathbf{g} & -\mathbf{G}' \\ -\mathbf{g}^T & a & -\mathbf{p}^T \\ -\mathbf{G}'^T & -\mathbf{p} & diag(\mathbf{G}^T\mathbf{e}_m) \end{vmatrix} \times \begin{vmatrix} \mathbf{C}'_j \\ C^* \\ \mathbf{0} \end{vmatrix} = \begin{vmatrix} \mathbf{0} \\ F^* \\ \mathbf{F}'_e \end{vmatrix} \tag{4.281}$$

The flux F^* into the internal node i, the concentrations $\mathbf{C}'_j$ of the other internal nodes, and the individual fluxes $\mathbf{F}'_e$ out of each of the external nodes are then given by:

$$\mathbf{C}'_j = \mathbf{A}'^{-1}\mathbf{g}C^* \tag{4.282}$$

$$F^* = \left(a - \mathbf{g}^T\mathbf{A}'^{-1}\mathbf{g}\right)C^* \tag{4.283}$$

$$\mathbf{F}'_e = \left(-\mathbf{p} - \mathbf{G}'^T\mathbf{A}'^{-1}\mathbf{g}\right)C^* \tag{4.284}$$

In parallel, the partitioned matrix $\begin{vmatrix} \mathbf{A}' & -\mathbf{g} \\ -\mathbf{g}^T & a \end{vmatrix}$ can easily be inverted using standard methods, yielding:

$$\begin{vmatrix} \mathbf{A}' & -\mathbf{g} \\ -\mathbf{g}^T & a \end{vmatrix}^{-1} = \begin{vmatrix} (\mathbf{A}' - \mathbf{g}a^{-1}\mathbf{g}^T)^{-1} & \mathbf{A}'^{-1}\mathbf{g}(a - \mathbf{g}^T\mathbf{A}'^{-1}\mathbf{g})^{-1} \\ a^{-1}\mathbf{g}^T(\mathbf{A}' - \mathbf{g}a^{-1}\mathbf{g}^T)^{-1} & (a - \mathbf{g}^T\mathbf{A}'^{-1}\mathbf{g})^{-1} \end{vmatrix} \tag{4.285}$$

A comparison of the inversion in Eq. (4.285) with the solution in Eqs. (4.282)–(4.284) yields the earlier claimed results at the beginning of this section in Eqs. (4.276) and (4.277).

The matrix $(\mathbf{P} + diag\,(\mathbf{Ge}_n))^{-1}$ can therefore be decomposed as a product of two simple matrices:

$$(\mathbf{P} + diag(\mathbf{Ge}_n))^{-1} = \mathbf{F}^{*-1} \times \mathbf{\Gamma}^* \tag{4.286}$$

where $\mathbf{F}^*$ is a diagonal matrix that contains the fluxes in the chosen internal nodes while $\mathbf{\Gamma}^*$ is a matrix with unity on the diagonal, while the other elements on that row contain the normalized concentrations for that particular case. Again, each row contains a separate solution of the diffusion problem specific for the node identified with that row.

4.3.4.3.6.1 Worked Example for a Diffusional Point Source Calculation in a Discrete Network The calculations in this section have been performed using Mathcad 15 software. Consider the network in Figure 4.9 with four internal nodes (1, 2, 3, and 4 in blue) and three external nodes (5, 6, and 7 in white). The numbers in red between the nodes are the mass transfer coefficients between those nodes. Simple verification then shows the matrices in Appendix 4.1 along with the numerical solutions. To verify the point source solution, take, for example, row 4 of the matrix $(\mathbf{P} + diag\,(\mathbf{Ge}_n))^{-1}$. This means that node 4 is considered open

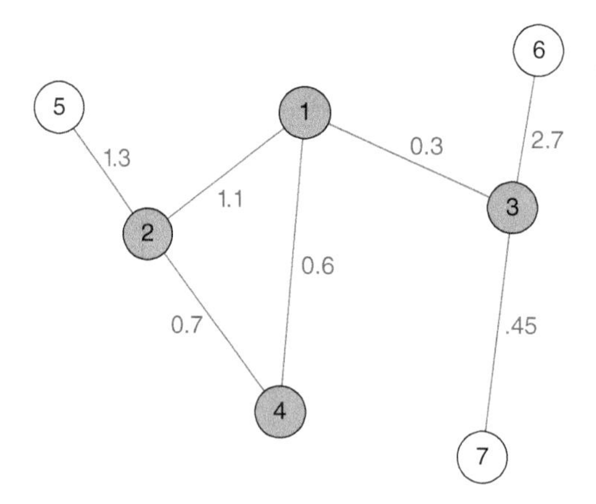

Figure 4.9 Trial network for diffusion only.

and its concentration is set to unity. The other internal nodes (1, 2, and 3) are closed, while all external nodes (5, 6, and 7) are set at a zero concentration. The diagonal element of row 4 contains the inverse of the flux entering node 4, hence:

$$Flux = 1/1.4547 = 0.6874$$

The concentrations in nodes 1, 2, 3, and 4 are then given by dividing each row element by the diagonal element:

$$C_1 = 0.7793/1.4547 = 0.5357$$

$$C_2 = 0.6050/1.4547 = 0.4159$$

$$C_3 = 0.0678/1.4547 = 0.0466$$

$$C_4 = 1.4547/1.4547 = 1$$

The material balance in node 4 then verifies:

$$Flux = 0.6874 = 0.7 \times (1 - 0.4159) + 0.6 \times (1 - 0.5357)$$

This can then easily be further verified for all the nodes.

4.3.4.3.7 Diffusion from a Point Source in the Network

The meaning that the matrix $(\mathbf{P} + \, diag\,(\mathbf{Ge}_n))^{-1}$ is a comprehensive point source solution for each of the internal nodes when all external nodes are put at a zero concentration leads one to suspect that perhaps it can assist in finding a methodology for obtaining an overall mass transfer coefficient. For this, consider a node located at the "center" of a three-dimensional discrete network and assume that this network is large enough that it can also be approximated reasonably well by the classical continuum approach. Let the concentration in this center node be designated by C^*. The flux F^* entering into the network at this node is then given by:

$$F^* = C^*/(\mathbf{P} + diag(\mathbf{Ge}_n))^{-1}_{*,*} \tag{4.287}$$

The average concentration of the network can be found from:

$$C_{avg} = \frac{C^*}{m+n} \sum_{\substack{j=1,m \\ j \neq *}} (\mathbf{P} + diag(\mathbf{Ge}_n))^{-1}_{*,j} / (P + diag(\mathbf{Ge}_n))^{-1}_{*,*} \tag{4.288}$$

where $(m + n)$ is the total number of nodes in the network, internal plus external. Hence, we notice that the ratio F^*/C_{avg} is independent of C^* and is given by:

$$F^*/C_{avg} = (m+n) / \sum_{\substack{j=1,m \\ j \neq *}} (\mathbf{P} + diag(\mathbf{Ge}_n))^{-1}_{*,j} \tag{4.289}$$

This formulation looks very plausible when the "center" node has a very average value for the local mass transfer coefficient compared to the rest of the network. But what happens when, for instance, we happen to pick a "center" node with, by chance, a set of connecting mass transfer coefficients that happen to be very small compared to the entire distribution of mass transfer coefficients? In that case, the flux F^* will become very small as well. However, the rest of the network concentrations will also become very small, and the ratio of the flux over the average concentration therefore is likely not very sensitive to the local properties of the "center node."

4.3.4.3.8 Diffusion from a Point Source along Classic Lines

The mathematics of diffusion from a point source in space have been well covered in the open literature. Here, I only show essential results for reasons of continuity for the reader.

Let D_e be the effective diffusion coefficient for mass transport in three-dimensional space. The DE that governs mass transport for diffusion in a sphere of radius R is given by:

$$\frac{d}{dr}\left(4\pi D_e r^2 \frac{dc}{dr}\right) = 0 \tag{4.290}$$

Imposing a mass flux F^* at the point source or the total mass flux at the perimeter R allows one to solve for the concentration profile:

$$c = \frac{F^*}{4\pi D_e}\left(\frac{1}{r} - \frac{1}{R}\right) \tag{4.291}$$

Calculating the volume average concentration C_{avg} is done from:

$$C_{avg} = \frac{\int_0^R 4\pi r^2 c dr}{\int_0^R 4\pi r^2 dr} \tag{4.292}$$

and yields:

$$\frac{F^*}{C_{avg}} = 8\pi D_e R \tag{4.293}$$

From this, notice that again the ratio F^*/C_{avg} is independent of the imposed flux itself and is only a function of the effective diffusion coefficient D_e and the geometry (here spherical), as in the discrete pore network case in Section 4.3.4.3.7.

4.3.4.3.9 Merging the Network Approach and the Classic Approach

Equations (4.289) and (4.293) allow one to calculate the effective diffusion coefficient from the individual mass transfer coefficients in the discrete network as:

$$D_e = \frac{1}{8\pi R}\left((m+n)/\sum_{\substack{j=1,m \\ j\neq *}} (\mathbf{P} + diag(\mathbf{Ge}_n))^{-1}_{*,j}\right) \tag{4.294}$$

It is obvious that a very interesting question and treatment now arises regarding the expected value of the matrix $(\mathbf{P} + diag\,(\mathbf{Ge}_n))^{-1}$ when the individual mass transfer coefficients and connections have a certain statistical distribution.

4.3.4.3.10 A Property of the Diffusion Matrix P

The diffusion matrix **P** is symmetrical, has negative off-diagonal elements, and each diagonal element contains the negative sum of all the off-diagonal elements on that row. The matrix **P** is obviously singular, hence having no inverse. Their does exist a pseudo-inverse matrix $\mathbf{P}^+$ called the Moore–Penrose inverse that is applicable to any matrix, including singular and rectangular matrices. The pseudo-inverse $\mathbf{P}^+$ is unique, always exists, and must satisfy the following four conditions to be called the Moore–Penrose inverse:

$$\mathbf{PP}^+\mathbf{P} = \mathbf{P} \tag{4.295}$$

$$\mathbf{P}^+\mathbf{PP}^+ = \mathbf{P}^+ \tag{4.296}$$

$$(\mathbf{PP}^+)^* = \mathbf{PP}^+ \tag{4.297}$$

$$(\mathbf{P}^+\mathbf{P})^* = \mathbf{P}^+\mathbf{P} \tag{4.298}$$

where the subscript $*$ stands for the Hermitian transpose, also known as the conjugate transpose. The Moore–Penrose inverse can be conveniently obtained via the well-known singular-value decomposition of a matrix. There is also a limit definition of the Moore–Penrose inverse, and this can be found in Albert [19]:

$$\mathbf{P}^{+} = \lim_{\varepsilon \to 0}\left[\left(\mathbf{P}^{T}\mathbf{P} + \varepsilon\mathbf{I}\right)^{-1}\mathbf{P}^{T}\right] = \lim_{\varepsilon \to 0}\left[\mathbf{P}^{T}\left(\mathbf{P}\mathbf{P}^{T} + \varepsilon\mathbf{I}\right)^{-1}\right]$$

An interesting special case exists for matrices of the diffusional kind as described here. Let $\mathbf{\Lambda}$ and $\mathbf{V}$ be the matrix of eigenvalues with respect to the eigenvectors of a square matrix $\mathbf{P}$. It is well known that:

$$\mathbf{P} = \mathbf{V}\mathbf{\Lambda}\mathbf{V}^{T} \tag{4.299}$$

When the matrix $\mathbf{P}$ is real, singular, and of the diffusional kind with the properties as described in Eq. (4.254), then it is shown herein without formal proof that the following is obtained: define $\mathbf{\Lambda}^{\mathbf{o}}$ as a diagonal matrix containing, in the same place, the element inverses of $\mathbf{\Lambda}$, except for the zero-eigenvalue element, which is kept at zero in $\mathbf{\Lambda}^{\mathbf{o}}$. A semi-inverse $\mathbf{P}^{\mathbf{o}}$ is obtained by applying:

$$\mathbf{P}^{\mathbf{o}} = \mathbf{V}\mathbf{\Lambda}^{\mathbf{o}}\mathbf{V}^{T} \tag{4.300}$$

such that:

$$\mathbf{P}^{\mathbf{o}}\mathbf{P} = \mathbf{P}\mathbf{P}^{\mathbf{o}} = \mathbf{I} - \mathbf{U}/n \tag{4.301}$$

where $\mathbf{I}$ is the identity matrix, $\mathbf{U}$ is the unit matrix (i.e. each element of $\mathbf{U}$ is equal to unity), and n is the dimension of the matrix $\mathbf{P}$. This relationship becomes clear when one realizes that not only the eigenvectors given as the columns of $\mathbf{V}$ are normalized and orthogonal with each other, but also the rows are normalized and orthogonal with each other.

Thus, Eq. (4.301) shows that, when the singular matrix $\mathbf{P}$ becomes arbitrarily large, the semi-inverse becomes arbitrarily close to the identity matrix.

4.3.4.3.10.1 Worked Example Appendix 4.2 shows the Mathcad 15 software calculations of the semi-inverse properties of the matrix $\mathbf{P}$ for the case in Figure 4.9. Mathcad readily gives the eigenvalues and the eigenvectors of the singular matrix $\mathbf{P}$. The eigenvalues for the semi-inverse are evaluated in the column vector **EVAL0** in Appendix 4.2. The bottom of Appendix 4.2 then shows the expected relationship between $\mathbf{P}$ and $\mathbf{P}^{\mathbf{0}}$

4.3.4.4 Diffusion and First-order Reaction in an Arbitrary Network

4.3.4.4.1 The Material Balance Matrix for Reaction and Diffusion

Consider now that a first-order reaction occurs in the nodes of the network (i.e. each node acts as a small continuous stirred tank reactor for material fluxes that are entering from neighboring nodes). The rate of reaction r in each node obeys the rate equation kSC and is proportional to a particular surface area S assigned to that node. Hence, the rate of reaction in the node is given by:

$$r = kSC \tag{4.302}$$

The same also applies for the external nodes. The overall material balance matrix including a reaction now becomes:

$$\mathbf{M} = \begin{vmatrix} \mathbf{P} + diag(\mathbf{G}\mathbf{e}_n + k\mathbf{S}_i) & -\mathbf{G} \\ -\mathbf{G}^T & \mathbf{Q} + diag\left(\mathbf{G}^T\mathbf{e}_m + k\mathbf{S}_e\right) \end{vmatrix} \tag{4.303}$$

4.3.4.4.2 General Solution for Reaction and Diffusion

The overall matrix equation for the network with reaction is given by:

$$\begin{vmatrix} \mathbf{P} + diag(\mathbf{G}\mathbf{e}_n + k\mathbf{S}_i) & -\mathbf{G} \\ -\mathbf{G}^T & \mathbf{Q} + diag\left(\mathbf{G}^T\mathbf{e}_m + k\mathbf{S}_e\right) \end{vmatrix} \times \begin{vmatrix} \mathbf{C}_i \\ \mathbf{C}_e \end{vmatrix} = \begin{vmatrix} \mathbf{0} \\ \mathbf{F}_e \end{vmatrix} \tag{4.304}$$

The matrix equation is easily solved and yields:

$$\mathbf{\Omega}_i = (\mathbf{P} + diag(\mathbf{G}\mathbf{e}_n + k\mathbf{S}_i))^{-1}\mathbf{G} \tag{4.305}$$

$$\mathbf{H}_e = \mathbf{Q} + diag\left(\mathbf{G}^T\mathbf{e}_m + k\mathbf{S}_e\right) - \mathbf{G}^T(\mathbf{P} + diag(\mathbf{G}\mathbf{e}_n + k\mathbf{S}_i))^{-1}\mathbf{G} \tag{4.306}$$

with:

$$\mathbf{C}_i = \mathbf{\Omega}_i\mathbf{C}_e \tag{4.307}$$

$$\mathbf{F}_e = \mathbf{H}_e\mathbf{C}_e \tag{4.308}$$

As in the case of diffusion only, $\mathbf{\Omega}_i$ is a rectangular matrix of weight factors that shows how the concentrations in the internal nodes depend on the concentrations arbitrarily set in the external nodes. Notice that due to loss of reagent resulting from the chemical reaction, the sum of coefficients of each row of $\mathbf{\Omega}_i$ no longer adds up to unity. The square matrix $\mathbf{H}_e$ shows how the fluxes going into the network depend again on the concentrations arbitrarily set in the external nodes. Here, again, due to reaction, the flux entering the network (diagonal element) is balanced with the fluxes exiting the network (off-diagonal elements) and the amount of material reacted away in the network. The following relationship in Eq. (4.309) is easily derived from Eq. (4.305):

$$\mathbf{\Omega}_i = \left(\mathbf{I} + (\mathbf{P} + diag(\mathbf{G}\mathbf{e}_n))^{-1}diag(k\mathbf{S}_i)\right)^{-1}(\mathbf{P} + diag(\mathbf{G}\mathbf{e}_n))^{-1}\mathbf{G} \tag{4.309}$$

Let:

$$\mathbf{Z} = (\mathbf{P} + diag(\mathbf{G}\mathbf{e}_n))^{-1}diag(k\mathbf{S}_i) \tag{4.310}$$

hence:

$$\mathbf{\Omega}_i = (\mathbf{I} + \mathbf{Z})^{-1}\mathbf{\Omega}_{i,D} \tag{4.311}$$

where $\mathbf{\Omega}_{i,\ D}$ contains the diffusional weight factors earlier defined. The matrix $(\mathbf{I} + \mathbf{Z})^{-1}$ can be called an effectiveness matrix $\mathbf{E}$ and can be written symbolically as:

$$\mathbf{E} = \frac{\mathbf{I}}{\mathbf{I} + \dfrac{diag(k\mathbf{S}_i)}{\mathbf{P} + diag(\mathbf{Ge}_n)}} = \frac{\mathbf{I}}{\mathbf{I} + \mathbf{Z}} \tag{4.312}$$

The expression for the matrix $\mathbf{E}$ is valid for an arbitrary network and hence also for any catalyst particle shape. The matrix $\mathbf{Z}$ is nonsymmetrical; hence, in the general case, a canonical representation is only possible through:

$$\mathbf{Z} = \mathbf{U}\mathbf{\Lambda}\mathbf{U}^{-1} \tag{4.313}$$

where $\mathbf{\Lambda}$ and $\mathbf{U}$ are, respectively, the eigenvalues and eigenvectors of the matrix $\mathbf{Z}$. The expression of $\mathbf{E}$ is possible in the Taylor series form:

$$\mathbf{E} = \mathbf{I} - \mathbf{Z} + \mathbf{Z}^2 - \mathbf{Z}^3 + \ldots \tag{4.314}$$

This is valid on the condition that the absolute value of each of the eigenvalues of $\mathbf{Z}$ is less than unity in order to yield convergence of the series. This is the case for small enough values of the rate coefficient k. For values of k that are too large, one has to proceed directly with the inverse.

From Eq. (4.305), we also obtain:

$$\mathbf{\Omega}_i\mathbf{e}_n = (\mathbf{P} + diag(\mathbf{Ge}_n + k\mathbf{S}_i))^{-1}\mathbf{Ge}_n \tag{4.315}$$

Hence, with the help of the corollary established earlier in Section 4.3.4.2, one can easily show the following relationship:

$$\mathbf{\Omega}_i\mathbf{e}_n = \mathbf{e}_m - (\mathbf{P} + diag(\mathbf{Ge}_n + k\mathbf{S}_i))^{-1}k\mathbf{S}_i \tag{4.316}$$

Thus, the sum of the elements of each row of $\mathbf{\Omega}_i$, which is essentially the normalized concentration in each internal node, is either equal to unity for diffusion only or less than unity when a reaction is also considered. For a very high rate of reaction, it is clear that the concentration approaches zero for each internal node, as it should. The near-zero concentration in the internal nodes does not mean that the chemical rate is near zero, since the rate constant is exceedingly high. This is as it should be, since a finite rate of reactant is diffusing into the network from the external nodes to the immediate subsurface nodes. It will be shown later that for networks that represent real catalysts, the reaction rate in a node is only a small perturbation of the diffusion rate. Yet, large and small effectiveness factors are easily obtained in this case by considering networks of a size that is a realistic representation of real catalyst particles.

4.3.4.4.3 The Bosanquet Average Reaction–Diffusion Matrix

The column vector containing the rates of reaction in each of the internal nodes is now obtained from:

$$\mathbf{R}_i = diag(k\mathbf{S}_i)\mathbf{C}_i = \left((diag(k\mathbf{S}_i))^{-1} + (\mathbf{P} + diag(\mathbf{G}\mathbf{e}_n))^{-1}\right)^{-1}\boldsymbol{\Omega}_{i,D}\mathbf{C}_e \tag{4.317}$$

or:

$$\mathbf{R}_i = \mathbf{B}_o\boldsymbol{\Omega}_{i,D}\mathbf{C}_e \tag{4.318}$$

The total rate of reaction in the network is then obtained from:

$$R_\Sigma = \mathbf{e}_m^T\mathbf{B}_o\boldsymbol{\Omega}_{i,D}\mathbf{C}_e \tag{4.319}$$

Hence, reaction and diffusion are properly combined for an arbitrary network when using the Bosanquet average of the purely diffusion quantity $(\mathbf{P} + diag(\mathbf{G}\mathbf{e}_n))^{-1}$ of the network and the purely reaction quantity $(diag(k\mathbf{S}_i))^{-1}$ of the network. It is interesting to note that even in the fully arbitrary case the Bosanquet average matrix $\mathbf{B}_0$ remains fully symmetrical. Hence, the canonical representation can be written as:

$$\mathbf{B}_o = \mathbf{W}\boldsymbol{\Lambda}\mathbf{W}^T \tag{4.320}$$

where $\boldsymbol{\Lambda}$ and $\mathbf{W}$ are, respectively, the eigenvalues and eigenvectors of the matrix $\mathbf{B}_0$. The following can also be derived for the fluxes at the perimeter of the network:

$$\mathbf{H}_e = \mathbf{Q} + diag\left(\mathbf{G}^T\mathbf{e}_m + k\mathbf{S}_e\right) - \boldsymbol{\Gamma}_e\mathbf{B}_o\boldsymbol{\Omega}_{i,D}\mathbf{C}_e \tag{4.321}$$

with:

$$\boldsymbol{\Gamma}_e = \mathbf{G}^T(diag(k\mathbf{S}_i))^{-1} \tag{4.322}$$

The elements of the matrix $\boldsymbol{\Gamma}_e$ are related to the individual fluxes from each reacting internal node to each of the external nodes when only that one connection is available.

4.3.4.4.4 The Meaning of $(\mathbf{P} + diag(\mathbf{G}\mathbf{e}_n + k\mathbf{S}))^{-1}$

I expect this to be similar to the case of a point source for diffusion without reaction. Again, beyond the external nodes, select an arbitrary internal node, call it node i, and declare it open. Define the concentration in that internal node to be equal to C^* and set all other external node concentrations equal to zero. Using the same method as used in Section 4.3.4.3.6, it can be shown that the flux going into that internal node is given by:

$$flux_i = C^*/(\mathbf{P} + diag(\mathbf{G}\mathbf{e}_n + k\mathbf{S}_i))_{i,i}^{-1} \tag{4.323}$$

The concentration in the other internal nodes j will be shown to be:

$$Conc_j = \ flux_i \times (\mathbf{P} + diag(\mathbf{Ge}_n + k\mathbf{S}_i))^{-1}_{i,j} \tag{4.324}$$

Hence, the matrix $(\mathbf{P} + \ diag\,(\mathbf{Ge}_n + k\mathbf{S}_i))^{-1}$ fully solves as many internal diffusional reaction problems as there are internal nodes. The proof goes along the same lines as for the case of diffusion only.

4.3.4.4.4.1 An Example Calculation of Reaction and Diffusion Consider as in the previous network of Figure 4.9 that each *internal* node now also has a first-order reaction coefficient k and that the surface area in the nodes is collected in the column vector **S**. Mathcad 15 software readily calculates the matrix of interest $(\mathbf{P} + diag\,(\mathbf{Ge}_n + k\mathbf{S}_i))^{-1}$ shown in Appendix 4.3. Consider, again, node 4 as an example. The flux entering node 4 is given by:

$$Flux = 1/0.4396 = 2.2748$$

The concentrations in nodes 1, 2, 3, and 4 are then given by dividing each row element by the diagonal element of that row:

$$C_1 = 0.0992/0.4396 = 0.2257$$

$$C_2 = 0.1191/0.4396 = 0.2709$$

$$C_3 = 0.0072/0.4396 = 0.0164$$

$$C_4 = 0.4396/0.4396 = 1$$

The material balance in node 4 then verifies:

$$Flux = 2.2748 = 0.7 \times (1 - 0.2709) + 0.6 \times (1 - 0.2257) + 1.3$$

This can then easily be verified for all other nodes as well.

As already mentioned in Section 4.1, I do not expect to see any more interesting relationships between network performance properties and matrix properties beyond the actual matrix equations, or at least I could not find any. Appendix 4.4 shows numerical calculations that illustrate the point.

4.3.4.4.4.2 An Example Calculation for a Perturbation in a Small Network Consider again the case of a simple network as shown in Figure 4.9, but now with a 1000-fold smaller rate coefficient than in Appendix 4.3. The Mathcad 15 calculations in Appendix 4.5 show that the smallest eigenvalue of $\mathbf{H}_e$ is very close to the sum of all elements of $\mathbf{H}_e$ (i.e. the total chemical rate in the network) divided by the number of external nodes. Although the network is arbitrary, it is very small, and hence it has an effectiveness factor near unity. Nonetheless, keeping the chemical rate coefficient a perturbation compared to the mass transfer coefficients retains the relationship of the minimum eigenvalue of $\mathbf{H}_e$, with the total chemical rate in the network interestingly just as it occurs in regular networks.

It is then that I realized that in order to investigate a network with a low effectiveness factor, instead of increasing the rate constant, a better approach is to keep the chemical rate constant at the level of a perturbation and instead increase the depth of the network. This is how Section 4.3.5 on very deep networks came into being. In addition, take a second look at the similarities with Section 4.3.3.3.3, starting with the paragraph prior to Eq. (4.83).

4.3.4.4.5 Diffusion and Reaction from a Point Source along Classic Lines

The mathematics of diffusion and reaction from a point source in a porous structure that is homogeneous and isotropic has also been covered in the open literature, but perhaps not as frequently as the case of diffusion only. Here, I only show essential results for reasons of continuity for the reader. Let D_e be the effective diffusion coefficient for diffusion in three-dimensional space. The DE that governs mass transport for diffusion and first-order reaction in a sphere of radius R around the source point is given by:

$$\frac{d}{dr}\left(D_e r^2 \frac{dc}{dr}\right) = r^2 kSc \tag{4.325}$$

or:

$$\frac{d}{d\xi}\left(\xi^2 \frac{dc}{d\xi}\right) = \left(\frac{R^2 kS}{D_e}\right)\xi^2 c = \phi^2 \xi^2 c \tag{4.326}$$

where S is the specific surface area per unit volume while k is the first-order rate constant expressed in moles reacted per unit surface area and per unit time. The symbol ϕ stands for the well-known Thiele modulus. The symbol ξ stands for the dimensionless distance from the center of the sphere. Again, imposing a mass flux F^* at the point source and assuming a zero concentration along the surface of the sphere allows one to solve for the unknown concentration profile:

$$c(\xi) = \frac{F^*}{4\pi R D_e}\frac{sinh\left[\phi(1-\xi)\right]}{\xi sinh\phi} \tag{4.327}$$

Calculating the volume average concentration C_{avg} then yields:

$$c_{avg} = \frac{F^*}{8\pi R D_e} 6\left[\frac{1}{\phi^2} - \frac{1}{\phi \sinh(\phi)}\right] \tag{4.328}$$

The continuum approach and the discrete approach then allows the following relationship:

$$\frac{D_e}{6\left[\frac{1}{\phi^2} - \frac{1}{\phi \sinh(\phi)}\right]} = \frac{1}{8\pi R}\left((n+m)/\sum_{\substack{j=1,m \\ j\neq *}} (\mathbf{P} + diag(\mathbf{Ge}_n + k\mathbf{S}_i))^{-1}_{*,j}\right) \tag{4.329}$$

To preserve the same value of the rate coefficient in the continuum approach and the network approach, be aware of the need to balance the total surface area for either:

$$(4\pi R^3/3)S = n\sum_i S_i \tag{4.330}$$

Inserting the expression for D_e from Eq. (4.289) into Eq. (4.329), I get:

$$6\left[\frac{1}{\phi^2} - \frac{1}{\phi \sinh(\phi)}\right] = \frac{\sum_{\substack{j=1,m \\ j\neq *}} (\mathbf{P} + diag(\mathbf{Ge}_n + k\mathbf{S}_i))^{-1}_{*,j}}{\sum_{\substack{j=1,m \\ j\neq *}} (\mathbf{P} + diag(\mathbf{Ge}_n))^{-1}_{*,j}} \tag{4.331}$$

The right-hand side of Eq. (4.331), which is determined by the network properties, is now linked to the classic Thiele modulus used in the function on the left-hand side of the same equation.

4.3.4.4.6 A "Black Body" Catalyst: A Closed Sphere with a Pinhole Access Leading to the Center

Consider that the catalytic sphere has a point opening located on the surface with a narrow channel leading to the center of the sphere. Apart from the pinhole, the rest of the surface of the sphere is closed off. Again, we will assume a first-order reaction. Because of the closure of the surface, the concentration gradient on the inside of the sphere at the surface is zero. The solution method is the same as what is used in Section 4.3.4.4.5 and leads to the following results:

$$c(\xi) = \frac{F^*}{4\pi RD_e} \times \frac{\phi cosh[\phi(1-\xi)] - sinh[\phi(1-\xi)]}{\phi cosh\phi - sinh\phi} \times \frac{1}{\xi} \tag{4.332}$$

and:

$$\frac{F^*}{c_{avg}} = \frac{4\pi RD_e}{3} \times \phi^2 \tag{4.333}$$

4.3.4.4.7 Alternative Solution Method

There is an alternative solution method available for a first-order reaction. This can be accomplished by adding a single virtual node to the existing network. By definition, this virtual node is considered open (i.e. it allows for flux with the bulk fluid). Let this virtual node have a mass transfer connection to each internal node as well as to each external node. As we will see, the use of the virtual node allows one to transform the problem of diffusion and reaction into a problem of diffusion only. The mass transfer coefficient from each node to the virtual node is selected to be equal to the corresponding element of kS_i for each internal node and to the corresponding element of kS_e for each external node. Chemical reaction in each of the nodes is set to zero. In addition, let the concentration in the virtual node be set to zero. With this arrangement, the diffusion flux from each network node to the virtual node plays the role of the first-order reaction in that node. With this virtual node, the problem of diffusion and reaction has now been turned into a pure diffusion problem, albeit with one extra node. The total flux exiting the virtual node is then also identical to the total amount of reagent reacted away in the network. The mass transfer matrix dimension has now been augmented by one, and the overall matrix equation is now given by:

$$\begin{vmatrix} \mathbf{P} + diag(\mathbf{G}\mathbf{e}_n + k\mathbf{S}_i) & -\mathbf{G} & -k\mathbf{S}_i \\ -\mathbf{G}^T & \mathbf{Q} + diag(\mathbf{G}^T\mathbf{e}_m + k\mathbf{S}_e) & -k\mathbf{S}_e \\ -k{\mathbf{S}_i}^T & -k{\mathbf{S}_e}^T & kS_N \end{vmatrix} \times \begin{vmatrix} \mathbf{C}_i \\ \mathbf{C}_e \\ 0 \end{vmatrix} = \begin{vmatrix} \mathbf{0} \\ \mathbf{F}_e \\ F_v \end{vmatrix} \tag{4.334}$$

where S_N represents the total surface area of the network:

$$S_N = {\mathbf{S}_i}^T\mathbf{e}_i + {\mathbf{S}_e}^T\mathbf{e}_e \tag{4.335}$$

and F_v represents the total flux exiting the network via the virtual node. For ease of notation, let:

$$\mathbf{A} = \mathbf{P} + diag(\mathbf{G}\mathbf{e}_n + k\mathbf{S}_i) \tag{4.336}$$

and:

$$\mathbf{B} = \mathbf{Q} + diag(\mathbf{G}^T\mathbf{e}_m + k\mathbf{S}_e) \tag{4.337}$$

Partitioning the material balance matrix just below and just to the right of the matrix $\mathbf{A}$ and writing out the equation leads to the following expected relationship for $\mathbf{C}_i$:

$$\mathbf{A}\mathbf{C}_i - |\mathbf{G} \quad k\mathbf{S}_i| \begin{vmatrix} \mathbf{C}_e \\ 0 \end{vmatrix} = \mathbf{0} \tag{4.338}$$

or:

$$\mathbf{C}_i = \mathbf{A}^{-1}|\mathbf{G} \quad k\mathbf{S}_i| \begin{vmatrix} \mathbf{C}_e \\ 0 \end{vmatrix} = \mathbf{\Omega}_i\mathbf{C}_e \tag{4.339}$$

The perimeter fluxes and the flux from the virtual node are obtained from:

$$\begin{vmatrix} \mathbf{F}_e \\ F_v \end{vmatrix} = - \begin{vmatrix} \mathbf{G}^T \\ k\mathbf{S}_i^T \end{vmatrix} \mathbf{C}_i + \begin{vmatrix} \mathbf{B} & -k\mathbf{S}_e \\ -k\mathbf{S}_e^T & kS_N \end{vmatrix} \begin{vmatrix} \mathbf{C}_e \\ 0 \end{vmatrix} \tag{4.340}$$

or:

$$\begin{vmatrix} \mathbf{F}_e \\ F_v \end{vmatrix} = \left\{ - \begin{vmatrix} \mathbf{G}^T \\ k\mathbf{S}_i^T \end{vmatrix} \mathbf{A}^{-1} |\mathbf{G} \quad k\mathbf{S}_i| + \begin{vmatrix} \mathbf{B} & -k\mathbf{S}_e \\ -k\mathbf{S}_e^T & kS_N \end{vmatrix} \right\} \begin{vmatrix} \mathbf{C}_e \\ 0 \end{vmatrix} \tag{4.341}$$

and:

$$\begin{vmatrix} \mathbf{F}_e \\ F_v \end{vmatrix} = \left\{ - \begin{vmatrix} \mathbf{G}^T\mathbf{A}^{-1}\mathbf{G} & \mathbf{G}^T\mathbf{A}^{-1}k\mathbf{S}_i \\ k\mathbf{S}_i^T\mathbf{A}^{-1}\mathbf{G} & k\mathbf{S}_i^T\mathbf{A}^{-1}k\mathbf{S}_i \end{vmatrix} + \begin{vmatrix} \mathbf{B} & -k\mathbf{S}_e \\ -k\mathbf{S}_e^T & kS_N \end{vmatrix} \right\} \begin{vmatrix} \mathbf{C}_e \\ 0 \end{vmatrix} \tag{4.342}$$

or:

$$\begin{vmatrix} \mathbf{F}_e \\ F_v \end{vmatrix} = \begin{vmatrix} \mathbf{H}_e & -\mathbf{\Omega}_i^T k\mathbf{S}_i - k\mathbf{S}_e \\ -k\mathbf{S}_i^T\mathbf{\Omega}_i - k\mathbf{S}_e^T & kS_N - k\mathbf{S}_i^T\mathbf{A}^{-1}k\mathbf{S}_i \end{vmatrix} \begin{vmatrix} \mathbf{C}_e \\ 0 \end{vmatrix} = \mathbf{H}_v \begin{vmatrix} \mathbf{C}_e \\ 0 \end{vmatrix} \tag{4.343}$$

The matrix $\mathbf{H}_v$ is thus augmented in size from the matrix $\mathbf{H}_e$ by one extra column and one extra row. The off-diagonal elements in the far right column contain the fluxes toward the virtual node and hence represent the actual chemical rates in each node in the presence of the diffusional limitations in the network. The term $(kS_N - k\mathbf{S}_i^T\mathbf{A}^{-1}k\mathbf{S}_i)$ represents the sum total of all of the chemical rates. This can be more easily understood by realizing that the problem of diffusion only with the virtual node at zero concentration and all other external node concentrations at unity is the same as putting the virtual node concentration equal to unity and keeping all of the other external nodes at zero. This allows one to obtain the effectiveness factor η of the network from:

$$\eta = 1 - k\mathbf{S}_i^T\mathbf{A}^{-1}\mathbf{S}_i/S_N \tag{4.344}$$

or more specifically:

$$\eta = 1 - k\mathbf{S}_i^T[\mathbf{P} + diag(\mathbf{G}\mathbf{e}_n + k\mathbf{S}_i)]^{-1}\mathbf{S}_i/S_N \tag{4.345}$$

From this, it is clear that as k approaches zero the effectiveness factor approaches unity, as is to be expected. One can perform the following manipulation:

$$\eta = 1 - k\mathbf{S}_i^T(diag(k\mathbf{S}_i))^{-1}\left[(\mathbf{P} + diag(\mathbf{G}\mathbf{e}_n))(diag(k\mathbf{S}_i))^{-1} + \mathbf{I}\right]^{-1}\mathbf{S}_i/S_N \tag{4.346}$$

which simplifies to:

$$\eta = 1 - \mathbf{e}_i^T \left[(\mathbf{P} + diag(\mathbf{Ge}_n))(diag(k\mathbf{S}_i))^{-1} + \mathbf{I} \right]^{-1} \mathbf{S}_i / S_N \tag{4.347}$$

For large values of the rate constant k, the matrix $(\mathbf{P} + diag\,(\mathbf{Ge}_n))\,diag\,(k\mathbf{S}_i)^{-1}$ becomes small; hence, we can approximate the inverse above as:

$$\left[(\mathbf{P} + diag(\mathbf{Ge}_n))(diag(k\mathbf{S}_i))^{-1} + \mathbf{I} \right]^{-1} \cong \mathbf{I} - (\mathbf{P} + diag(\mathbf{Ge}_n))(diag(k\mathbf{S}_i))^{-1} \tag{4.348}$$

For large k, one therefore obtains:

$$\eta(k \gg 1) = 1 - \mathbf{e}_i^T \left[\mathbf{I} - (\mathbf{P} + diag(\mathbf{Ge}_n))(diag(k\mathbf{S}_i))^{-1} \right] \mathbf{S}_i / S_N \tag{4.349}$$

or:

$$\eta(k \gg 1) = 1 - \frac{\mathbf{e}_i^T \mathbf{S}_i}{S_N} + \mathbf{e}_i^T (\mathbf{P} + diag(\mathbf{Ge}_n))(diag(k\mathbf{S}_i))^{-1} \mathbf{S}_i / S_N \tag{4.350}$$

The first two terms in the expression above, $\left(1 - \mathbf{e}_i^T \mathbf{S}_i / S_N\right)$, represent the fraction of the total surface that is located in the external nodes. Since the external nodes are not considered mass transfer limited, the chemical rates in these nodes will never be hampered, and hence the contribution to the overall effectiveness factor will, for a large enough k, become equal to the fraction of the total surface area that is located in the perimeter nodes, as it should. The expression above can be further simplified to:

$$\eta(k \gg 1) = 1 - \frac{\mathbf{e}_i^T \mathbf{S}_i}{S_N} + \mathbf{e}_i^T (\mathbf{P} + diag(\mathbf{Ge}_n)) \mathbf{e}_i / k S_N \tag{4.351}$$

Since $\mathbf{P}$ is a diffusional matrix, we have:

$$\mathbf{e}_i^T \mathbf{P} \mathbf{e}_i = 0 \tag{4.352}$$

and hence:

$$\eta(k \gg 1) = 1 - \frac{\mathbf{e}_i^T \mathbf{S}_i}{S_N} + \mathbf{e}_i^T diag(\mathbf{Ge}_n) \mathbf{e}_i / k S_N \tag{4.353}$$

The expression $\mathbf{e}_i^T diag(\mathbf{Ge}_n)\mathbf{e}_i$ represents the sum of all mass transfer coefficients from the external nodes or "perimeter of the network" into the network, and it also represents the highest possible mass transfer rate into the network when the node activities are assumed to be very high and hence when the concentrations in the subsurface nodes can be considered zero. Dividing this highest possible rate by the chemical activity kS_N of the entire network yields the logical contributing expression for the overall effectiveness factor.

4.3.4.4.8 Network Perimeter Blocking

The two equations that allow one to calculate the internal concentrations and the fluxes going into the network at the perimeter when all external nodes are open are:

$$\mathbf{C}_i = \mathbf{\Omega}_i \mathbf{C}_e \tag{4.354}$$

$$\mathbf{F}_e = \mathbf{H}_e \mathbf{C}_e \tag{4.355}$$

Assume now that of those external nodes, a number of them have been blocked off for flow and reaction. These blocked nodes seen from the inside form a dead end. Hence, overall mass transfer in and out of the network is becoming hampered and therefore will lead to a smaller effectiveness factor. This perimeter blocking issue is easily handled with the current matrix approach. Rearrange the matrix $\mathbf{H}_e$ by switching columns and respective rows such that blocked external nodes are in one group and the non-blocked external nodes are in the other group. Then the matrix can be partitioned and rewritten as:

$$\begin{vmatrix} \mathbf{0} \\ \mathbf{F}_{nb} \end{vmatrix} = \begin{vmatrix} \mathbf{H}_1 & \mathbf{H}_2 \\ \mathbf{H}_3 & \mathbf{H}_4 \end{vmatrix} \begin{vmatrix} \mathbf{C}_b \\ \mathbf{C}_{nb} \end{vmatrix} \tag{4.356}$$

This then yields:

$$\mathbf{0} = \mathbf{H}_1 \mathbf{C}_b + \mathbf{H}_2 \mathbf{C}_{nb} \tag{4.357}$$

$$\mathbf{F}_{nb} = \mathbf{H}_3 \mathbf{C}_b + \mathbf{H}_4 \mathbf{C}_{nb} \tag{4.358}$$

Solving leads to:

$$\mathbf{C}_b = -\mathbf{H}_1^{-1} \mathbf{H}_2 \mathbf{C}_{nb} \tag{4.359}$$

$$\mathbf{F}_{nb} = \left(-\mathbf{H}_3 \mathbf{H}_1^{-1} \mathbf{H}_2 + \mathbf{H}_4\right) \mathbf{C}_{nb} \tag{4.360}$$

The matrix $\left(-\mathbf{H}_3 \mathbf{H}_1^{-1} \mathbf{H}_2 + \mathbf{H}_4\right)$ is the Schur complement of the matrix $\mathbf{H}_1$ in the rearranged matrix $\mathbf{H}_e$.

4.3.4.4.9 Considering Surface and Subsurface Nodes

Consider the following mass balance matrix:

$$\mathbf{M} = \begin{vmatrix} \mathbf{P} + diag(\mathbf{G}\mathbf{e}_n + k\mathbf{S}_i) & -\mathbf{G} \\ -\mathbf{G}^T & \mathbf{Q} + diag\left(\mathbf{G}^T \mathbf{e}_m + k\mathbf{S}_e\right) \end{vmatrix} \tag{4.361}$$

For this system, we get as the solution:

$$\mathbf{H}_e = \mathbf{Q} + diag\left(\mathbf{G}^T \mathbf{e}_m + k\mathbf{S}_e\right) - \mathbf{G}^T (\mathbf{P} + diag(\mathbf{G}\mathbf{e}_n + k\mathbf{S}_i))^{-1} \mathbf{G} \tag{4.362}$$

$$\mathbf{\Omega}_i = (\mathbf{P} + diag(\mathbf{G}\mathbf{e}_n + k\mathbf{S}_i))^{-1} \mathbf{G} \tag{4.363}$$

The set of internal nodes $\mathbf{C}_i$ and external nodes $\mathbf{C}_e$ can be augmented with a new set of nodes that only connect to the previous external nodes and have no connections with any of the internal nodes. The new nodes effectively take over from the old external nodes and reduce the old external nodes to subsurface nodes at a concentration C_{ss}. The new material balance matrix then becomes:

$$\mathbf{M}' = \begin{vmatrix} \mathbf{P} + diag(\mathbf{G}\mathbf{e}_n + k\mathbf{S}_i) & -\mathbf{G} & \mathbf{0} \\ -\mathbf{G}^T & \mathbf{Q} + diag\left(\mathbf{G}^T\mathbf{e}_m + k\mathbf{S}_e + \mathbf{U}\mathbf{e}_k\right) & -\mathbf{U} \\ \mathbf{0}^T & -\mathbf{U}^T & \mathbf{R} + diag\left(\mathbf{U}^T\mathbf{e}_k + k\mathbf{S}_e{}'\right) \end{vmatrix} \tag{4.364}$$

One obtains the following matrix equations:

$$(\mathbf{P} + diag(\mathbf{G}\mathbf{e}_n + k\mathbf{S}_i))\mathbf{C}_i - \mathbf{G}\mathbf{C}_{ss} = \mathbf{0} \tag{4.365}$$

$$-\mathbf{G}^T\mathbf{C}_i + \left(\mathbf{Q} + diag\left(\mathbf{G}^T\mathbf{e}_m + k\mathbf{S}_e + \mathbf{U}\mathbf{e}_k\right)\right)\mathbf{C}_{ss} - \mathbf{U}\mathbf{C}_e = \mathbf{0} \tag{4.366}$$

$$-\mathbf{U}^T\mathbf{C}_{ss} + \left(\mathbf{R} + diag\left(\mathbf{U}^T\mathbf{e}_k + k\mathbf{S}_e{}'\right)\right)\mathbf{C}_e = \mathbf{F}_e' \tag{4.367}$$

The first equation yields:

$$\mathbf{C}_i = (\mathbf{P} + diag(\mathbf{G}\mathbf{e}_n + k\mathbf{S}_i))^{-1}\mathbf{G}\mathbf{C}_{ss} = \mathbf{\Omega}_i\mathbf{C}_{ss} \tag{4.368}$$

The second equation yields:

$$\left(-\mathbf{G}^T\mathbf{\Omega}_i + \mathbf{Q} + diag\left(\mathbf{G}^T\mathbf{e}_m + k\mathbf{S}_e + \mathbf{U}\mathbf{e}_k\right)\right)\mathbf{C}_{ss} - \mathbf{U}\mathbf{C}_e = 0 \tag{4.369}$$

or:

$$(\mathbf{H}_e + diag(\mathbf{U}\mathbf{e}_k))\mathbf{C}_{ss} - \mathbf{U}\mathbf{C}_e = 0 \tag{4.370}$$

Hence:

$$\mathbf{C}_{ss} = (\mathbf{H}_e + diag(\mathbf{U}\mathbf{e}_k))^{-1}\mathbf{U}\mathbf{C}_e \tag{4.371}$$

and then finally:

$$\mathbf{H}_e' = -\mathbf{U}^T(\mathbf{H}_e + diag(\mathbf{U}\mathbf{e}_k))^{-1}\mathbf{U} + \left(\mathbf{R} + diag\left(\mathbf{U}^T\mathbf{e}_k + k\mathbf{S}_e{}'\right)\right) \tag{4.372}$$

This then gives the new matrix $\mathbf{H}_e'$ that allows one to calculate the fluxes from the arbitrarily chosen external concentrations based on the old matrix $\mathbf{H}_e$. This relationship is of the Riccati type.

The method shown here is preferred for solving large networks numerically because of its high speed and minimized memory usage for large sparse matrices.

4.3.4.4.10 Non-first-order Reaction

Here, we consider reaction rates that are not first order but still have a monotonous response to their dependence on concentration. The case of a zero-order reaction is not considered since it requires very careful inventory of concentrations in nodes due to the limited supply of reactant. Let the quantity $\mathcal{R}(c)$ represent the actual rate expression in the node and let it include the surface area dependence. We can still apply the virtual node approach by considering a mass transfer coefficient that is set by the reaction rate:

$$k = \mathcal{R}(c)/c \tag{4.373}$$

The material balance equation with the k-values column-vectorized is:

$$\begin{vmatrix} \mathbf{P}+ diag(\mathbf{G}\mathbf{e}_n + \mathcal{R}(C_i)/C_i) & -\mathbf{G} & -\mathcal{R}(C_i)/C_i \\ -\mathbf{G}^T & \mathbf{Q} + diag\left(\mathbf{G}^T\mathbf{e}_m + \mathcal{R}(C_e)/C_e\right) & -\mathcal{R}(C_e)/C_e \\ -(\mathcal{R}(C_i)/C_i)^T & -(\mathcal{R}(C_e)/C_e)^T & \Sigma \end{vmatrix} \times \begin{vmatrix} \mathbf{C_i} \\ \mathbf{C_e} \\ 0 \end{vmatrix} = \begin{vmatrix} 0 \\ \mathbf{F}_e \\ F_v \end{vmatrix} \tag{4.374}$$

with:

$$\Sigma = \Sigma\mathcal{R}(C_i)/C_i + \Sigma\mathcal{R}(C_e)/C_e \tag{4.375}$$

The procedure can be started by using an initial estimate of the concentrations and calculating from that the virtual mass transfer coefficients. This then allows one to calculate the hopefully improved node concentrations, which then iteratively can be used again to calculate the improved mass transfer coefficients. This procedure is continued until the concentration values in the internal nodes converge. Other numerical scenarios may need to be explored if convergence fails.

4.3.5 Treatise on Very Deep Networks

So far, I have considered general networks with a first-order reaction in the nodes and mass transfer by diffusion from node to node. Closing all of the material balances in each node allows one to solve the network for the concentrations in all of the internal nodes. Except for a zero minimum eigenvalue of the Schur complement in the case of diffusion only, I have no other specific statements on general networks beyond the actual solutions in matrix form.

Most real commercial catalyst extrudates, pellets, and granules range in size from around 30 μm up to 3 mm. Catalyst pellets are made up of tiny constituent particulates, and the interstitial space between them creates a massive porous network. The voids (pores, channels, nodes) in the structure of an amorphous catalyst are typically ~2–50 nm in size, but they can range widely from 1 to 2000 nm. The void space is created during the preparation or the manufacture of the raw materials that make up the catalyst or during the shaping and forming of the catalyst itself. The constituent solid particles can be viewed as stacked on top of each other in disordered layers, and they expose their surface chemistry to the reactants. The stacking of the particles can be categorized as belonging to layers perpendicular to the direction of interest, just like the layers of an onion. The networks that are created in this way are typically ~1000 to 100 000 layers deep, with many thousands of nodes in each layer. I mention here a similarity with standing on a bed of pebbles. One can clearly see the first layer of pebbles. The voids between the pebbles allow one to see the layer underneath to some degree, but this second layer now also effectively blocks further viewing straight down. The second layer of pebbles requires a redirection for further viewing to proceed. As such, the voids form an irregular network of nodes and channels for mass transfer through the bed of pebbles. Similarly, in a catalyst, a molecule can only follow the irregular open paths in the void space back and forth and sideways as it penetrates the catalyst structure (see cover for an example).

To give a pictorial of how large such a massive network is in a single catalyst pellet, consider a commercial cylindrical extrudate of 1.6 mm in diameter. Multiply the size by 1000 so that the extrudate diameter now becomes 1.6 m or approximately the height of an average person. The length of an extrudate is typically about three to five times the diameter. If the density were 1 g/cm^3, then the weight of the scaled extrudate would be 8000 kg, which is rather substantial. The scaled diameter of a 10 nm pore opening on the catalyst external surface then becomes 10 μm, or approximately half the diameter of the thinnest human hair. These scaled "pore openings" are then densely located next to each other, separated by only a very few tens of microns. A back-of-the-envelope calculation shows that the areal density of the pore openings on the scaled surface is approximately 10 000 times higher than the average areal density of hair on a human scalp. Entering an opening on the scaled surface would, within a few tens of microns of the scaled extrudate, lead to a junction in the void space formed by neighboring particulates with multiple possible routes to follow. Each route shows abrupt differences at that very location from the other routes, and at every junction the process repeats over the many thousands of nodes and layers. This simple picture hopefully gives an idea of the enormous size and complexity of a real catalyst network.

In this last section, I call the catalyst node network a VDN, standing for "very deep network." The word "deep" is chosen over the word "large" because "deep" has the intuitive connotation that the network extends far (deep) in the direction of interest (i.e. in the direction leading from the surface of the particle to its core). A large network, on the other hand, does not necessarily need to extend deep.

As with regular networks, the irregular network is solved by closing the material balance in every node. With this approach, local mass balance closure in each node is guaranteed. This task – initially at least – appears to be monumental, and that is indeed the case for truly arbitrary networks (in the mathematical sense).

To describe the irregular void space of the catalyst, I will use "white noise" variability to quantify the abrupt changes in the local structure of a catalyst. White noise variability here means that the value of a variable (e.g. a mass transfer coefficient) from one point to a neighboring point along a coordinate changes abruptly and yields a spiked pattern of values of that variable along the coordinate. The changes are random but always occur according to a given unchanging distribution. For only a few steps along the coordinate, the pattern is very irregular and random. For many hundreds or thousands of steps, however, the value of the variable acts like a well-behaved smooth variable within bounds set by the distribution. The number of steps required to get a smooth behavior depends on the width of the distribution: the wider the distribution, the more steps required.

The concentration of reagent or product along the path varies and is determined by the abrupt behavior of the variables at play and by the local mass balance closures along the path. Locally, the value of the concentration changes abruptly from node to neighboring node. With a large enough number of nodes (i.e. a small enough step size), the pattern of the concentration becomes well behaved. Be aware that the pattern of the concentration is well behaved for a VDN (i.e. the concentration varies locally within certain small variability bounds), but the "derivative" of the concentration remains highly irregular (i.e. the derivative is discontinuous and the concentration is "not differentiable" along the path). This paragraph outlines what I meant in Section 4.1 by solutions that are "analytical but of a special kind." The concentration converges to a solution within any preset level of variability bounds but is not differentiable.

For real catalyst porous networks, I will demonstrate that the VDN concept allows one to obtain a very specific solution to mass transfer. The solution applies to a VDN that has "white noise" variability in the node-to-node mass transfer coefficient v. The solution of the discrete network converges to the *mathematical form* of the classic continuum solution in the case of a VDN. With the VDN concept, there is closure of the material balance in each node and no need for a tortuosity factor.

Finally, in order to deal with a first-order reaction in a VDN, I introduce perturbation induced by chemical reaction, here called VDN perturbation (VDNP). On top of the white noise variability in mass transfer from node to node, I allow white noise variability of the surface area S allocated to the nodes, as well as white noise variability of the rate coefficient k. The white noise variability for the mass transfer, surface area, and rate coefficient is always according to its own specific distribution and is free to be chosen by the reader. The distribution of the rate coefficient allows one to introduce a variety of chemistries: an acid site strength distribution, a metal cluster size distribution, or simply active site density variations from node to node. Surprisingly, with this approach, here I will also show convergence to the *mathematical form* of the classic continuum solutions, as was done with the regular networks. The convergence to the mathematical form of the classic solutions is a major advantage for VDNP. The parameter in the mathematical form is a variant of the classic Thiele modulus. With VDNP, the parameter is now based on the distribution of the variables and the architecture of the network. VDNP imposes explicit closure of all material balances in all of the nodes but nevertheless benefits from the classic solutions already established over nearly a century. Here, again, no tortuosity factor is required.

The VDN and VDNP approach can take any distribution of properties of interest, but the wider the distribution, the deeper the network that is required. What I mean by this is that the network needs to be large enough such that the local irregularities have a chance to be evened out over the long-range architecture of the network. However, it will be shown that these networks are still very manageable in terms of size, even when handling a very broad distribution. The specific statements I will make later on apply to the long-range architecture of the catalyst and not to the local irregularities.

In all of the examples in this book, I have used distributions for k, S and v that are independent of each other. I do not think that they necessarily have to be independent. As long as the three parameters are selected on a consistent basis and do not lead to long-range directional pattern variations, I expect VDNP to work satisfactorily.

The "onion layer"-like structure of real catalysts allows one to solve the VDN and VDNP approach very efficiently and requires one to deal with matrices that are only a small fraction of the size of the overall material balance matrix.

To clarify regarding perturbation: the "P" in VDNP means that here it is considered that the reaction component (i.e. the reactivity in a node) is a mere perturbation imposed on the diffusion field (i.e. the mass transfer coefficients linking the node to the neighboring nodes). The actual distance between the nodes is exceedingly small, and Beeckman [15, 20] has shown that this distance typically amounts to only a few pore or channel diameters. Because of this short distance, the mass transfer coefficients become exceedingly large. For real catalysts, the chemical reactivity is distributed over all of the nodes. For a VDN representing a single catalyst extrudate or a catalyst bead, the more nodes that are chosen, the lower the reactivity per node has to be. The reason for this is that the total chemical reactivity should be constant for a pellet (think of the metal content, the total acid site count, or the active site count for the total chemical reactivity in a catalyst pellet). Hence, for a VDN, the chemical rate coefficient kS in a node becomes exceedingly small compared to the mass transfer coefficients, and therefore reaction can be viewed as a mere perturbation of diffusion in a single-node arrangement.

The term "perturbation" points to a small contribution of the reaction to the overall material balance closure by diffusion in each node. This would at first sight imply little mass transfer limitation, and this statement would be true if the network were small. However, here I consider VDNs with nodes that extend deep and far into the particle. This way, VDNP allows for networks with large and small effectiveness factors and hence catalyst particles with weak and strong mass transfer limitations.

For arbitrary networks one can consider any distribution of properties, but a Gaussian or also a uniform distribution appears to make a lot of sense. For a random reaction perturbation introduced on top of the diffusion, the mathematics draws parallels with the field of stochastic DEs, except of course that here there is no time dependence, since we are dealing with steady-state conditions. *Overall, the important point is that even though the distribution of properties is allowed to be wide and randomized from node to node, the numerical solutions converge functionally to the classic continuum solutions in the case of VDNP.*

A point I want to stress is that the *distribution of properties* should not change over the entire size of the network (i.e. the catalyst pellet). Locally, at the node level, properties do of course change, and they do so abruptly, as one would expect for a real catalyst. Changes in the distribution would only occur for catalyst particles that have long-range inhomogeneity in their makeup. The issue is not that the VDNP approach would not work – on the contrary, it would. The issue is that typical commercial catalysts are prepared with great care to avoid such inhomogeneity and hence can be considered to have a *constant distribution* of properties throughout the entire catalyst pellet. For applications, I think it is most important to consider networks of such a size that local irregularities are evened out for the depth considered. How exactly to accomplish this will become clear later on when we are dealing with the minimum eigenvalues of specific matrices and the specific numerical accuracy the reader imposes.

I refer to Chen, Degnan, and Smith [9], who point to two drawbacks of a purely stochastic approach. The authors duly highlight first that the numerical task is substantial in order to get to a meaningful average for a particular property, and second that it is difficult to obtain generalizations. It is in both of these areas that I hope to make a contribution to the literature with this last section.

The challenge with random variables, stochastic DEs, and high local variability combined with smooth long-range behavior for catalysts lies in identifying the appropriate mathematical framework to use with the problem at hand.

For catalyst VDN structures, and for VDNP to include reaction, it appears to me that the matrix approach here presented is very capable of dealing with the intricacies of real catalyst porous networks. *It will be demonstrated that the minimum eigenvalue of a specific matrix is uniquely linked to the correct average of certain properties, such as the overall mass transfer coefficient through a network and the overall rate of reaction in a network that displays white noise variability. This linkage is crucial in order to define the correct parameters to use in the classic Thiele modulus definition. For VDNP, the Thiele modulus, the effectiveness factor, and the concentration profile along the direction of interest merge with the classic continuum approach to any degree of accuracy desired. The greater the degree of accuracy that is chosen, the deeper the network and the tinier the reaction perturbation have to be.*

Lastly, I point to an advantage of the network approach over the continuum approach in that the network approach allows one to treat arbitrary "point sources" with great ease, while the continuum approach quickly becomes quite difficult to solve analytically, even with the advantage of homogeneity.

4.3.5.1 Random Variables

I want to now introduce a random variable and its distribution. The random variable in the context of this book can be many things, such as a distance in terms of step size, a surface area in a node, a rate constant in a node, a mass transfer coefficient between two nodes, etc. Each time the random variable is sampled, it is done so according to a distribution that is free to be chosen and defined by the reader. In the worked examples in this book, I use a uniform distribution for ease of calculation. A fundamental method for generating the sample values of a random variable according to a given arbitrary distribution is called "inverse transform sampling," and I refer the reader to the general literature for more on this method. I prefer one particular write-up on this subject by Boucher [21].

4.3.5.1.1 The Distribution of Random Variables

The distribution of a random variable x is called $g(x)$ and the probability that a random sample of the variable ends up in the interval $(x, x + dx)$ is given by $g(x)dx$. The distribution is normalized, which means that:

$$\int_{-\infty}^{+\infty} g(x)dx = 1 \tag{4.376}$$

The mean μ of the variable x is defined as:

$$\mu = \int_{-\infty}^{+\infty} xg(x)dx \tag{4.377}$$

The harmonic mean μ_h of the variable will come in handy and is obtained from:

$$\mu_h^{-1} = \int_{-\infty}^{+\infty} x^{-1}g(x)dx \tag{4.378}$$

For a uniform distribution within the two positive limits a and $a + b$ (with $b > 0$) and zero everywhere else, I obtain:

$$g(x) = \frac{1}{b} \text{ for } a \leq x \leq a + b \tag{4.379}$$

$$\mu = a + b/2 \tag{4.380}$$

$$\mu_h = \frac{b}{\ln\left(\frac{a+b}{a}\right)} \tag{4.381}$$

while a random sample of the variable is obtained with the Mathcad 15 software as:

$$x = a + rnd(b) \tag{4.382}$$

where $rnd(b)$ is a random number generated between the limits (0, b).

For a Gaussian distribution with mean μ and variance σ^2, the literature yields:

$$g(x) = \frac{1}{\sqrt{2\pi\sigma^2}} e^{-\frac{(x-\mu)^2}{2\sigma^2}} \tag{4.383}$$

The Mathcad 15 software here also conveniently generates a random sample of the variable x conforming to the Gaussian distribution with a mean μ and a variance σ^2. I refer the reader to the literature for the harmonic mean of the variable.

4.3.5.2 Convergence for the Network Approach

The convergence of the actual value of the discrete variable (but not any of the derivatives) to the exact limit case is what I mean by convergence of a discrete variable to the classic solution of DEs. I will show that discrete networks of nodes converge as close as one wants to the mathematical form of the classic continuum models by using VDN and VDNP. The solution with the network approach retains all of the material balance closures in the nodes, no matter the size of the network.

As an example of convergence, consider the distance traveled going left to right as a summation of small discrete steps. Each step is considered to be a sample of the random variable and each step is sampled from a constant distribution. Here, I use a step size s_i sampled from a uniform distribution between the two limits a and $a + b$. The step size s_i for a single step can then be written as per Eq. (4.384):

$$s_i = a + rnd(b) \tag{4.384}$$

Starting from the origin, the coordinate x_i reached after step i is obtained from Eq. (4.385):

$$x_i = x_{i-1} + s_i \tag{4.385}$$

with:

$$x_0 = 0 \tag{4.386}$$

Normalize all of the coordinates by dividing each coordinate by the final coordinate x_n after n steps as per Eq. (4.387):

$$y_i = x_i/x_n \tag{4.387}$$

Thus, for any number of steps, I travel from the origin to the point with a coordinate equal to unity in exactly n steps randomly chosen from a distribution. Figure 4.10 shows the trace of the travel for three trials each with a 10-fold increase in the number of steps from the previous one. Appendix 4.6 shows the Mathcad code used to generate the traces. For a large number of steps, the trace or the y-coordinate increases almost linearly with the x-coordinate and "acts" as a continuous variable. However, on close examination, the trace is not continuously differentiable and remains so no matter how large the number of steps chosen. In the sense of "convergence," the value of the trace or the y-coordinate at every point can be brought as close as one wants to the limit case $y = i/n$ by using a large enough number of steps.

4.3.5.3 Diffusion in Deep Networks

4.3.5.3.1 Diffusion along a Deep String of Nodes

Consider a string of nodes as shown in Figure 4.1 that are connected by channels with a mass transfer coefficient ν that is a random variable. The string has N internal nodes and two open nodes at each end of the string. At the far left open node of

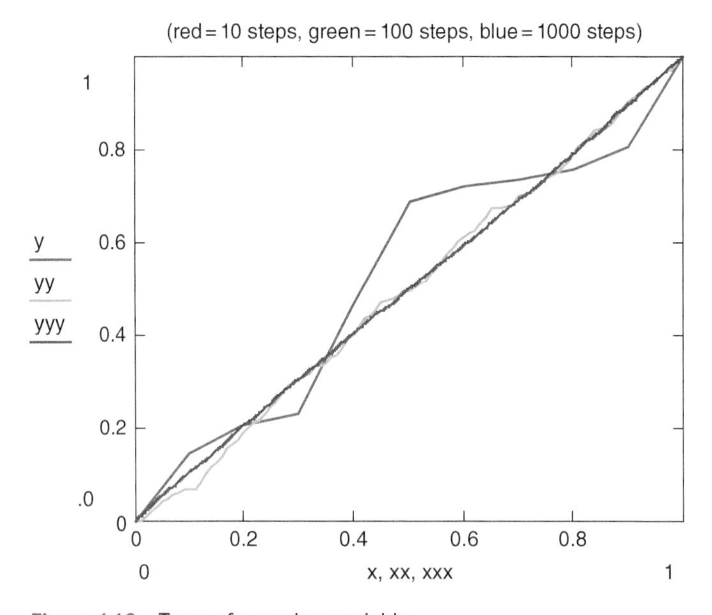

Figure 4.10 Trace of a random variable.

the string, the concentration of reagent is set at C_A, while at location $N+1$ on the far right, the concentration is set at C_B. Making a material balance in each node leads to the matrix in Eq. (4.388):

$$\begin{vmatrix} \nu_0 & -\nu_0 & 0 & 0 & 0 \\ -\nu_0 & \nu_0+\nu_1 & -\nu_1 & 0 & 0 \\ 0 & -\nu_{i-1} & \nu_{i-1}+\nu_i & -\nu_i & 0 \\ 0 & 0 & -\nu_{N-1} & \nu_{N-1}+\nu_N & -\nu_N \\ 0 & 0 & 0 & -\nu_N & \nu_N \end{vmatrix} \begin{vmatrix} C_A \\ C_1 \\ C_i \\ C_N \\ C_B \end{vmatrix} = \begin{vmatrix} f_A \\ 0 \\ 0 \\ 0 \\ f_B \end{vmatrix} \tag{4.388}$$

and can be written succinctly as:

$$\begin{vmatrix} \nu_0 & -\mathbf{g} & 0 \\ -\mathbf{g}^T & \mathbf{Z} & -\mathbf{q} \\ 0 & -\mathbf{q}^T & \nu_N \end{vmatrix} \begin{vmatrix} C_A \\ \mathbf{C}_i \\ C_B \end{vmatrix} = \begin{vmatrix} f_A \\ \mathbf{0} \\ f_B \end{vmatrix} \tag{4.389}$$

Equation (4.389) leads to:

$$\nu_0 C_A - \mathbf{g}\mathbf{C}_i = f_A \tag{4.390}$$

$$-\mathbf{g}^T C_A + \mathbf{Z}\mathbf{C}_i - \mathbf{q}C_B = \mathbf{0} \tag{4.391}$$

$$-\mathbf{q}^T\mathbf{C}_i + \nu_N C_B = f_B \tag{4.392}$$

and leads to the solution:

$$\mathbf{C}_i = \mathbf{Z}^{-1}\left(\mathbf{g}^T C_A + \mathbf{q}C_B\right) \tag{4.393}$$

$$f_A = \left(\nu_0 - \mathbf{g}\mathbf{Z}^{-1}\mathbf{g}^T\right)C_A - \mathbf{g}\mathbf{Z}^{-1}\mathbf{q}C_B \tag{4.394}$$

$$f_B = -\mathbf{q}^T\mathbf{Z}^{-1}\mathbf{g}^T C_A + \left(\nu_N - \mathbf{q}^T\mathbf{Z}^{-1}\mathbf{q}\right)\mathbf{C}_B \tag{4.395}$$

Since there is no chemical reaction, the four coefficients in Eqs. (4.394) and (4.395) are the same in absolute value.

As far as determining the overall flux between the two open nodes is concerned, one can proceed more efficiently. Consider, for instance, two channels with individual mass transfer coefficients ν_1 and ν_2. A material balance in the single closed node yields:

$$\nu_1(C_A - C) + \nu_2(C_B - C) = 0 \tag{4.396}$$

Solving Eq. (4.396) for the unknown concentration C yields:

$$C = \frac{\nu_1 C_A + \nu_2 C_B}{\nu_1 + \nu_2} \tag{4.397}$$

and the overall flux f through the channels is then given by:

$$f = \frac{1}{\frac{1}{\nu_1} + \frac{1}{\nu_2}}(C_A - C_B) = \nu_e(C_A - C_B) \tag{4.398}$$

The effective average mass transfer coefficient ν_e that replaces two channels in series by a single channel is thus obtained as the Bosanquet average of the two individual channels. The extension to the effective mass transfer coefficient ν_e replacing a string of $N+1$ channels in series by a single channel is obtained from Eq. (4.399):

$$\nu_e = \frac{1}{\sum_{i=1}^{N+1} \frac{1}{\nu_i}} = \frac{1/(N+1)}{\frac{1}{\mu_h}} \tag{4.399}$$

A set of $N+1$ arbitrary channels can be replaced by a set of $N+1$ identical channels each with a mass transfer coefficient ν_h obtained from Eq. (4.400):

$$\nu_h = \nu_e(N+1) = \mu_h \tag{4.400}$$

where μ_h is the harmonic mean of the set of $N+1$ channels with individual mass transfer coefficients ν_i. Notice that the order in which the channels are combined has no impact on the effective mass transfer coefficient. For any number of individual channels with mass transfer coefficients taken randomly according to a distribution, the effective mass transfer coefficient equals the harmonic mean divided by the number of channels.

Appendix 4.7 shows the Mathcad 15 code for solving the string concentrations. The mass transfer coefficient ν_i is sampled from a uniform distribution per Eq. (4.401):

$$\nu_i = 0.1 + rnd(2) \tag{4.401}$$

Figure 4.11 shows the trace of the concentration for a string of 100 nodes with a concentration C_A set at a value of two at the far left and a value for C_B set at unity at the far right. Except for the bumps and kinks due to picking extremities in the random sampling of the distribution and the closure of the material balances in each node, the trace in Figure 4.11 appears to be linear along the x-coordinate, as expected. A larger number of steps would make the trace look smoother, but the derivative would remain spiked. The classic continuum model yields a continuously differentiable linear trace between the two end-point concentrations. The discrete node solution here converges to the classic solution to within any small error bound $\pm\varepsilon$ chosen at will by employing a large enough number of nodes.

The harmonic mean of the mass transfer coefficients is obtained in an analytical way from Eq. (4.402):

$$\nu_h = \left(\frac{1}{2.1-0.1}\ln\left(\frac{2.1}{0.1}\right)\right)^{-1} \simeq 0.6569 \tag{4.402}$$

Sample calculations of the mass transfer flux as a function of the number of nodes are shown in Table 4.4 and support the comments made by Chen, Degnan

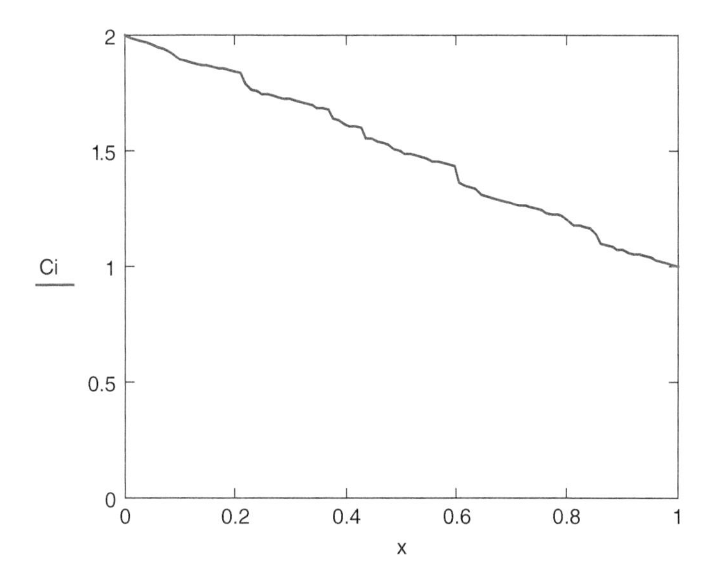

Figure 4.11 Concentration profile for diffusion only along a string.

and Smith [9] on a purely stochastic approach. Guidance based on a theoretical foundation is required in order to extract the full benefit from purely stochastic work.

Table 4.4 shows that a reasonable agreement is obtained between the stochastic approach and the theoretical harmonic mean method when applying upward of 500 nodes. Physically, 500 nodes, each, say, 10 nm in size and about 30 nm apart, translates to a linear distance of 20 μm or about 1% of the diameter of a 2 mm extrudate. Hence, the 500 node network is still small compared to the overall size of the catalyst particle, as one might expect.

Table 4.4 Comparison of the stochastic flux versus the analytical solution for diffusion in a string of N nodes ($v = 0.1 + rnd(2)$, consistent defined units).

N	$f = \frac{v_h}{N+1}(C_A - C_B)$ (analytical, $v_h \simeq 0.6569\ldots$)	Flux by stochastic simulation
10	0.0597	0.0695
100	0.00650	0.00572
500	0.001311	0.001315
1000	0.000656	0.0006731
2000	0.0003283	0.0003287
3000	0.0002189	0.0002204

4.3.5.3.2 Diffusion in a Deep Slab

Consider now a deep network that connects a set of n external nodes contained in the concentration vector C_A to a set of m external nodes contained in the concentration vector C_B. The network is considered to have a flat slab geometry in either two or three dimensions. The dimensions of both vectors are not necessarily the same. Consider the following material balance matrix **M** for the network:

$$\mathbf{M} = \begin{vmatrix} diag(\mathbf{Ge}_n) & -\mathbf{G} & \mathbf{0} \\ -\mathbf{G}^T & \mathbf{S} + diag(\mathbf{G}^T\mathbf{e}_n) + diag(\mathbf{Qe}_m) & -\mathbf{Q} \\ \mathbf{0} & -\mathbf{Q}^T & diag(\mathbf{Q}^T\mathbf{e}_m) \end{vmatrix} \quad (4.403)$$

This leads to:

$$\begin{vmatrix} diag(\mathbf{Ge}_n) & -\mathbf{G} & \mathbf{0} \\ -\mathbf{G}^T & \mathbf{S} + diag(\mathbf{G}^T\mathbf{e}_n) + diag(\mathbf{Qe}_m) & -\mathbf{Q} \\ \mathbf{0} & -\mathbf{Q}^T & diag(\mathbf{Q}^T\mathbf{e}_m) \end{vmatrix} \times \begin{vmatrix} \mathbf{C}_A \\ \mathbf{C}_i \\ \mathbf{C}_B \end{vmatrix} = \begin{vmatrix} \mathbf{F}_A \\ \mathbf{0} \\ \mathbf{F}_B \end{vmatrix} \quad (4.404)$$

where both $\mathbf{F}_A$ and $\mathbf{F}_B$ are counted positive when the flux is going into the network. There are n external nodes on the A-side and m external nodes on the B-side, and there are k internal nodes. I obtain the following equations:

$$diag(\mathbf{Ge}_n)\mathbf{C}_A - \mathbf{GC}_i = \mathbf{F}_A \quad (4.405)$$

$$-\mathbf{G}^T\mathbf{C}_A + \left[\mathbf{S} + diag(\mathbf{G}^T\mathbf{e}_n) + diag(\mathbf{Qe}_m)\right]\mathbf{C}_i - \mathbf{QC}_B = \mathbf{0} \quad (4.406)$$

$$-\mathbf{Q}^T\mathbf{C}_i + diag(\mathbf{Q}^T\mathbf{e}_m)\mathbf{C}_B = \mathbf{F}_B \quad (4.407)$$

Let:

$$\mathbf{Z} = \mathbf{S} + diag(\mathbf{G}^T\mathbf{e}_n) + diag(\mathbf{Qe}_m) \quad (4.408)$$

We then get as solution:

$$\mathbf{C}_i = \mathbf{Z}^{-1}(\mathbf{G}^T\mathbf{C}_A + \mathbf{QC}_B) \quad (4.409)$$

$$\mathbf{F}_A = \left[diag(\mathbf{Ge}_n) - \mathbf{GZ}^{-1}\mathbf{G}^T\right]\mathbf{C}_A - \mathbf{GZ}^{-1}\mathbf{QC}_B \quad (4.410)$$

$$\mathbf{F}_B = -\mathbf{Q}^T\mathbf{Z}^{-1}\mathbf{G}^T\mathbf{C}_A + \left[diag(\mathbf{Q}^T\mathbf{e}_m) - \mathbf{Q}^T\mathbf{Z}^{-1}\mathbf{Q}\right]\mathbf{C}_B \quad (4.411)$$

Since there is no reaction, the sum of all of the fluxes on the one side and the sum of all of the fluxes on the other side must be equal to zero, and this leads to a set of identities:

$$\mathbf{e}_n^T\left[diag(\mathbf{G}\mathbf{e}_n)-\mathbf{G}\mathbf{Z}^{-1}\mathbf{G}^T\right]\mathbf{e}_n=\mathbf{e}_m^T\mathbf{Q}^T\mathbf{Z}^{-1}\mathbf{G}^T\mathbf{e}_n=\mathbf{e}_m^T\left[diag\left(\mathbf{Q}^T\mathbf{e}_m\right)-\mathbf{Q}^T\mathbf{Z}^{-1}\mathbf{Q}\right]\mathbf{e}_m=\mathbf{e}_n^T\mathbf{G}\mathbf{Z}^{-1}\mathbf{Q}\mathbf{e}_m \tag{4.412}$$

VDNs have the property that the sum of the fluxes from one side can be expressed as the number of nodes from that side times the minimum eigenvalue of the corresponding matrix. This is not the case for networks that are shallow even when they have many nodes from that side. The proof for the minimum eigenvalue relationship with the total flux is simple when one realizes that the sum of all of the elements in any row of these matrices is positive and very small compared to the elements of that row. In fact, the deeper the network, the smaller the sum, since the resistance to flow is getting larger and larger.

However, perhaps I would now like to invite any student in mathematics to help with the next task, which is to find the minimum eigenvalue of the Schur complement expressed as a function of the distribution of the random elements for a VDN.

Appendix 4.8 shows an example calculation that illustrates this for a network that is 50 layers deep and with five nodes per layer in an otherwise square arrangement, as can be seen in Figure 4.3. Let the reader be aware that all of the mass transfer coefficients ν are different and are now sampled randomly from a uniform distribution:

$$\nu = 1 + rnd(1) \tag{4.413}$$

so the individual mass transfer coefficients all have values between 1 and 2. Let λ_A be the minimum eigenvalue of the matrix $\mathbf{H}_A = [diag(\mathbf{G}\mathbf{e}_n) - \mathbf{G}\mathbf{Z}^{-1}\mathbf{G}^T]$ and let λ_B be the minimum eigenvalue of the matrix $\mathbf{H}_B = [diag(\mathbf{Q}^T\mathbf{e}_m) - \mathbf{Q}^T\mathbf{Z}^{-1}\mathbf{Q}]$.

Appendix 4.9 shows the results when the nodes on either side are different in number.

Based on numerous example calculations with deep networks as shown in Appendix 4.8 and Appendix 4.9, I can make the specific statement that the total mass flux through the network can be written as:

$$\mathbf{e}_n^T\mathbf{H}_A\mathbf{e}_n = \mathbf{e}_m^T\mathbf{H}_B\mathbf{e}_m = n\lambda_A = m\lambda_B \tag{4.414}$$

Equation (4.414) yields a good method for knowing that one has enough depth in the network to call it a VDN because this estimate of the smallest eigenvalue of $\mathbf{H}_A$ and $\mathbf{H}_B$ can be compared to the numerical value obtained via the Mathcad "eigenvals" software, and this should be within the error bounds that have been considered as acceptable. If they are not, one needs to increase the depth.

Let:

$$\mathbf{C}_A = C_A\mathbf{e}_n \text{ and } \mathbf{C}_B = C_B\mathbf{e}_m \tag{4.415}$$

where both C_A and C_B are scalars. We can write the overall mass transfer flux as:

$$\mathbf{e}_n^T\mathbf{F}_A = \mathbf{e}_m^T\mathbf{F}_B = n\lambda_A(C_A - C_B) = m\lambda_B(C_A - C_B) \tag{4.416}$$

The overall mass transfer coefficient v_e and the overall mass transfer flux F through the deep network is very simply stated as:

$$v_e = n\lambda_A = m\lambda_B \tag{4.417}$$

$$F = v_e(C_A - C_B) \tag{4.418}$$

In addition, when m equals n, let λ_C be the *maximum* eigenvalue of the matrix $\mathbf{GZ}^{-1}\mathbf{Q}$, then:

$$\lambda_C = \lambda_A = \lambda_B \tag{4.419}$$

These relationships with the eigenvalues are only valid for deep networks or VDNs, in which case the values converge very well. However, experience shows that the convergence for λ_C in Eq. (4.419) is more difficult to obtain than for λ_A and λ_B in the same equation.

I believe that the reason for this simplicity is due to the fact that the total number of internal mass transfer equations that have to be zero in the overall material balance matrix is very numerous and tends to even out the very few non-zero sides of the network where the fluxes come into play (i.e. the sum of the coefficients of the matrix **M** internally are all zero while the sum of the few coefficients of **M** on the sides of the network [the corners of the matrix] are not). This then also explains why a shallow network, albeit a wide one, gives poor convergence with respect to the minimum eigenvalue. The calculations also show that for a deep network the eigenvalues of interest are easily estimated from:

$$n\lambda_A = \mathbf{e}_n^T\left[diag(\mathbf{G}\mathbf{e}_n) - \mathbf{GZ}^{-1}\mathbf{G}^T\right]\mathbf{e}_n \tag{4.420}$$

$$m\lambda_B = \mathbf{e}_m^T\left[diag\left(\mathbf{Q}^T\mathbf{e}_m\right) - \mathbf{Q}^T\mathbf{Z}^{-1}\mathbf{Q}\right]\mathbf{e}_m \tag{4.421}$$

$$n\lambda_C = \mathbf{e}_n^T\mathbf{GZ}^{-1}\mathbf{Q}\mathbf{e}_m \text{ for } n = m \tag{4.422}$$

Lastly, a rectangular network as used in Appendix 4.8 with white noise mass transfer coefficients can be replaced with an equivalent network of the same size and arrangement, but with all of the mass transfer coefficients of the same value ν_{eq}. For a VDN, the equivalent network has the same total mass flux as a network with random mass transfer coefficients. For the equivalent network, the mass transfer coefficient ν_{eq} is obtained from Eq. (4.423):

$$\nu_{eq} = (L-1)\lambda_A \tag{4.423}$$

where L stands for the total number of layers of nodes (i.e. the two external [A-side an B-side] and the $L-2$ internal layers). The term $L-1$ therefore portrays the sum of the number of segments or the "total length" of the network.

4.3.5.3.3 The Equivalent Network Efficacy

Consider a network as in Appendix 4.8 with four nodes and 500 layers deep. The uniform distribution of mass transfer coefficients is taken as:

$$v = a + rnd(b) \tag{4.424}$$

The value of a is kept at unity and I change the random part b over several orders of magnitude. I compare the random network equivalent v_{eq} to the arithmetic mean v_{am} of the distribution. The results are shown in Table 4.5, and they reveal that the equivalent network (or the actual random network) for a highly random part has lost about 25% of its efficacy for mass transport compared to a network of the same size and arrangement that has the same v_{am} for every channel. The reason for this is that channels with a low mass transfer coefficient effectively act as blockages for mass transfer and the molar flux has to follow a corrugated and percolating path around the blockages.

Table 4.6 shows the results for slabs with more nodes per layer. Similar results hold for the corrugated percolating path as in the previous case.

4.3.5.4 First-order Reaction and Diffusion in Very Deep Networks

4.3.5.4.1 The Minimum Eigenvalue of H_e

For real catalysts, the depth of the network and the number of pore mouths on the surface are exceedingly large. For a VDN, the minimum eigenvalue of the matrix $\mathbf{H}_e$ in Eq. (4.306) is intimately linked to the performance of the network (the catalyst). Each diagonal element of $\mathbf{H}_e$ is larger than the sum of all of the absolute values of the non-diagonal elements of that row of $\mathbf{H}_e$, hence the matrix $\mathbf{H}_e$ is diagonally dominant (even for the case of diffusion where the diagonal element equals the sum of absolute values of the non-diagonal row elements). The arithmetic

Table 4.5 Diffusion through a slab.

(500 layers, 4 nodes/layer)				
a	b	$v_{am} = a + b/2$	$v_{eq} = (L-1)\lambda_{min}(H_A)$	v_{eq}/v_{am}
1	1	1.5	1.47	0.98
1	5	3.5	3.10	0.88
1	10	6	5.00	0.83
1	20	11	8.58	0.78
1	100	51	39.1	0.77
1	1000	501	365	0.73
1	10 000	5001	3737	0.75

Table 4.6 Diffusion through a broad slab ($L = 200$, $a = 1$).

b	N	v_{eq}/v_{am}
1	4	0.965
10	4	0.851
100	4	0.767
1000	4	0.736
10 000	4	0.711
100 000	4	0.761
1	8	0.973
10	8	0.860
100	8	0.784
1000	8	0.791
10 000	8	0.753
100 000	8	0.755
1	16	0.981
10	16	0.855
100	16	0.784
1000	16	0.778
10 000	16	0.782
100 000	16	0.790

average over all of the rows (or pore mouths) now can be shown to become an excellent estimate for the minimum eigenvalue λ_R of $\mathbf{H}_e$:

$$\lambda_R = \frac{\mathbf{e}_N^T \mathbf{H}_e \mathbf{e}_N}{N} \tag{4.425}$$

where N is the number of pore mouths on the periphery of the network. Equation (4.425) gives a good method for determining whether there is enough depth to the network and a small enough perturbation to call it a VDNP because this estimate of the smallest eigenvalue of $\mathbf{H}_e$ can be compared to the numerical value obtained via the Mathcad "eigenvals" software, and they should be within the error bounds that have been set as acceptable.

The eigenvector $\mathbf{x}$ corresponding to the minimum eigenvalue converges to a unit vector $\mathbf{e}_N$ divided by the square root of the number of pore mouths:

$$\mathbf{x} = \frac{1}{\sqrt{N}} \mathbf{e}_N \tag{4.426}$$

Hence, one can write for the overall rate of reaction R of a VDN, including the mass transfer limitations:

$$R = N\lambda_R \tag{4.427}$$

To summarize what is worked and stated here:

- For very deep networks, such as those present in real catalyst particles, the rate coefficient in a node can be considered to be a perturbation of the mass transfer coefficients between nodes.
- Very strong mass transfer limitations can be obtained in the case of perturbation by considering deep enough networks.
- The mass transfer coefficients, the rate coefficients, and the surface areas allocated to channels and nodes can be arbitrary distributions, and the network node connections can be arbitrary.
- The methodology presented here for solving the network guarantees that the material balance in each node of the network is satisfied.
- Although none of the rows of $\mathbf{H}_e$ add up to the same value, the minimum eigenvalue estimate of $\mathbf{H}_e$ equals the arithmetic average of all of the sums of each row.
- The network solution converges to the classic continuum solution as close as one wants by using an appropriate VDN.
- The effectiveness factor η of the VDN is obtained from:

$$\eta = \frac{N\lambda_R}{\sum\limits_i kS_i} = \frac{\lambda_R}{\frac{1}{N}\sum\limits_i kS_i} \tag{4.428}$$

- The effectiveness factor is thus the ratio of the minimum eigenvalue of $\mathbf{H}_e$ divided by the arithmetic mean of the network rate coefficients. Hence, λ_R also represents the average mass transfer limited chemical rate coefficient of a node in a network.
- The solution of a VDN converges to the mathematical formulation of the classic continuum model. This allows one to conveniently define the Thiele modulus for a network. For instance, for a flat slab:
  - I base the Thiele modulus ϕ_η for a VDN on it's effectiveness factor, and ϕ_η for a flat slab is then conveniently obtained from the implicit equation:

$$\eta = \frac{\tanh\left(\phi_\eta\right)}{\phi_\eta} \tag{4.429}$$

  - The concentration profile along the dimensionless depth ξ of the VDN is then calculated from:

$$C = C_L \frac{\cosh\left(\xi\phi_\eta\right)}{\cosh\left(\phi_\eta\right)} \tag{4.430}$$

4.3.5.4.1.1 Worked Example Perhaps it is easiest to start with an example calculation that shows some of the results. Consider a network as shown in Figure 4.4 used in Section 4.3.3.3.3 on regular networks. The network will be considered open on one side only. However, we will now entertain the case where the mass transfer coefficients between the nodes and the surface areas in the nodes are no longer constant. For both individual mass transfer coefficients and individual surface areas, I assume a uniform random distribution:

For the mass transfer coefficient, I assume:

$$v = 1 + rnd(1) \tag{4.431}$$

where $rnd(1)$ is a uniform random number generated in the interval (0, 1) by the Mathcad 15 software. The flux between two nodes at concentrations C_1 and C_2 is then written as:

$$f = v(C_1 - C_2) \tag{4.432}$$

For the surface area in each node, I assume the same distribution:

$$S = 1 + rnd(1) \tag{4.433}$$

The rate of reaction in a node is then written as:

$$r = kSC \tag{4.434}$$

where k is here considered to be a constant and represents the first-order rate constant. For the rate of reaction to be a perturbation of the diffusion rate, it is clear that the value of k has to be small enough such that kS/v is small compared to unity. As an example, consider a network that has five external nodes and is 500 layers deep (i.e. five interconnected parallel strings). To appreciate the randomness of both distributions, Figure 4.12 shows the surface area located in the

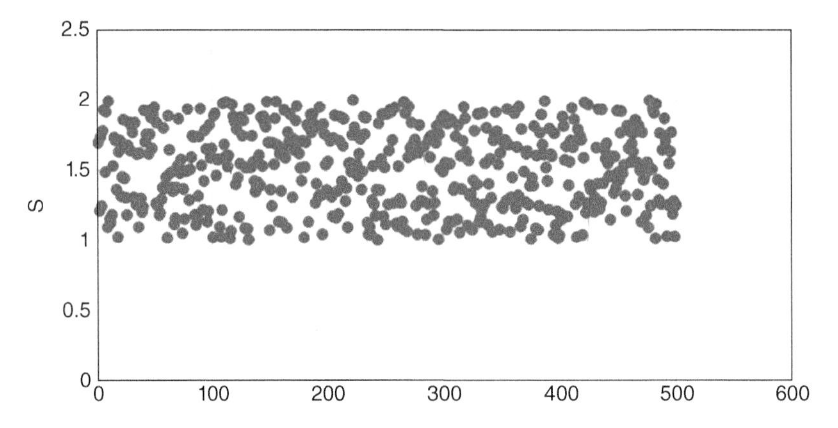

Figure 4.12 Distribution of surface areas in the nodes.

500 nodes of one particular string. The distribution of the mass transfer coefficients has the same appearance (although different in every location) because of the definition. Consider a rate constant value of 0.01 and hence the chemical rate coefficient is a small perturbation of the mass transfer coefficients. The results are obtained with the Mathcad 15 software and they are listed in Appendix 4.10. The sums of each row of the matrix $\mathbf{H}_e$ are substantially smaller that the respective diagonal elements. The exact total rate of reaction over the entire network calculates at 0.7041, while the kinetic rate is $2500 \times 0.01 \times 1.5 = 37.5$. The effectiveness factor for the network therefore calculates at about 1.9%. Hence, the network is strongly mass transfer limited, notwithstanding the fact that, as mentioned, the chemical rate is slow compared to the local diffusion rates. The product of the minimum eigenvalue of $\mathbf{H}_e$ and the number of external pore mouths calculates at 0.7025. Hence, the agreement with the total rate clearly demonstrates the essence of what was said in the previous section on the minimum eigenvalue λ_R of the matrix $\mathbf{H}_e$. Numerically, this is quite fascinating: $\mathbf{H}_e$ is symmetrical, but none of the rows (and hence the columns) add up to the same value; in fact, they are all substantially different. Yet, with VDNP, the sum of all of the elements of $\mathbf{H}_e$ equals its minimum eigenvalue times the number of rows (or pore mouths) of the network. This is even more fascinating because all of the elements of the material balance matrix are random, or they can be any other distribution for that matter. This is very different from the regular network cases considered so far, where all of the rows indeed add up to exactly the same value, and hence, evidently, this sum is then also a minimum eigenvalue of the matrix.

It is interesting to note the following: the matrix $\mathbf{H}_{e,\,D}$ for diffusion in a VDN and the matrix $\mathbf{H}_{e,\,R}$ for diffusion and reaction via VDNP are different, but the difference is not great due to the nature of a perturbation. However, the network is still substantially mass transfer limited, and this is not something that is intuitively clear from the classic continuum approach.

4.3.5.4.2 *Perturbation in a Very Deep Network Case*

My approach introduces a carving of the reaction–diffusion–architecture parameter space where reaction is looked upon as a perturbation on top of the diffusion field for all cases of very high and very low effectiveness factors. The carving brings about certain links between the catalyst performance and the matrix solutions in a very natural way and in the same way as for regular networks. Sometimes, the literature presents statements such as "the rate of mass transfer is slow compared to the rate of reaction and hence leads to strong mass transfer limitations in the catalyst particle." This can be misleading to the novice or student in the area because at the node level mass transfer in real catalysts is always very fast, much faster than reaction in the node. Mass transfer limitations in real catalysts are simply due to the VDN nature of these materials. Both the mass transfer and the reaction can be and realistically always are like white noise-like and irregular, but the reaction is always a small perturbation of the mass transfer, and this is the case for most commercial catalysts. The example in Section 4.3.5.4.1.1 already gave a flavor of a reaction perturbation of a VDN with diffusion and showed some interesting results. Let us now take a deeper look, investigate the concentration profile, and search for links with the classic theory. For this, consider a single string of nodes (like a string of pearls) and allow each node to have a widely variable surface area:

$$S = 0.5 + rnd(1.5) \tag{4.435}$$

Hence, the surface allocated to a node ranges from 0.5 to 2.0. The profile of surface areas in the 500 nodes along the string considered can be observed in Figure 4.13. The signal is highly irregular and looks similar to what one might expect in a real case. I allow a rate coefficient:

$$k = 0.000045 \tag{4.436}$$

For the mass transfer coefficient between the nodes, I take:

$$v = 0.5 + rnd(1.5) \tag{4.437}$$

In this section I always use mass transfer and surface area from the same distribution. It is clear that the k value indicated makes the rate constant kS a small perturbation of the field of mass transfer coefficients. As we will see, the mass transfer limitation will be substantial, and this is a consequence of the deep network (here a string) that is being considered. Solving this network is easily done with the methodology outlined and yields the results for the concentration of

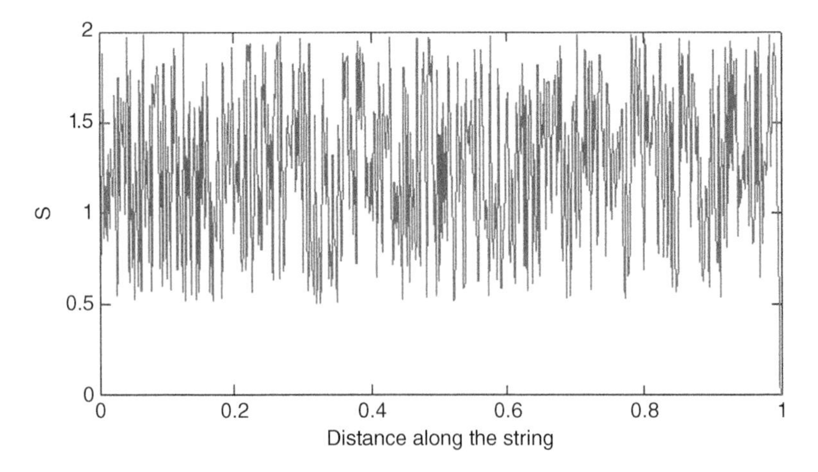

Figure 4.13 Surface area distribution.

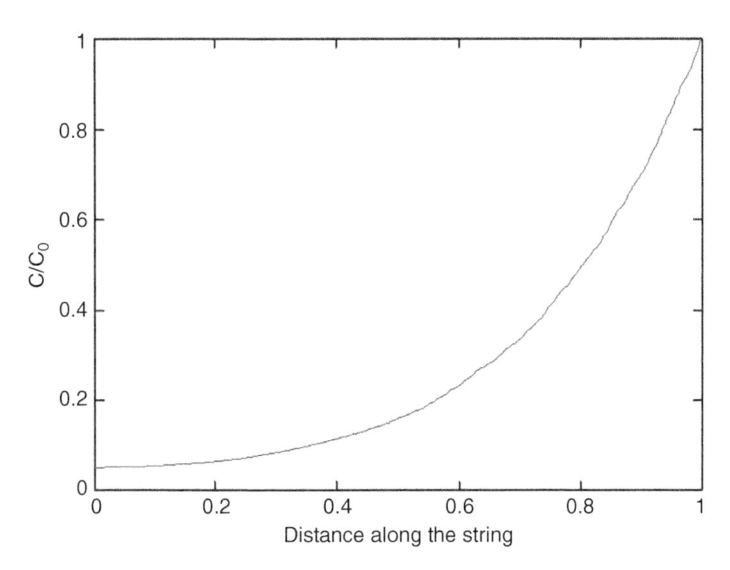

Figure 4.14 Diffusion and reaction in a multi-segment string.

reactant (the string is closed on the left and open on the right) shown in Figure 4.14. Upon inspection of the profile, it (i) has a fairly normal shape and (ii) still shows some small bumps and kinks as one traverses the concentration values. The smooth profile, notwithstanding the wide variability of inputs of surface areas and mass transfer coefficients, is quite remarkable. The semi-smoothness of the profile is due to the perturbation considered. The bumps and kinks, on the other hand, are due to closure of the material balances in each node as required based on the very local values of the surface areas and mass transfer coefficients. Furthermore, the profile also looks quite like what one would expect from the classic approach where one has the well-known profile:

$$c(x) = \frac{\cosh(\phi x)}{\cosh(\phi)} \tag{4.438}$$

with:

$$\phi = L\sqrt{\frac{2k'}{rD_e}} \tag{4.439}$$

To obtain an estimate of the Thiele modulus from the string properties, we let the string be represented by a hollow pore, which then yields for the flux through a slice:

$$D_e \pi r^2 \Delta c_i / \Delta x_i = \nu_i \Delta c_i \tag{4.440}$$

while for the surface it yields:

$$2\pi r k' \Delta x_i = kS_i \tag{4.441}$$

Hence, the Thiele modulus estimate is then obtained from the average string properties as:

$$\phi = N\sqrt{\frac{kS_{avg}}{\nu_{avg}}} \tag{4.442}$$

where N stands for the total number of nodes while the node properties here have been simply arithmetically averaged as an estimate. Inserting the appropriate values then yields:

$$\phi = 500\sqrt{\frac{0.000045 \times 1.25}{1.25}} \simeq 3.354 \tag{4.443}$$

The concentration of reagent at the core of the string is then calculated as:

$$C_0 = \frac{1}{\cosh(\phi)} \simeq \frac{1}{\cosh(3.354)} \simeq 0.0698 \tag{4.444}$$

Figure 4.15 shows the concentration profile for this continuum case for a value of the Thiele modulus of 3.354. This profile is of course totally smooth and fits fairly well with the network profile shown in Figure 4.14. In Section 4.3.5.3.1,

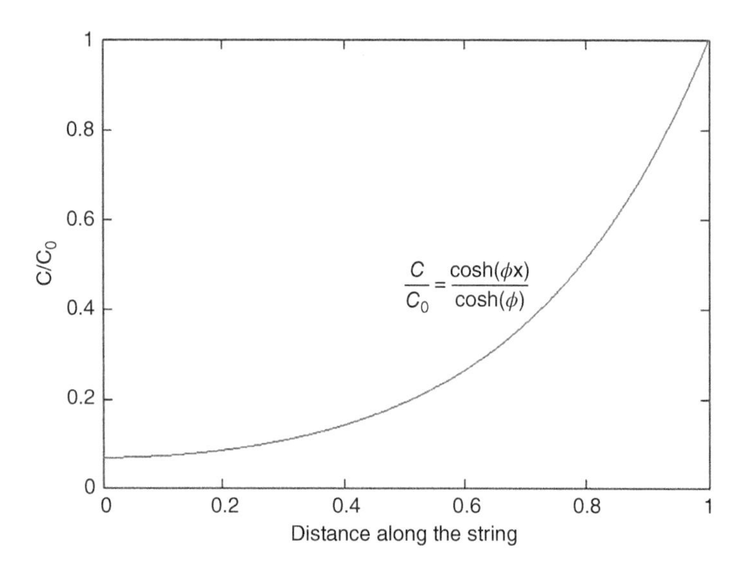

Figure 4.15 The continuum approach concentration profile.

I have elaborated that an arithmetic average for the v_i is not optimal; rather, a harmonic mean should be used for a string as per Eq. (4.400):

$$v_h = 1/\left(\frac{1}{2-0.5}\int_{0.5}^{2}\frac{1}{v}dv\right) = 1.5/[\ln(2) - \ln(0.5)] \simeq 1.082 \tag{4.445}$$

Applying the harmonic mean of the mass transfer coefficients, the value for the Thiele modulus becomes:

$$\varphi = 500\sqrt{\frac{0.000045 \times 1.25}{1.082}} \simeq 3.605 \tag{4.446}$$

The concentration of reagent at the core of the string is now calculated as:

$$C_0 = \frac{1}{\cosh(\phi)} \simeq \frac{1}{\cosh(3.605)} \simeq 0.0543 \tag{4.447}$$

Applying the harmonic mean for the mass transfer coefficients is clearly an improvement, and the calculated value in Eq. (4.447) fits the concentration at the core of Figure 4.14 nearly perfectly.

Let us now take a look at the range of concentration profiles obtained from the network approach as a function of the width of the variability band in surface area and mass transfer coefficients. I will keep the values of N and k the same so that the assumption on the perturbation is met, and I will simply show the variability band and the impact on the profile.

In Figure 4.16, I have a wide variability, but I omitted the very low values of surface areas and mass transfer coefficients by using a 0.1 minimum value. The concentration profile shown in Figure 4.17 for a large maximum/minimum range, $S = 0.1 + rand(1)$, is fairly well behaved, but one can observe a very realistic snaking of the profile due to the local irregularities and consistent mass balance closures. In Figure 4.18, I have the same bandwidth, but the ratio of highest values to lowest values is much smaller, with a small maximum/minimum range of $S = 1 + rand(1)$. Hence, the concentration profile as shown in Figure 4.19 is much better behaved. In fact, the profile looks nearly exactly as would be expected from the classic theory (I calculate a 0.0653 core concentration). Widening the variability more but staying above an arbitrary cutoff of 0.1 of course yields a wilder concentration profile, as is shown in Figures 4.20 and 4.21 for a surface area distribution with an average maximum/minimum range of $S = 0.1 + rand(2)$. The concentration profile becomes a little wilder yet, as is shown in Figures 4.22 and 4.23, for a surface area distribution with an average maximum/minimum range of $S = 0.1 + rand(3)$. As one might expect, lowering the cutoff to zero makes the concentration profile very wild, yet surprisingly, at least to myself, it seems to have a realistic appearance, as can be seen in Figures 4.24 and 4.25 for a surface

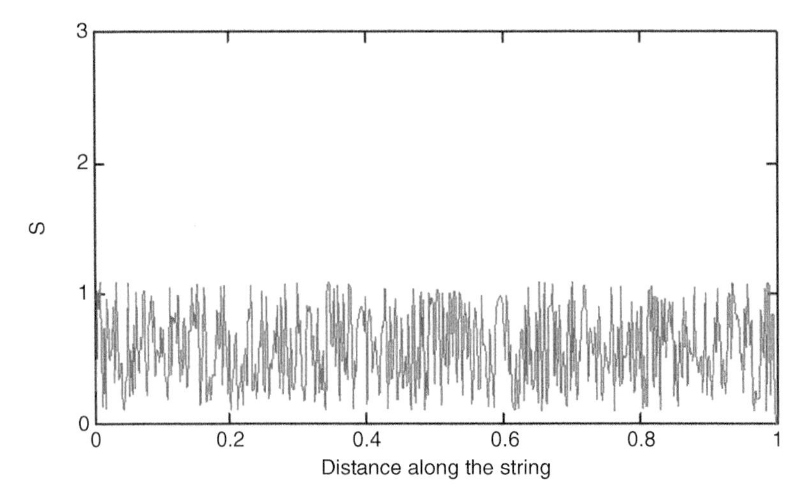

Figure 4.16 Surface area distribution for $S = 0.1 + rnd(1)$.

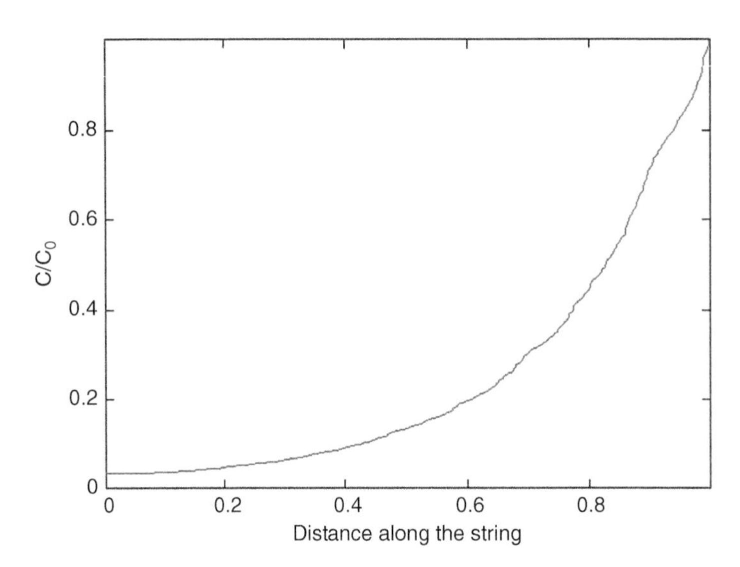

Figure 4.17 Concentration profile for $S = 0.1 + rnd(1)$.

area distribution with an average maximum/minimum range of $S = rand(1)$. From this and the previous band example, it is clear that the concentration profile stays well behaved when the VDNP assumption is kept valid. In fact, it is not just well behaved; in the limit, it goes to what I believe is exactly the classic continuum solution. The difference is that the network approach closes the material balance in every node based always on the local values of surface areas and mass transfer

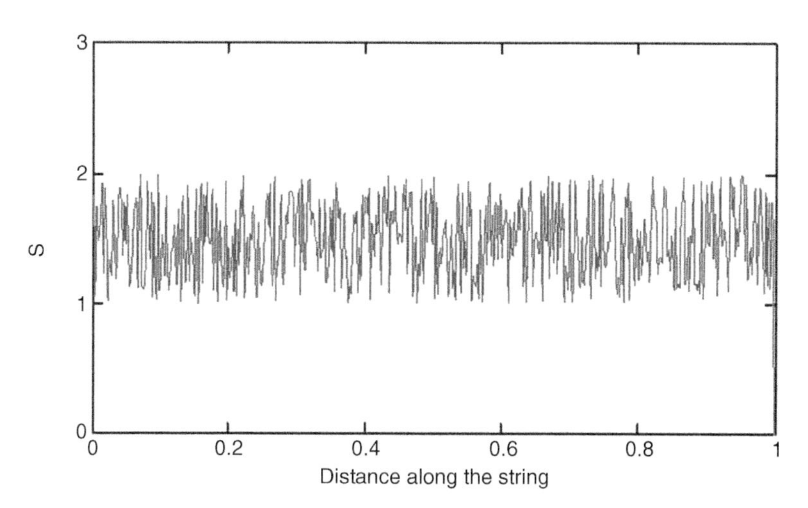

Figure 4.18 Surface area distribution for $S = 1 + rnd(1)$.

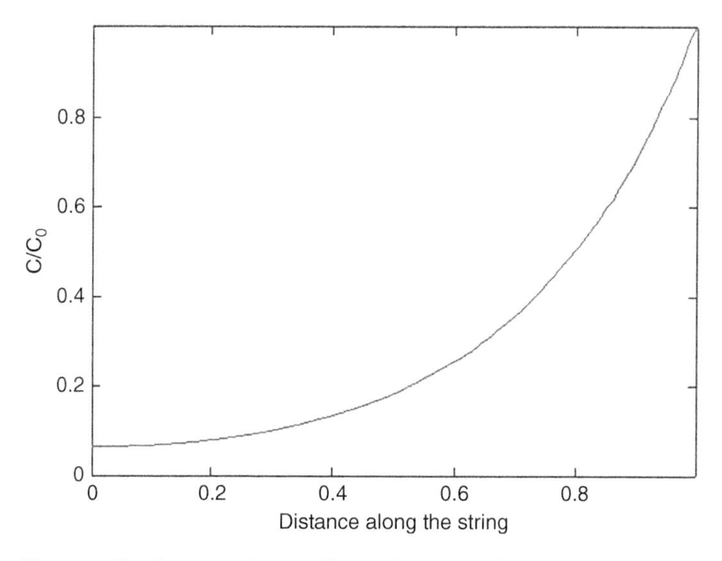

Figure 4.19 Concentration profile for $S = 1 + rnd(1)$.

coefficients, which of course can be highly variable. These profiles can always be brought closer to the classic continuum case by assuming deeper VDNs, as is illustrated in Figures 4.26–4.29. When adjusting for a higher number of nodes N, be aware to take into account the change required in k from Eq. (4.442) (i.e. doubling the value of N requires a reduction in the value of k by a factor of four to keep the same Thiele modulus).

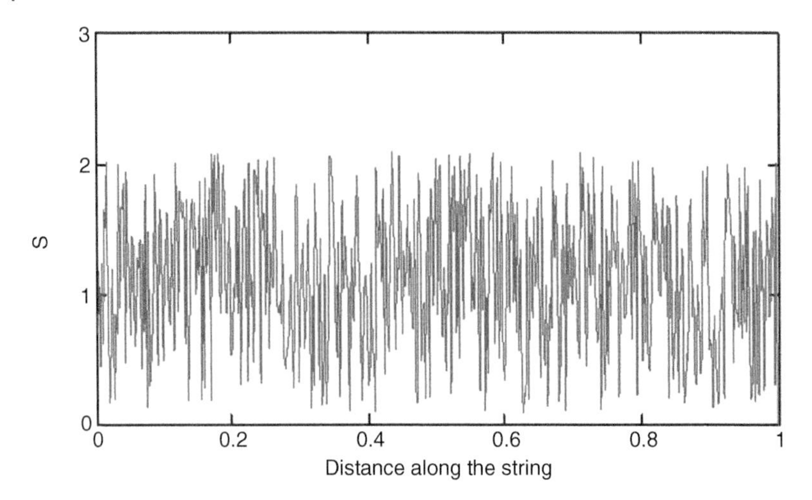

Figure 4.20 Surface area distribution for $S = 0.1 + rnd(2)$.

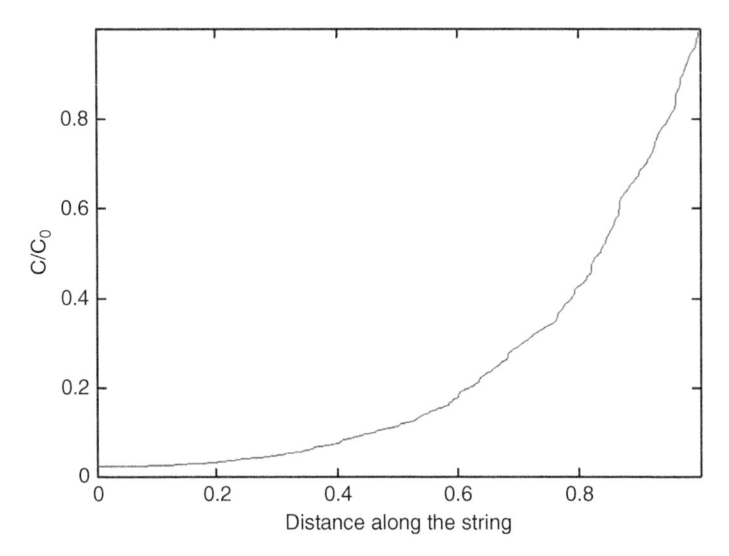

Figure 4.21 Concentration profile for $S = 0.1 + rnd(2)$.

4.3.5.4.3 More on the Local Average of Node Properties

Consider a string of $i = 1, \ldots, N$ nodes and let node N be external with an arbitrary concentration C_N. Let the node on the far left start with the index $i = 1$ and have increasing index values toward the right, ending up with the index $i = N$. Represent the molar flux leaving node i to the left by F_i. Define the flux coefficient ϕ_i as:

$$F_i = \phi_i C_i \tag{4.448}$$

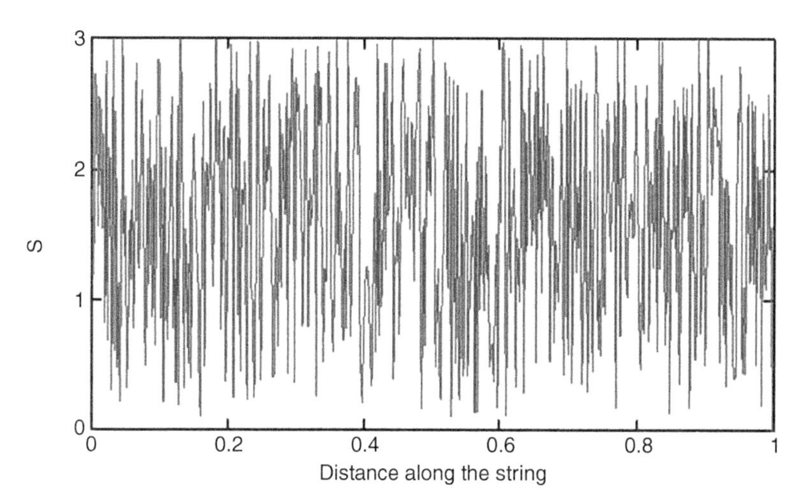

Figure 4.22 Surface area distribution for $S = 0.1 + rnd(3)$.

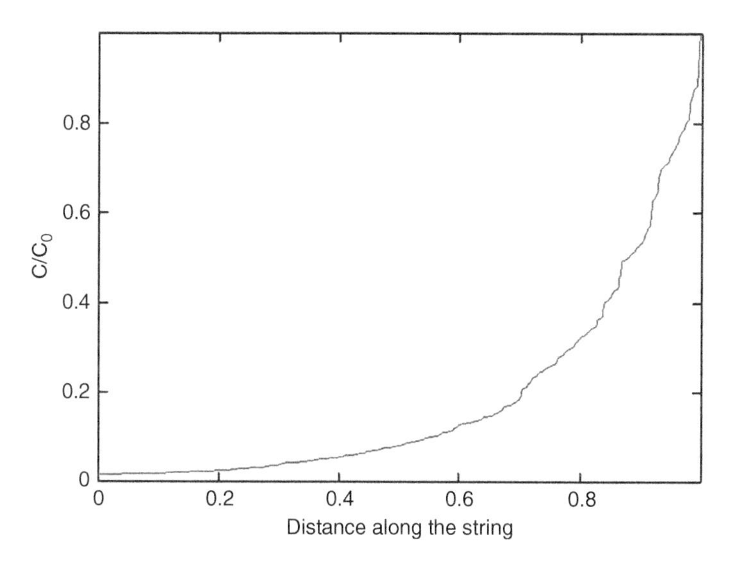

Figure 4.23 Concentration profile for $S = 0.1 + rnd(3)$.

Let the surface area in the nodes and mass transfer coefficients be distributed as:

$$S_i = 1 + rnd(1) \tag{4.449}$$

and let the mass transfer coefficient from node $i - 1$ to node i be represented by v_i:

$$v_i = 1 + rnd(1) \tag{4.450}$$

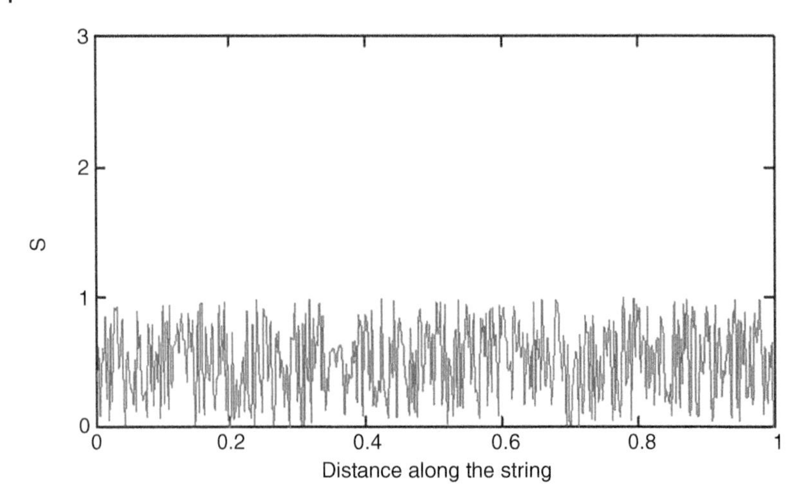

Figure 4.24 Surface area distribution for $S = rnd(1)$.

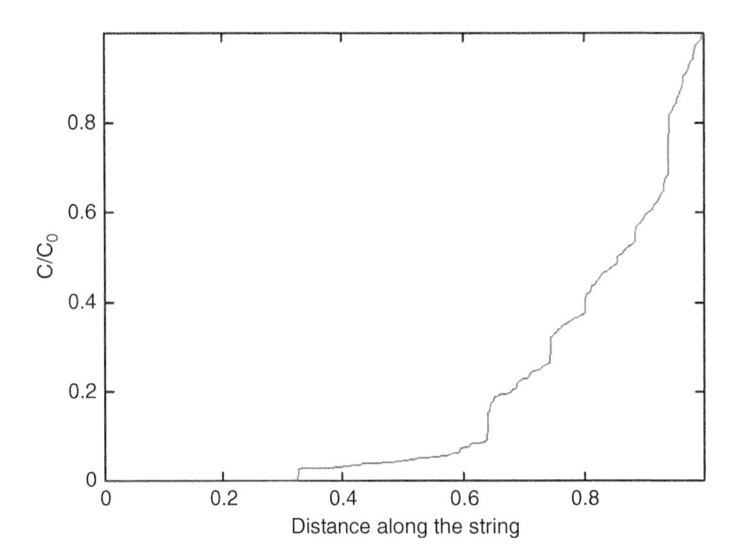

Figure 4.25 Concentration profile for $S = rnd(1)$.

where $rnd(1)$ stands for a random number uniformly distributed between zero and unity. Both properties are then uniformly distributed between one and two. The starting string is assumed to be a single node and no molar flux is leaving that node, hence:

$$\phi_1 = 0 \tag{4.451}$$

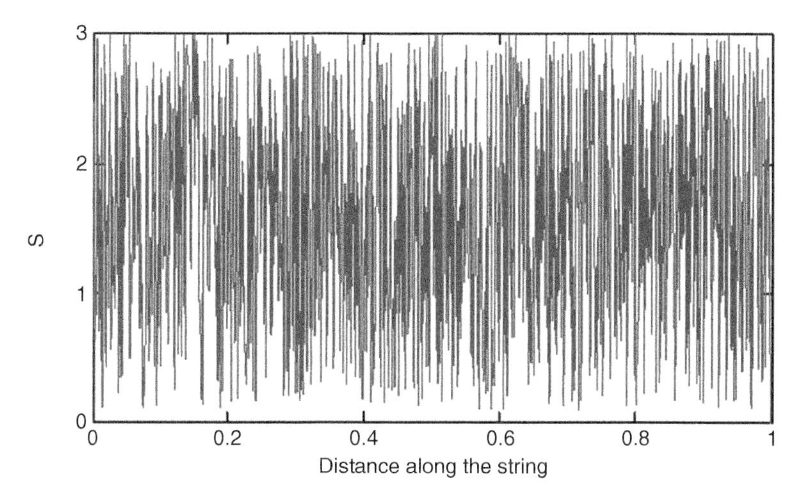

Figure 4.26 Surface area distribution for $S = 0.1 + rnd(3)$, $N = 1000$, $k = 0.00001125$.

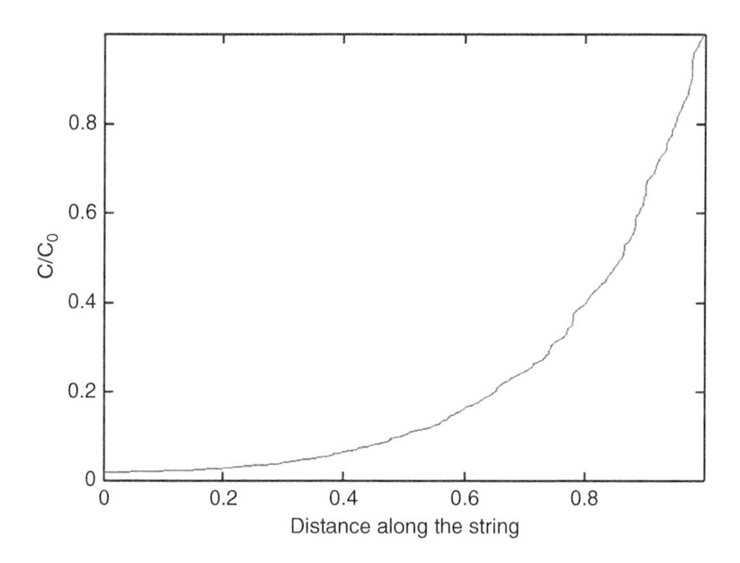

Figure 4.27 Concentration profile for $S = 0.1 + rnd(3)$, $N = 1000$, $k = 0.00001125$.

A mass transfer channel with value v_1 connects the starting node to node 2, and so on.

Add a single node (with a chemical rate coefficient equal to k) to this arrangement. This allows one to write the material balance as:

$$F_N + kS_NC_N = (C_{N+1} - C_N)v_{N+1} = F_{N+1} \tag{4.452}$$

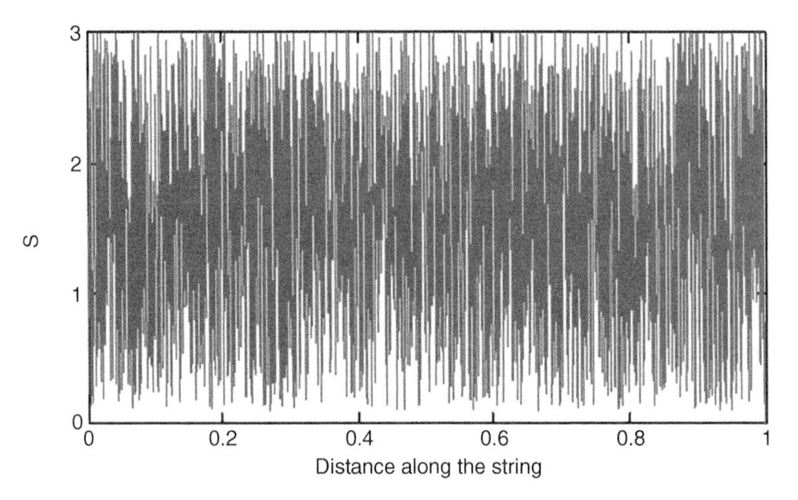

Figure 4.28 Surface area distribution for $S = 0.1 + rnd(3)$, $N = 2000$, $k = 0.0000028125$.

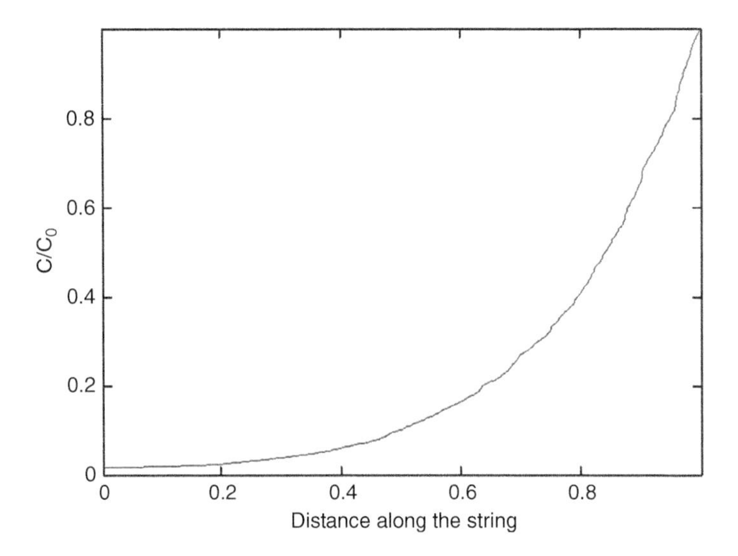

Figure 4.29 Concentration profile for $S = 0.1 + rnd(3)$, $N = 2000$, $k = 0.0000028125$.

Solving for C_N leads to:

$$C_N = \frac{v_{N+1}}{\phi_N + kS_N + v_{N+1}} C_{N+1} \tag{4.453}$$

which then leads to the expression for ϕ_{N+1}:

$$\phi_{N+1} = \frac{(\phi_N + kS_N)v_{N+1}}{\phi_N + kS_N + v_{N+1}} \tag{4.454}$$

This equation is of the Riccati type and can be rearranged to read:

$$\phi_{N+1}\phi_N + \phi_{N+1}(kS_N + v_{N+1}) - \phi_N v_{N+1} - kS_N v_{N+1} = 0 \qquad (4.455)$$

This equation is a second-order finite-difference equation with variable coefficients. The solution can be obtained via the standard available methods or by numerical means. The S_N, v_{N+1} and hence ϕ_{N+1} are all distributed variables and functions. Because of VDN perturbation, we do assume that kS_N is small compared to the other terms. It is this simple assumption that allows the variable ϕ_{N+1} to line out to a constant value with perhaps a small variation, as is shown in Figure 4.30 for $k = 10^{-4}$. The "lined-out" value is approximately 0.0146. From the graph, it can be observed that about 250 nodes are required to reach this maximum ϕ value. Of course, for $k = 0$, the value of ϕ_1 starts at zero and remains at zero because there is no reaction. Choosing a smaller value for k increases the number of nodes required to line out, but also of course reduces the jitter of the curve. For example, choosing $k = 10^{-5}$ yields Figure 4.31. Be aware that this so-called smoother curve of course has only to be looked at with a magnifying glass, as is shown in Figure 4.32, to see that the underlying jagged behavior due to the material balance closures is still prevalent. Hence, convergence to the classic continuum solution is to be understood as the difference between both solutions becoming smaller and smaller. This is not the case for the local gradient, as one can observe, because this gradient will swing widely from closure to closure, as can be seen in Figure 4.32. Increasing the value of k to 10^{-3} in Figure 4.33 leads to the opposite behavior, and it is a lot of fun to play around with this. The lined-out value for ϕ_∞ for a large number of nodes, assuming averages for the properties, can be obtained from the quadratic equation in Eq. (4.456):

$$\phi_\infty^2 + \phi_\infty kS_{avg} - kS_{avg}v_{avg} = 0 \qquad (4.456)$$

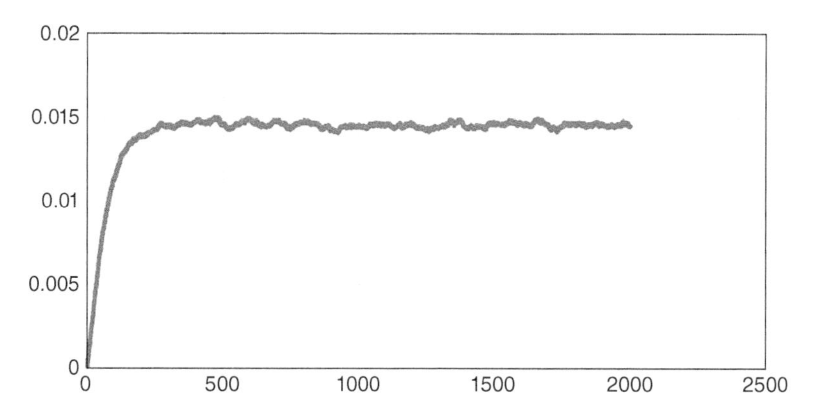

Figure 4.30 Flux coefficient as a function of the number of nodes.

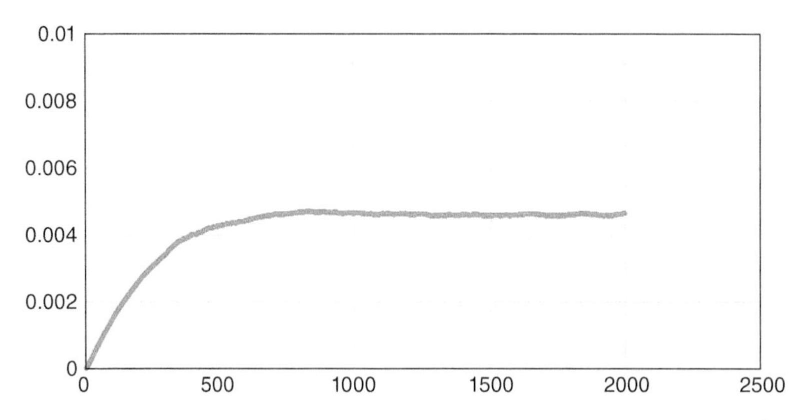

Figure 4.31 Flux coefficient as a function of the number of nodes for $k = 10^{-5}$.

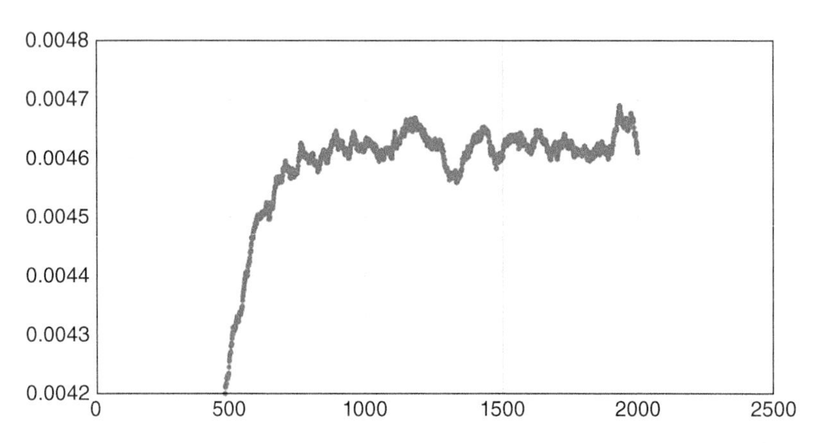

Figure 4.32 Flux coefficient as a function of the number of nodes for $k = 10^{-5}$ (reduced axis range).

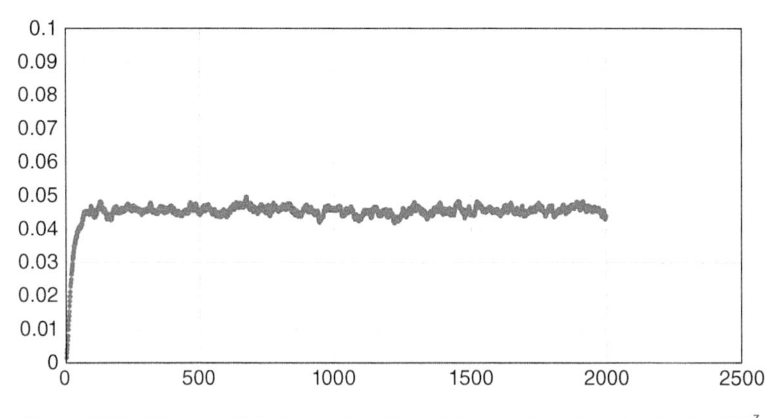

Figure 4.33 Flux coefficient as a function of the number of nodes for $k = 10^{-3}$.

The solution is given by:

$$\phi_{\infty} = \frac{1}{2}\left[-kS_{avg} + \sqrt{\left(kS_{avg}\right)^2 + 4kS_{avg}v_{avg}}\right] \tag{4.457}$$

When using the arithmetic averages for both S and v, the numerical solution for Eq. (4.411) with $k = 0.0001$ leads to $\phi_\infty = 0.0149$. The 0.0149 value is close to the actual numerical value of 0.0146. However, using the arithmetic mean for both S and v is not necessarily optimal, as already mentioned in Section 4.3.5.4.2.

For instance, in Section 4.3.5.3.1, it became clear that using the harmonic mean for v leads to a much better prediction of the diffusivity. The harmonic mean can be obtained via Excel from:

$$v_h = \frac{N}{\sum_{i=1}^{N} \frac{1}{v_i}} = \frac{N}{\sum_{i=1}^{N} \frac{1}{[1 + rnd(1)]_i}} \tag{4.458}$$

I decided to use an analytical method for the case at hand:

$$v_h = 1/\left(\frac{1}{2-1}\int_1^2 \frac{1}{v}dv\right) = 1/\ln(2) \simeq 1.443 \tag{4.459}$$

Applying the harmonic mean obtained from Eq. (4.459) in Eq. (4.457) leads to a value of $\phi_\infty = 0.01463$ for $k = 10^{-4}$ and a value of $\phi_\infty = 0.004645$ for $k = 10^{-5}$. Clearly, the lined-out values of ϕ_∞ using the harmonic mean are very satisfactory, as can be observed in Figures 4.30 and 4.33.

4.3.5.4.4 Diffusion and First-order Reaction in a Network of Interconnected Parallel Strings

In Section 4.3.5.4.2, I have shown that a harmonic mean for the average mass transfer coefficient is the proper value to use in calculating the Thiele modulus for the VDNP of a string. Interconnected parallel strings as shown in Figure 4.4 are different, and in Figures 4.34–4.37 I show the traces of the concentration profiles along each of the strings. By manipulating the depth of the network, it is clear that a VDNP is required in order to get the concentration profiles to converge to a single line of the classic continuum form. Appendix 4.11 shows the Mathcad 15 code I used to generate the profiles. I applied the same uniform distribution of mass transfer coefficients and nodal surface areas, and I draw the reader's attention to the fact that the distribution is very wide, as per Eq. (4.460):

$$v = 1 + rnd(100) \tag{4.460}$$

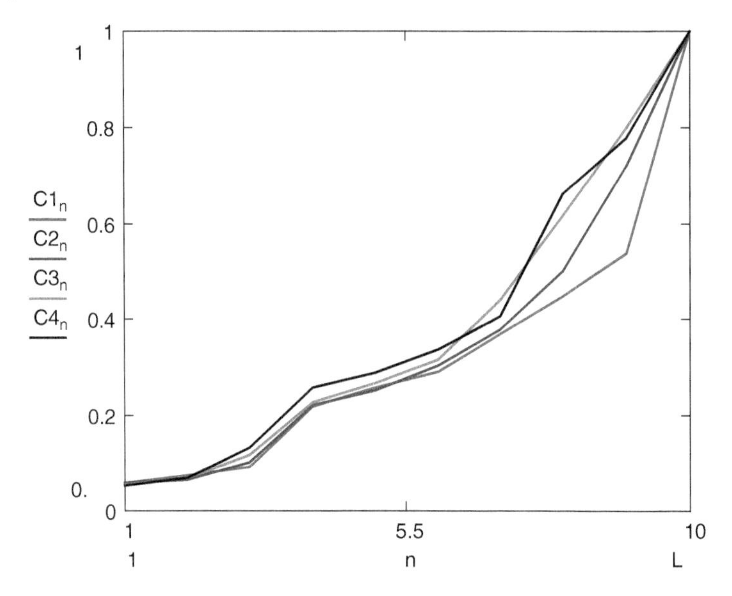

Figure 4.34 Concentration profile for 10 layers, four nodes per layer.

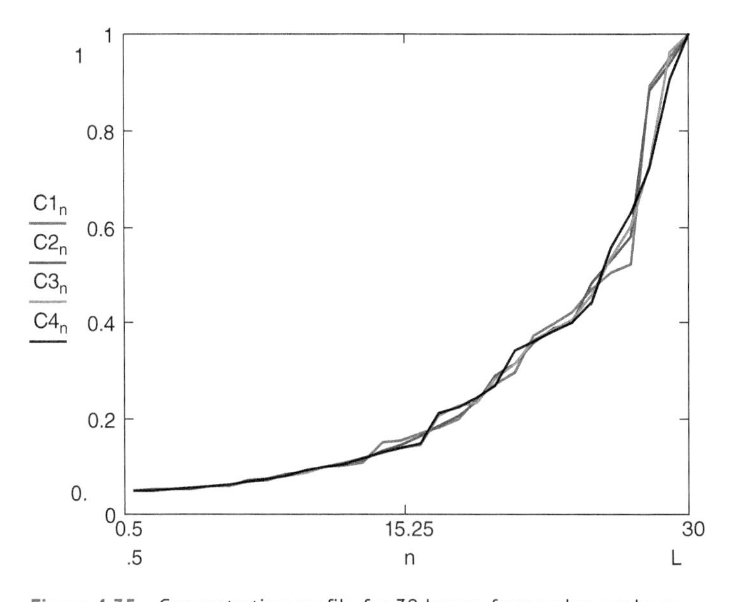

Figure 4.35 Concentration profile for 30 layers, four nodes per layer.

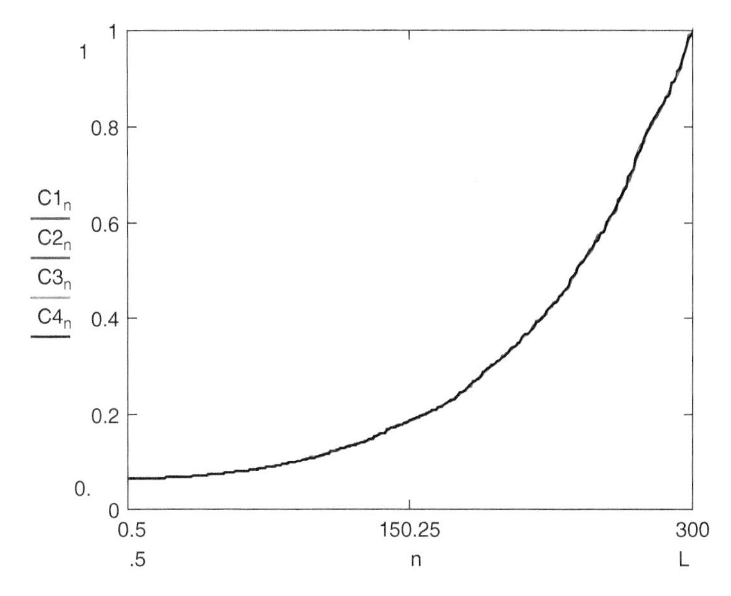

Figure 4.36 Concentration profile for 300 layers, four nodes per layer.

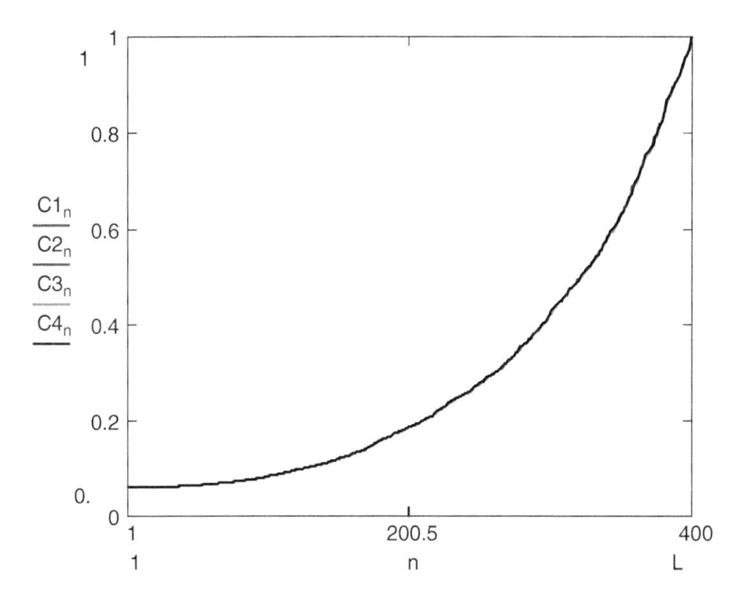

Figure 4.37 Concentration profile for 400 layers, four nodes per layer.

Table 4.7 Diffusion and first-order reaction as a function of the number of layers with four nodes per layer.

L	k	η	ϕ_η	ϕ_{am} (v = arithmetic mean)	ϕ_{hm} (v = harmonic mean)
10	0.15	0.277	3.60	4.06	6.24
30	0.01	0.243	4.12	3.05	4.68
300	0.0001	0.299	3.34	3.00	4.61
400	0.00006	0.286	3.49	3.12	4.76

In Table 4.7, I adapted the rate constant as I changed the depth of the network to keep the effectiveness factor just below about 30%. I calculated the Thiele modulus ϕ_η of the ensemble of strings based on the calculated effectiveness factor η of the network and by applying the convergence of VDNP to the classic solutions shown per the implicit equation in Eq. (4.461):

$$\eta = \frac{\tanh\left(\phi_\eta\right)}{\phi_\eta} \tag{4.461}$$

Based on calculations not further detailed here, I found the match of the concentration profile based on ϕ_η to be very satisfactory. In Table 4.7, I show the Thiele modulus based on either the arithmetic mean (ϕ_{am}) or the harmonic mean (ϕ_{hm}) of the distribution of the mass transfer coefficients, but neither is satisfactory. This is not surprising since neither takes the interconnectivity of the parallel string into consideration.

During my study, it became obvious that I should use the equivalent mass transfer coefficient v_{eq} defined in Eq. (4.423). The reason for this is simply because the value of the equivalent mass transfer coefficient is indeed dependent on all of the random coefficients and also takes the interconnectivity of the network into consideration. The Thiele modulus for the random network is then calculated as per Eq. (4.462):

$$\phi_{eq} = (N + 1/2)\sqrt{\frac{kS_{eq}}{v_{eq}}} \tag{4.462}$$

where I take S_{eq} as the arithmetic mean of the surface area distribution. Table 4.8 shows that the agreement between the Thiele modulus from Eq. (4.462) and the Thiele modulus based on the effectiveness factor is excellent.

Table 4.8 Diffusion and first-order reaction in random networks.

$(L = 300, N = 4, S = a_s + rnd(b_s), v = a_v + rnd(b_v))$ $(S_{eq} = a_s + b_s/2, v_{eq} = (L-1)\lambda_{min}(H_D))$									
a_S	b_S	a_v	b_v	k	v_{eq}	S_{eq}	η	ϕ_η	ϕ_{eq}
1	1	1	1	0.0001	1.46	1.50	0.327	3.04	3.04
1	1	1	1	0.001	1.46	1.50	0.104	9.65	9.61
1	2	1	2	0.00001	1.89	2.00	0.770	0.97	0.98
1	2	1	2	0.0001	1.89	2.00	0.324	3.07	3.09
1	2	1	2	0.001	1.89	2.00	0.103	9.72	9.77
1	10	1	10	0.00001	5.03	6.00	0.750	1.03	1.04
1	10	1	10	0.0001	5.03	6.00	0.306	3.26	3.28
5	3	1	1	0.00001	1.46	6.50	0.481	2.00	2.00
5	3	1	1	0.0001	1.46	6.50	0.157	6.38	6.33

4.3.5.4.5 Taking Greater Advantage of Very Deep Network Perturbation and the Classic Continuum Approach Equivalence

Taking advantage of the convergence of VDNP to the classic continuum approach can also be achieved via:

1) Determining the effectiveness factor experimentally based on small and large catalyst particle data.
2) Calculating the Thiele modulus from the implicit equation in Eq. (4.461).
3) Calculating the dimensionless concentration υ anywhere in the flat slab from:

$$\upsilon = \frac{\cosh(\xi\phi_\eta)}{\cosh(\phi_\eta)} \tag{4.463}$$

4) Obtaining the classic ratio of the kinetic rate constant over the effective diffusion coefficient from:

$$\frac{k}{D_{eff}} = \frac{\phi_\eta^2}{S_{BET}L^2} \tag{4.464}$$

4.3.5.4.6 A Concise Formulation for H_R Based on H_D in the Case of Very Deep Network Perturbation

Consider an arbitrary VDN and let the mass transfer matrix equation be given by Eq. (4.465) for diffusion:

$$\begin{vmatrix} \mathbf{P} + diag(\mathbf{G}\mathbf{e}_n) & -\mathbf{G} \\ -\mathbf{G}^T & \mathbf{Q} + diag(\mathbf{G}^T\mathbf{e}_m) \end{vmatrix} \begin{vmatrix} \mathbf{C}_i \\ \mathbf{C}_e \end{vmatrix} = \begin{vmatrix} 0 \\ \mathbf{F}_e \end{vmatrix} \tag{4.465}$$

and by Eq. (4.66) for diffusion and a first-order reaction:

$$\begin{vmatrix} \mathbf{P} + diag(\mathbf{G}\mathbf{e}_n + k\mathbf{S}_i) & -\mathbf{G} \\ -\mathbf{G}^T & \mathbf{Q} + diag(\mathbf{G}^T\mathbf{e}_m + k\mathbf{S}_e) \end{vmatrix} \begin{vmatrix} \mathbf{C}_i \\ \mathbf{C}_e \end{vmatrix} = \begin{vmatrix} \mathbf{0} \\ \mathbf{F}_e \end{vmatrix} \tag{4.466}$$

The matrix equations are easily solved, yielding:

$$\mathbf{\Omega}_D = (\mathbf{P} + diag(\mathbf{G}\mathbf{e}_n))^{-1}\mathbf{G} \tag{4.467}$$

$$\mathbf{H}_D = \mathbf{Q} + diag(\mathbf{G}^T\mathbf{e}_m) - \mathbf{G}^T(\mathbf{P} + diag(\mathbf{G}\mathbf{e}_n))^{-1}\mathbf{G} \tag{4.468}$$

$$\mathbf{\Omega}_R = (\mathbf{P} + diag(\mathbf{G}\mathbf{e}_n) + diag(k\mathbf{S}_i)))^{-1}\mathbf{G} \tag{4.469}$$

$$\mathbf{H}_R = \mathbf{Q} + diag(\mathbf{G}^T\mathbf{e}_m) + diag(k\mathbf{S}_e) - \mathbf{G}^T(\mathbf{P} + diag(\mathbf{G}\mathbf{e}_n) + diag(k\mathbf{S}_i)))^{-1}\mathbf{G} \tag{4.470}$$

The matrix $\mathbf{H}_D$ is singular and symmetrical and the canonical decomposition is given by:

$$\mathbf{H}_D = \mathbf{V}_D^T\mathbf{\Lambda}_D\mathbf{V}_D \tag{4.471}$$

The minimum eigenvalue of $\mathbf{H}_D$ is zero because the matrix is singular. The matrix of eigenvectors $\mathbf{V}_D$ is orthogonal. The eigenvector $\mathbf{v}_0$ associated with the minimum eigenvalue is given by:

$$\mathbf{v}_0 = \frac{1}{\sqrt{n}}\mathbf{e}_n \tag{4.472}$$

For the case with reaction:

$$\mathbf{H}_R = \mathbf{V}_R^T\mathbf{\Lambda}_R\mathbf{V}_R \tag{4.473}$$

For a VDNP, the matrix $\mathbf{H}_R$ is close to $\mathbf{H}_D$ (i.e. for the fluxes entering and exiting the network periphery because the fluxes are large compared to the loss of reagent due to reaction). The eigenvector matrices $\mathbf{V}_R$ and $\mathbf{V}_D$ respectively of $\mathbf{H}_R$ and $\mathbf{H}_D$ converge to each other. In addition, the eigenvalues of the same converge to each other, except for the minimum eigenvalue. In the case of reaction, the minimum eigenvalue of $\mathbf{H}_R$ is equal to the total consumption of reagent over the entire network R divided by the number of nodes on the periphery. We thus get:

$$\mathbf{V}_R \simeq \mathbf{V}_D \tag{4.474}$$

$$\mathbf{\Lambda}_R \simeq \mathbf{\Lambda}_D + \begin{vmatrix} R/N & \mathbf{0} \\ \mathbf{0} & \mathbf{0} \end{vmatrix} = \mathbf{\Lambda}_D + \begin{vmatrix} \lambda_{\min}(\mathbf{H}_R) & \mathbf{0} \\ \mathbf{0} & \mathbf{0} \end{vmatrix} = \mathbf{\Lambda}_D + \begin{vmatrix} (\mathbf{\Lambda}_R)_{1,1} & \mathbf{0} \\ \mathbf{0} & \mathbf{0} \end{vmatrix} \tag{4.475}$$

For a VDNP, we thus get a very simple expression for $\mathbf{H}_R$ based in large part on the properties of $\mathbf{H}_D$:

$$\mathbf{H}_R \simeq \mathbf{H}_D + \mathbf{V}_D \begin{vmatrix} \lambda_{\min}(\mathbf{H}_R) & \mathbf{0} \\ \mathbf{0} & \mathbf{0} \end{vmatrix} \mathbf{V}_D^T \tag{4.476}$$

This reminded me of the alternative method of solving reaction and diffusion in an arbitrary network through the addition of just one virtual node in Section 4.3.4.4.7.

4.3.5.4.7 Summary and Conclusion

I have tried to condense the content of this chapter into a single image (Figure 4.38), and hopefully I have been able to at least give a flavor of it. The classic continuum approach is based on the assumption that the catalyst porous structure is homogeneous and isotropic. This assumption is almost always fulfilled in commercial catalysts that have been carefully prepared starting from extruded powder mixtures or compressed powder mixtures. Very locally, however, any close-up picture of a catalyst, such as with scanning electron microscopy, makes it obvious that the porous structure is highly variable and complex. Yet, there are so many layers of particulates on top of more layers of particulates that averaging the properties is a valuable and practical approach. Thinking of a catalyst as a network of nodes allows one to write material balances in every node. Closing all of the material balances yields a very detailed description of mass transfer and reaction. These networks would likely be well suited to describe the class of materials called "amorphous catalysts". In this book, I concerned myself with networks that are made up of very large ensembles of particulates from the nano scale to the micron scale. Such networks, even in a 1 mm-sized particulate, would allow for many thousands of layers of particulates stacked on top of each other. Such networks I call VDNs or very large networks. Yet, as has already been demonstrated, the size of the VDN does not have to be so exorbitant that it would make them unwieldy. For evaluating networks, I highly recommend the "onion layer" approach described in Section 4.3.4.4.9 because it yields a very fast and efficient calculation. When considering VDNP, the networks yield and agree with the classic solutions, with the added feature and benefit that all material balances are closed in every node of the network. This in itself is a major advantage. A second advantage is that it allows one to calculate the Thiele modulus of the network based on the effectiveness factor of the network. The mass transfer characteristics of the network and the corrugated, winding, and percolating paths that reagents have to follow are properly taken into account. Even in the case of highly irregular properties and high node-to-node variability, the perturbation network approach allows one to take advantage of the existing classic solutions. I found

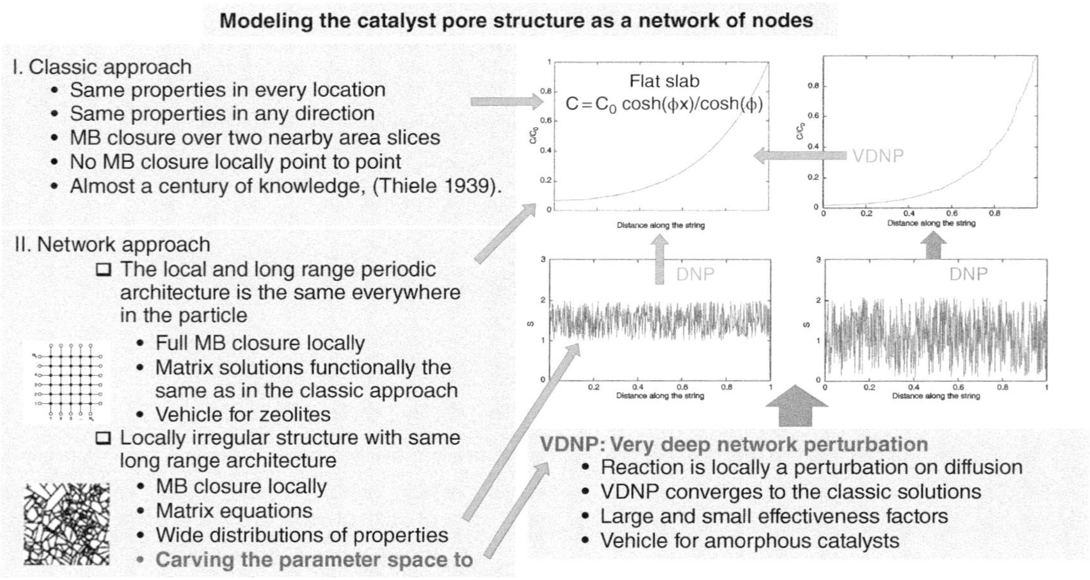

Figure 4.38 Comparison and convergence of a classic porous structure and a VDN architecture. MB = material balance.

the entire experience of VDN and VDNP very exciting, and it is similar to doing experiments on real catalysts in the laboratory on my PC. I expect that the perturbation network approach may also be applied advantageously to the case on non-first-order chemical reactions. Allow me to discuss perturbation a little more. The meaning of a catalytic activity perturbation on top of a diffusion field is that before a concentration gradient can establish itself, even in the slightest amount because of chemical reaction, the molecules have to diffuse through many tens or hundreds of layers of catalytic material or porous structure. Each layer of catalyst of just a few tens of nanometers thick is like a slice of Swiss cheese, containing many holes and apertures. Each molecule diffuses randomly across this gauntlet with many directional changes. Hence, a molecule of reagent travels through many hurdles of architecturally amorphous structure before it gets captured by a surface site and is reacted away. I think that catalysts that have extreme activities and where the perturbation approach may not hold are likely very rare and are controlled by external mass transfer limitations.

For a general arbitrary network, one is left with matrix equations, but there is not much else more specific to be said about this, or at least I am not able to. When I limit myself to VDNs with a reaction rate being a perturbation on top of the diffusion field, then I hope I have been able to have demonstrated that the following holds:

a) The VDN and VDNP approach converges to the classic continuum solutions, hence the classic solutions are special cases of the network solutions.
b) The network approach closes all material balances in each node, something the classic approach cannot do because it is not built in from the outset.
c) VDNP covers the range of effectiveness factors of commercial interest and is a good representation of actual catalysts.
d) The flexibility of networks now allows one to quantify how the range of individual mass transfer coefficients, surface areas, and rate constants in each node influence the overall performance of the catalyst. It is striking to see how these concentration profiles appear very real when they are calculated with broad, randomized ranges and how they neatly converge to the mathematical form of the classic solutions.

I believe that the approach outlined in this last section of the book can be applied to other fields that also have to deal with abrupt changes and irregular properties of materials from point to point. Examples include:

- Gas/liquid flow through a packing, as in:
  - A chemical reactor filled with catalyst pellets.
  - Liquid pressure drop through filter cake.
- Transport in disordered media.
- Electrical conductance in irregular media.

- Site–bond percolation.
- Light scattering.
- Traffic flow in large cities.
- Strength of disordered materials.

Nomenclature

Symbols in bold in this chapter always refer to vectors (lower case) or matrices (upper case). All symbols are defined in the text, and only the most common are mentioned here.

A	Component A
C	Concentration
D	Diffusion coefficient
i, j, k	Integers
f	Molar flux
F	Perimeter node mass transfer flux into the pore network
H	Specific mass transfer flux matrix
k^+	First-order rate constant in a node of a regular network
k	First-order rate coefficient in an irregular network
M	Normalized material balance matrix
N	Number of nodes along a principal direction
R	Rate of reaction in a node
S	Surface area allocated to a node
x, y, z	Principal directions

Greek Symbols

$\mathbf{\Psi}$	Matrix of adjacent layer concentration weight factors (–)
$\mathbf{\Lambda}$	Diagonal matrix of eigenvalues
ν	Mass transfer coefficient
ϕ	Thiele modulus (–)
$\mathbf{\Phi}$	Modulus matrix
η	Effectiveness factor
ξ	Dimensionless position
$\lambda_{\min}$	Minimum eigenvalue of a matrix (–)
$\mathbf{\Omega}$	Weight factor matrix (–)

References

1 Thiele, E.W. (1939). Relation between catalytic activity and size of particle. *Industrial and Engineering Chemistry* 31: 916–920.
2 Froment, G.F. and Bischoff, K.B. (1990). *Chemical Reactor Analysis and Design*, 2e. N. Y: Wiley.
3 Froment, G.F., Bischoff, K.B., and De Wilde, J. (2011). *Chemical Reactor Analysis and Design*, 3e, 195. Wiley.
4 Satterfield, C.N. (1980). *Mass Transfer in Heterogeneous Catalysis*. Malabar, FL: Robert Krieger Publishing Co. original edition (1970). Mass Transfer in Heterogeneous Catalysis. Cambridge, Massachusetts: MIT Press.
5 Aris, R. (1975). *The Mathematical Theory of Diffusion and Reaction in Permeable Catalysts*. Oxford University Press.
6 Hill, C. (1977). *An Introduction to Chemical Engineering and Reactor Design*, 440–446. Wiley.
7 Hegedus, L.L. and McCabe, R.W. (1984). *Catalyst Poisoning*. New York: Marcel Dekker.
8 Hegedus, L.L., Aris, R., Bell, A.T. et al. (1987). *Catalyst Design – Progress and Perspectives*. Wiley.
9 Chen, N.Y., Degnan, T.F. Jr., and Smith, C.M. (1994). *Molecular Transport and Reaction in Zeolites: Design and Application of Shape Selective Catalysts*. Wiley-VCH.
10 Becker, E.R. and Pereira, C.J. (1993). *Computer-Aided Design of Catalysts*, Chemical Industries Series, 51. Marcel Dekker.
11 Reyes, S.C. and Iglesia, E. (1993). Simulation techniques for the design and characterization of structural and transport properties of Mesoporous materials. In: *Computer Aided Innovation of New Materials II* (eds. M. Doyama, J. Kihara, M. Tanaka and R. Yamamoto), 1007. Elsevier Science Publishers B.V.
12 Kirkpatrick, S. (1973). Percolation and conduction. *Reviews of Modern Physics* 45: 574.
13 Mo, W.T. and Wei, J. (1986). Effective diffusivity in partially blocked zeolite catalyst. *Chemical Engineering Science* 41 (4): 703–710.
14 Beeckman, J.W. (1997). A rigorous matrix approach to site percolation for rectangular two-dimensional grids. *Industrial and Engineering Chemistry Research* 36 (8): 2964–2969.
15 Beeckman, J.W. (1991). Mathematical description of heterogeneous materials – effect of the branching direction. Symposium on Structure–Reactivity Relationships in Heterogeneous Catalysis, Boston, MA.
16 Beeckman, J.W. (1999). Diffusion, reaction and deactivation in pore networks. 8th International Symposium on Catalyst Deactivation, Bruges.

17 Levy, H. and Lessman, F. (1982). *Finite Difference Equations*. New York: Dover Publications.

18 Wei, J. (1982). A mathematical theory of enhanced para-xylene selectivity in molecular sieve catalysts. *Journal of Catalysis* 76: 433–439.

19 Albert, A. (1972). *Regression and the Moore–Penrose Pseudoinverse*. Academic Press.

20 Beeckman, J.W. (1990). Mathematical description of heterogeneous materials. *Chemical Engineering Science* 45: 2603–2610. Elsevier, doi:https://doi.org/10.1016/0009-2509(90)80148-8.

21 Boucher, C. (2016). Sampling Random Numbers from Probability Distribution Functions. https://uk.comsol.com/blogs/sampling-random-numbers-from-probability-distribution-functions (accessed 20 February 2020).

Appendix 4.1 Diffusion in a simple network

$$P = \begin{pmatrix} 2 & -1.1 & -0.3 & -0.6 \\ -1.1 & 1.8 & 0 & -0.7 \\ -0.3 & 0 & 0.3 & 0 \\ -0.6 & -0.7 & 0 & 1.3 \end{pmatrix} \quad G = \begin{pmatrix} 0 & 0 & 0 \\ 1.3 & 0 & 0 \\ 0 & 2.7 & 0.45 \\ 0 & 0 & 0 \end{pmatrix} \quad e3 = \begin{pmatrix} 1 \\ 1 \\ 1 \end{pmatrix}$$

$$P + \mathrm{diag}(G \cdot e3) = \begin{pmatrix} 2 & -1.1 & -0.3 & -0.6 \\ -1.1 & 3.1 & 0 & -0.7 \\ -0.3 & 0 & 3.45 & 0 \\ -0.6 & -0.7 & 0 & 1.3 \end{pmatrix}$$

$$(P + \mathrm{diag}(G \cdot e3))^{-1} = \begin{pmatrix} 1.0490 & 0.5482 & 0.0912 & 0.7793 \\ 0.5482 & 0.6537 & 0.0477 & 0.6050 \\ 0.0912 & 0.0477 & 0.2978 & 0.0678 \\ 0.7793 & 0.6050 & 0.0678 & 1.4547 \end{pmatrix}$$

Appendix 4.2 Property of the semi-inverse

$$P = \begin{pmatrix} 2 & -1.1 & -0.3 & -0.6 \\ -1.1 & 1.8 & 0 & -0.7 \\ -0.3 & 0 & 0.3 & 0 \\ -0.6 & -0.7 & 0 & 1.3 \end{pmatrix}$$

$$\text{eigenvals(P)} = \begin{pmatrix} 0.0000 \\ 0.3650 \\ 2.0120 \\ 3.0230 \end{pmatrix}$$

$$\text{eigenvals(P)} = \begin{pmatrix} 0.5000 & -0.1861 & -0.4163 & -0.7363 \\ 0.5000 & -0.3166 & -0.4458 & 0.6716 \\ 0.5000 & 0.8591 & 0.0729 & 0.0811 \\ 0.5000 & -0.3564 & 0.7891 & -0.0165 \end{pmatrix}$$

$$\text{EVAL0:=} \begin{pmatrix} 0 \\ \dfrac{1}{\text{eigenvals(P)}_2} \\ \dfrac{1}{\text{eigenvals(P)}_3} \\ \dfrac{1}{\text{eigenvals(P)}_4} \end{pmatrix} \quad \text{P0:=eigenvecs(P)}\cdot\text{diag(EVAL0)}\cdot\text{eigenvecs(P)}^{T}$$

$$\text{P}\cdot\text{P0} = \begin{pmatrix} 0.7500 & -0.2500 & -0.2500 & -0.2500 \\ -0.2500 & 0.7500 & -0.2500 & -0.2500 \\ -0.2500 & -0.2500 & 0.7500 & -0.2500 \\ -0.2500 & -0.2500 & -0.2500 & 0.7500 \end{pmatrix}$$

$$\text{P0}\cdot\text{P} = \begin{pmatrix} 0.7500 & -0.2500 & -0.2500 & -0.2500 \\ -0.2500 & 0.7500 & -0.2500 & -0.2500 \\ -0.2500 & -0.2500 & 0.7500 & -0.2500 \\ -0.2500 & -0.2500 & -0.2500 & 0.7500 \end{pmatrix}$$

Appendix 4.3 Diffusion and reaction in a simple network

$$\text{P} + \text{diag(G}\cdot\text{e3)} = \begin{pmatrix} 2 & -1.1 & -0.3 & -0.6 \\ -1.1 & 3.1 & 0 & -0.7 \\ -0.3 & 0 & 3.45 & 0 \\ -0.6 & -0.7 & 0 & 1.3 \end{pmatrix} \quad \text{S} = \begin{pmatrix} 2 \\ 0.4 \\ 0.7 \\ 1.3 \end{pmatrix} \quad \text{k} = 1$$

$$\text{P} + \text{diag(G}\cdot\text{e3)} + \text{diag(k}\cdot\text{S)} = \begin{pmatrix} 4 & -1.1 & -0.3 & -0.6 \\ -1.1 & 3.5 & 0 & -0.7 \\ -0.3 & 0 & 4.15 & 0 \\ -0.6 & -0.7 & 0 & 2.6 \end{pmatrix}$$

$$(\mathrm{P} + \mathrm{diag}(\mathrm{G}\cdot\mathrm{e3}) + \mathrm{diag}(\mathrm{k}\cdot\mathrm{S}))^{-1} = \begin{pmatrix} 0.29770 & 0.11340 & 0.02150 & 0.0992 \\ 0.11340 & 0.34520 & 0.00820 & 0.1191 \\ 0.02150 & 0.00820 & 0.24250 & 0.0072 \\ 0.09920 & 0.11910 & 0.00720 & 0.4396 \end{pmatrix}$$

Appendix 4.4 Matrix properties for diffusion and reaction in a simple network

$$\mathrm{e4} = \begin{pmatrix} 1 \\ 1 \\ 1 \\ 1 \end{pmatrix} \qquad \mathrm{He} := \mathrm{diag}\left(\mathrm{G}^{\mathrm{T}}\cdot\mathrm{e4}\right) - \mathrm{G}^{\mathrm{T}}\cdot(\mathrm{P} + \mathrm{diag}(\mathrm{G}\cdot\mathrm{e3}) + \mathrm{diag}(\mathrm{k}\cdot\mathrm{S}))^{-1}\cdot\mathrm{G}$$

e3 is defined in Appendix 4.1

$$\mathrm{He} = \begin{pmatrix} 0.7167 & -0.0288 & -0.0048 \\ -0.0288 & 0.9320 & -0.2947 \\ -0.0048 & -0.2947 & 0.4009 \end{pmatrix} \mathrm{He}\cdot\mathrm{e3} = \begin{pmatrix} 0.6831 \\ 0.6086 \\ 0.1014 \end{pmatrix} \mathrm{e3}^{\mathrm{T}}\cdot\mathrm{He}\cdot\mathrm{e3} = 1.3931$$

$$\mathrm{eigenvals(He)} = \begin{pmatrix} 0.2692 \\ 0.7155 \\ 1.0648 \end{pmatrix} \mathrm{eigenvecs(He)} = \begin{pmatrix} 0.0360 & -0.9969 & 0.0698 \\ 0.4073 & -0.0492 & -0.9120 \\ 0.9126 & 0.0612 & 0.4042 \end{pmatrix}$$

$$\mathrm{Effectiveness} := \frac{\left(\mathrm{e3}^{\mathrm{T}}\cdot\mathrm{He}\cdot\mathrm{e3}\right)}{\left(\mathrm{e4}^{\mathrm{T}}\cdot\mathrm{k}\cdot\mathrm{S}\right)} \quad \mathrm{Effectiveness} = 0.3166$$

Appendix 4.5 Perturbation in a simple network

$$\mathrm{P} + \mathrm{diag}(\mathrm{G}\cdot\mathrm{e3}) = \begin{pmatrix} 2 & -1.1 & -0.3 & -0.6 \\ -1.1 & 3.1 & 0 & -0.7 \\ -0.3 & 0 & 3.45 & 0 \\ -0.6 & -0.7 & 0 & 1.3 \end{pmatrix} \mathrm{S} = \begin{pmatrix} 2 \\ 0.4 \\ 0.7 \\ 1.3 \end{pmatrix} \mathrm{k} = 0.001$$

$$\mathrm{P} + \mathrm{diag}(\mathrm{G}\cdot\mathrm{e3}) + \mathrm{diag}(\mathrm{k}\cdot\mathrm{S}) = \begin{pmatrix} 2.002 & -1.1 & -0.3 & -0.6 \\ -1.1 & 3.1 & 0 & -0.7 \\ -0.3 & 0 & 3.451 & 0 \\ -0.6 & -0.7 & 0 & 1.301 \end{pmatrix} \mathrm{e4} = \begin{pmatrix} 1 \\ 1 \\ 1 \\ 1 \end{pmatrix}$$

$He:=diag\left(G^{T}\cdot e4\right)-G^{T}\cdot\left(P+diag(G\cdot e3)+diag(k\cdot S)\right)^{-1}\cdot G \quad \frac{\left(e3^{T}\cdot He\cdot e3\right)}{3}=0.00146$

$$He=\begin{pmatrix}0.1973 & -0.1667 & -0.0278\\ -0.1667 & 0.5298 & -0.3617\\ -0.0278 & -0.3617 & 0.3897\end{pmatrix} \quad He\cdot e3=\begin{pmatrix}0.0028\\0.0013\\0.0002\end{pmatrix} \quad eigenvals(He)=\begin{pmatrix}0.00146\\0.26761\\0.84772\end{pmatrix}$$

$Effectiveness:=\frac{\left(e3^{T}\cdot He\cdot e3\right)}{\left(e4^{T}\cdot k\cdot S\right)} \quad Effectiveness=0.9971 \quad min\,(eigenvals(He))\cdot 3=0.00437$

Appendix 4.6 A random variable

$a:=0 \quad b:=1$

$n:=10$

$i:=1..n+1 \quad j:=2..n+1$

$y_1:=0 \quad y_j:=y_{j-1}+a+rnd(b) \quad sum:=y_{n+1}$

$y_i:=\frac{y_i}{sum} \qquad x_1:=\frac{(i-1)}{n}$

$nn:=100$

$ii:=1..nn+1 \quad jj:=2..nn+1$

$yy_1:=0 \quad yy_{jj}:=yy_{jj-1}+a+rnd(b) \quad sum : yy_{nn+1}$

$yy_{ii}:=\frac{yy_{ii}}{sum} \quad xx_{ii}:=\frac{(ii-1)}{nn}$

$nnn:=1000$

$iii:=1..nn+1 \quad jjj:=2..nnn+1$

$yyy_1:=0 \quad yyy_{iii}:=yyy_{jjj-1}+a+rnd(b) \quad sum:=yyy_{nnn+1}$

$yyy_{iii}:=\frac{yyy_{iii}}{sum} \quad xxx_{iii}:=\frac{(iii-1)}{nnn}$

Appendix 4.7 Diffusion along a string of nodes

$N:=100 \; a:=0.1 \; b:=2 \; CA:=2 \; CB:=1 \; i=1..N \; k:=1..N+1 \; j:=1..N+2 \; eNp2_j:=1 \; eNp1_k:=1$

$M_{k,k+1}:=-(a+rnd(b)) \quad M_{N+2,N+2}:=0 \quad v_k:=-M_{k,k+1}$

$ve:=\left(eNp1^{T}\cdot v^{-1}\right)^{-1} \quad ve=0.0063559$

$M{:=}M + M^T - \mathrm{diag}\left[(M + M^T).eNp2\right]$ $\quad q{:=} - \mathrm{submatrix}(M, 2, N + 1, N + 2, N + 2,)$

$Z{:=}\mathrm{submatrix}(M, 2, N + 1, 2, N + 1)$ $\quad g{:=} - \mathrm{submatrix}(M, 1, 1, 2, N + 1)$ $\quad ZI{:=}Z^{-1}$

$C{:=}ZI\cdot\left(g^T\cdot CA + q\cdot CB\right)$

$M_{1,1} - g\cdot ZI\cdot g^T = 0.0063559$ $\quad g\cdot ZI\cdot q = 0.0063559$

$q^T\cdot ZI\cdot g^T = 0.0063559$ $\quad M_{N+2,N+2} - q^T\cdot ZI\cdot q = 0.0063559$

$Ci_1{:=}CA\ Ci_{N+2}{:=}CB\ Ci_{i+1}{:=}C_i\ x_1{:=}0\ x_{N+2}{:=}1\ x_{i+1}{:=}x_i + \dfrac{1}{N+1}$

Appendix 4.8 Diffusion in a rectangular strip with an equal number of nodes

Side-to-side diffusion in a network with square connectivity.
(L layers deep and N nodes per layer)
(Name: v2 overall mass transfer coefficient a.xmcd)

$L{:=}50\ N{:=}5\ M{:=}L\cdot N\ i{:=}1..M\ j{:=}1..M\ m{:=}1..N\ ii{:=}M - N + 1..M$

$kk{:=}1..N - 1\ eM_i{:=}1\ eN_m{:=}1\ iii{:=}1..M - N\ eMN_{iii}{:=}1$

$$P_{i,j}{:=}\mathrm{if}\left[j = i + N \vee \left[j = i + 1 \wedge \left(\frac{i}{N}\right) - \mathrm{trunc}\left(\frac{i}{N}\right) \neq 0\right], -1 - \mathrm{rnd}(1), 0\right]$$

$P_{ii,ii+1}{:=}0$ $\quad P_{kk,kk+1}{:=}0$ $\quad P{:=}\mathrm{submatrix}(P, 1, M, 1, M)$ $\quad MBD{:=}P + P^T - \mathrm{diag}\left[(P + P^T).eM\right]$

$A{:=}\mathrm{submatrix}(MBD, 1, N, 1, N)$ $\quad G{:=} - \mathrm{submatrix}(MBD, 1, N, N + 1, M - N)$

$Z{:=}\mathrm{submatrix}(MBD, N + 1, M - N, N + 1, M - N)$

$Q{:=} - \mathrm{submatrix}(MBD, N + 1, M - N, N + 1, M)$

$B{:=} - \mathrm{submatrix}(MBD, M - N + 1, M, M - N + 1, M)$

$HA{:=}A - G\cdot Z^{-1}\cdot G^T\ HB{:=}B - Q^T\cdot Z^{-1}\cdot Q$

$$HA = \begin{pmatrix} 0.6183 & -0.2678 & -0.1348 & -0.1089 & -0.0762 \\ -0.2678 & 0.7478 & -0.2136 & -0.1444 & -0.092 \\ -0.1348 & -0.2136 & 0.7404 & -0.2348 & -0.1284 \\ -0.1089 & -0.1444 & -0.2348 & 0.7636 & -0.2437 \\ -0.0762 & -0.092 & -0.1284 & -0.2437 & 0.5666 \end{pmatrix}$$

$$HA\cdot eN = \begin{pmatrix} 0.0306 \\ 0.0301 \\ 0.0288 \\ 0.0318 \\ 0.0264 \end{pmatrix} eN^T\cdot HA\cdot eN = 0.14760$$

$$N \cdot \min(\text{eigen vals}(HA)) = 0.14758$$

$$HB = \begin{pmatrix} 0.6594 & -0.3083 & -0.1539 & -0.1098 & -0.0616 \\ -0.3083 & 0.9127 & -0.2974 & -0.1808 & -0.9335 \\ -0.1539 & -0.2974 & 0.9547 & -0.3304 & -0.141 \\ -0.1098 & -0.1808 & -0.3304 & 0.9409 & -0.286 \\ -0.0616 & -0.0935 & -0.141 & -0.286 & 0.605 \end{pmatrix}$$

$$HB \cdot eN = \begin{pmatrix} 0.0258 \\ 0.0327 \\ 0.032 \\ 0.034 \\ 0.023 \end{pmatrix} \quad eN^T \cdot HB \cdot eN = 0.14760$$

$$N \cdot \min(\text{eigen vals}(HB)) = 0.14750$$

$$Q^T \cdot Z^{-1} \cdot G^T = \begin{pmatrix} 0.005341 & 0.005257 & 0.005036 & 0.005552 & 0.004605 \\ 0.006782 & 0.006675 & 0.006395 & 0.007050 & 0.005847 \\ 0.006636 & 0.006531 & 0.006257 & 0.006898 & 0.005721 \\ 0.007045 & 0.006934 & 0.006642 & 0.007323 & 0.06074 \\ 0.004764 & 0.004689 & 0.004492 & 0.004952 & 0.004107 \end{pmatrix}$$

$$eN^T \cdot Q^T \cdot Z^{-1} \cdot G^T \cdot eN = 0.14760$$

$$N \cdot \max\left(\text{eigen vals}\left(Q^T \cdot Z^{-1} \cdot G^T\right)\right) = 0.14852$$

Appendix 4.9 Diffusion in a rectangular strip with an unequal number of nodes

Side-to-side diffusion in a network with square connectivity but with unequal number of nodes.

N nodes on A-side, NN nodes on B-side.

A – side (L layer deep; N nodes per layer [N = 2 is minimum])

$L{:=}50 \; N{:=}3 \; M{:=}L \cdot N \; i{:=}1..M \; j{:=}1..M \; m{:=}1..N \; ii{:=}M - N + 1..M$

$kk{:=}1..N-1 \; eM_i{:=}1 eN_m{:=}1 \; iii{:=}1..M-N \; eMN_{iii}{:=}1 \; ik{:=}1..N$

$$P_{i,j}{:=}\text{if}\left[j = i + N \vee \left[j = i + 1 \wedge \left(\frac{i}{N}\right) - \text{trunc}\left(\frac{i}{N}\right) \neq 0\right], -1 - \text{rnd}(1), 0\right]$$

$P_{ii,ii+1} := 0 \ P_{kk,kk+1} := 0$

$P := \text{submatrix}(P, 1, M, 1, M) \ MBD := P + P^T - \text{diag}[(P + P^T).eM]$

B – side (LL layer deep; NN nodes per layer [NN = 2 is minimum])

$LL := 50 \ NN := 5 \ MM := LL \cdot NN \ i2 := 1..MM \ j2 := 1..MM \ m2 := 1..NN$

$ii2 := MM - NN + 1..MM \ kk2 := 1..NN - 1 \ eMM_{i2} := 1 \ eNN_{m2} := 1$

$iii2 := 1..MM - NN \ eMMNN_{iii2} := 1 \ ikk := 1..MM$

$$PP_{i2,j2} := \text{if}\left[j2 = i2 + NN \vee \left[j2 = i2 + 1 \wedge \left(\frac{i2}{NN}\right) - \text{trunc}\left(\frac{i2}{NN}\right) \neq 0\right], -1 - \text{rnd}(1), 0\right]$$

$PP_{ii2,ii2+1} := 0 \ PP_{kk2,kk2+1} := 0$

$Pp := \text{submatrix}(P, 1, M - N, 1, M - N)$

$Op_{ikk,iii} := 0 \ OOp_{iii,j2} := 0 \ PPP = \text{submatrix}(PP, 1, MM, 1, MM)$

$NEW := \text{augment}(\text{stack}(Pp, Op), \text{stack}(OOp, PPp))$

STITCHING

$NEW_{M-2N+ik,M-N+ik} := -1 - \text{rnd}(1) \ NEW_{kk2+M-N,kk2+M-N+1} := -1 - \text{rnd}(1)$

$jh := 1..MM + M - N \ eNEW_{jh} := 1 \ kz := MM + M - N$

$MBD := NEW + NEW^T - \mathit{diag}[(NEW + NEW^T) \cdot eNEW]$

$A := \text{submatrix}(MBD, 1, N, 1, N) \quad G := -\text{submatrix}(MBD, 1, N, N + 1, kz - NN)$

$Z := \text{submatrix}(MBD, N + 1, kz - NN, N + 1, kz - NN)$

$Q := -\text{submatrix}(MBD, N + 1, kz - NN, kz - NN + 1, kz)$

$B := -\text{submatrix}(MBD, kz - NN + 1, kz, kz - NN + 1, kz)$

$$HA = \begin{pmatrix} 0.5708 & -0.3084 & -0.2447 \\ -0.3084 & 0.7207 & -0.395 \\ -0.2447 & -0.395 & 0.6606 \end{pmatrix}$$

$$HA \cdot eN = \begin{pmatrix} 0.01761 \\ 0.01725 \\ 0.0209 \end{pmatrix} \quad eN^T \cdot HA \cdot eN = 0.05576$$

$N \cdot \min(\text{eigen vals}(HA)) = 0.05575$

$HB := B - Q^T \cdot Z^{-1} \cdot Q$

$$HB = \begin{pmatrix} 0.7022 & -0.2883 & -0.1996 & -0.1117 & -0.0904 \\ -0.2883 & -0.7199 & -0.2254 & -0.112 & -0.0845 \\ -0.1996 & -0.2254 & 0.9047 & -0.2854 & -0.1808 \\ -0.1117 & -0.112 & -0.2854 & 0.7967 & -0.2772 \\ -0.0904 & -0.0845 & -0.1808 & -0.2772 & 0.6426 \end{pmatrix}$$

$$HB{\cdot}eNN = \begin{pmatrix} 0.0122 \\ 0.00964 \\ 0.0135 \\ 0.01038 \\ 0.01004 \end{pmatrix}$$

$eNN^T{\cdot}HB{\cdot}eNN = 0.05576 \; NN{\cdot}\min(\text{eigen vals}(HB)) = 0.055754$

$eNN^T{\cdot}Q^T{\cdot}Z^{-1}{\cdot}G^T eN = 0.05576$

Appendix 4.10 Diffusion and first-order reaction in a very deep network of 500 layers deep and five nodes per layer

$$He = \begin{pmatrix} 0.8462 & -0.3354 & -0.172 & -0.1006 & -0.0707 \\ -0.3354 & 0.8922 & -0.2284 & -0.116 & -0.0758 \\ -0.712 & -0.2284 & 0.9107 & -0.2356 & -0.1337 \\ -0.1006 & -0.116 & -0.2356 & -0.8303 & -0.241 \\ -0.0707 & -0.0758 & -0.1337 & -0.241 & -0.6431 \end{pmatrix}$$

$$He{\cdot}eN = \begin{pmatrix} 0.1675 \\ 0.1366 \\ 0.141 \\ 0.1371 \\ 0.1219 \end{pmatrix}$$

$eN^T{\cdot}H{\cdot}eN = 0.7041$

$\min(\text{eigen vals}(H)){\cdot}N = 0.7025$

Appendix 4.11 Diffusion and first-order reaction

L layer deep, N notes per layer.

$$L:=300\ N:=4\ k:=0.0001\ a:=1\ b:=1\ M:=L\cdot N\ i:=1..M\ j:=1..M\ m:=1..N$$

$$ii:=M-N+1..M\ iii:=1..M-N\ kk:=1..L-1\ n:=1..L$$

$$eN_m:=1\ eM_i:=1\ eMN_{iii}:=1$$

$$S_i:=a+rnd(b)\ S_{ii}:=0\ SS=submatrix(k\cdot diag(S),1,M-N,1,M-N)$$

$$P_{i,j}:=if\left[j=i+N\vee\left[j=i+1\wedge\left(\frac{i}{N}\right)-trunc\left(\frac{i}{N}\right)\neq 0\right],-1-rnd(1),0\right]\ P_{ii,ii+1}:=0$$

$$nuavg:=a+\frac{b}{2}\qquad nus:=\frac{b}{\ln\left[\frac{(a+b)}{a}\right]}$$

$$nuavg=51\qquad nuavg=21.668$$

$$P:=submatrix(P,1,M,1,M)$$

$$MBD:=P+P^T-diag\left[\left(P+P^T\right).eM\right]\qquad G:=-submatrix(MBD,1,M-N,M-N+1,M)$$

$$AD:=submatrix(MBD,1,M-N,1,M-N)\quad BD:=submatrix(MBD,M-N+1,M,M-N+1,M)$$

$$OMD:=AD^{-1}\cdot G\qquad HD:=BD+G^T\cdot OMD$$

$$MBR:=P+P^T-diag\left[\left(P+P^T\right).eM\right]+k\cdot diag(S)$$

$$AR:=submatrix(MBR,1,M-N,1,M-N)\ BR:=submatrix(MBD,M-N+1,M,M-N+1,M)$$

$$OMR:=AR^{-1}\cdot G\qquad HR:=BR+G^T\cdot OMD$$

$$eM^T\cdot k\cdot S=6.1165\quad EFF:=\frac{eN^T\cdot HR\cdot eN}{eM^T\cdot k\cdot S}\quad EFF=0.288$$

$$\min(\text{eigen vals}(HR))=0.42051\qquad \frac{\left(eN^T\cdot HR\cdot eN\right)}{N\cdot\min(\text{eigen vals}(HR))}=1.0457$$

$$f(x):=EFF-\frac{\tanh(x)}{x}\ x:=0.5\ phief:=root(f(x),x)$$

$$phief=3.471\ phiaa:=\left(L+\frac{1}{2}\right)\cdot(k)^{0.5}\ phiaa=3.005]$$

$$phihm:=\left(L+\frac{1}{2}\right)\cdot\left(k\cdot\frac{a+\frac{b}{2}}{nus}\right)^{0.5}\quad phihm=4.61$$

Index

Catalyst Engineering Technology: Fundamentals and Applications,
First Edition. Jean W. L. Beeckman.